Energy at the Crossroads

Also by Vaclav Smil

China's Energy

Energy in the Developing World

Energy Analysis in Agriculture

Biomass Energies

The Bad Earth

Carbon-Nitrogen-Sulfur

Energy, Food, Environment

Energy in China's Modernization

General Energetics

China's Environmental Crisis

Global Ecology

Energy in World History

Cycles of Life

Energies

Feeding the World

Enriching the Earth

The Earth's Biosphere

Energy at the Crossroads

Global Perspectives and Uncertainties

Vaclav Smil

The MIT Press
Cambridge, Massachusetts
London, England

First MIT Press paperback edition, 2005

This book was set in Sabon by Achorn Graphic Services, Inc.
Printed on recycled paper and bound in the United States of America.

Library of Congress Cataloging-in-Publication Data

Smil, Vaclav.
Energy at the crossroads : global perspectives and uncertainties / Vaclav Smil.
p. cm.
Includes bibliographical references and index.
ISBN 0-262-19492-9 (hc. : alk. paper), 0-262-69324-0 (pb.)
1. Energy policy. 2. Energy policy—United States. 3. Energy policy—China. 4. Energy policy—Environmental aspects. 5. Energy development—Technological innovations. 6. Globalization—Environmental aspects. 7. Power resources—Political aspects—History. 8. Petroleum industry and trade—Political aspects—History. 9. History, Modern. I. Title.

HD9502.A2S543 2003
333.79—dc21

2002045222

10 9 8 7 6

Contents

Acknowledgments

This is my fifth book with the MIT Press, a fitting occasion to thank three people whose editorial efforts have seen these volumes from their conception to their publication: Larry Cohen, Deborah Cantor-Adams, and Clay Morgan. And, once again, thanks to Douglas Fast for preparing or reproducing nearly 140 images that illustrate this book.

Reflections on a Life of Energy Studies

I have always believed that books should be written only for truly compelling reasons. So why this book, and why now? The objective reason is simply the importance of its subject. Energy conversions are required for every process in the biosphere and for every human action, and our high-energy, predominantly fossil-fueled civilization is utterly dependent on unceasing flows of fuels and electricity. Frequent reexamination of these fundamental realities is imperative if we are to avoid fatal blunders and come closer to reconciling our need for abundant and reliable energy supplies with the existential requirement of maintaining the integrity of irreplaceable biospheric services.

It is the subjective reason for the book that I want to discuss here—the background that might help the reader understand my strong opinions. My goal in reviewing my long energy journey is to convince the reader that I have done my professional homework diligently and that my propensity to doubt and to judge is not based on momentary whims or fashionable trends.

I made my decision to begin systematic energy studies 40 years ago while an undergraduate at Prague's Carolinum University. No sudden impulse led to this decision—merely a slowly accumulating conviction that using first principles to understand the sources of energy, the techniques for harnessing these sources, and the enormous consequences of their conversions would help me understand the functioning of the biosphere and the rise and growing complexity of human civilization. I have never regretted my decision because it has given me endless opportunities for truly interdisciplinary research. I cannot imagine myself being engaged in any other form of inquiry, particularly not in any of those reductionist quests that are the hallmark of so much modern scientific work. Studies of energy systems also allowed me to work on a wide range of disparate problems that are, nevertheless, fundamentally related.

My earliest research looked at the environmental impacts of coal-fired electricity generation, particularly the effects of air pollution (SO_2 and particulates) on human health and ecosystems. This was the topic of my thesis, and a year later I published my first paper, "Energy Environment People," in *Vesmír* (*Universe*), the monthly journal of the Czech Academy of Sciences (Smil 1966). This publication breached the Communist taboo against open discussion of the environmental deterioration in the North Bohemian Brown Coal Basin, which later became internationally known as one apex of the infamous Sulfur Triangle. But from the very beginning of my studies I was also interested in grand patterns and was fascinated by disparate and shifting compositions of national energy supply and usage. I took the advantage of the abundant availability of Russian materials in Prague of the 1960s and found among them many impressive attempts at comprehensive analyses of entire energy systems.

When I moved to the United States in 1969, I found the country's energy research rich but very traditional, subdivided into standard extraction and conversion branches, with hardly any attention to final consumption patterns. At Penn State, one of the country's leading centers of energy research, I continued my work on air pollution while exploring other environmental impacts of energy production and consumption, with a particular emphasis on changes in principal extraction and conversion techniques. Although I have never been keen on economic reductionism, in this period I also learned to appreciate Richard Gordon's no-nonsense approach to energy economics.

My American thesis, completed in 1971 and published three years later (Smil 1974), was a long-term forecast of innovations in a wide range of energy extraction, conversion, and transportation techniques and the consequences of their adoption for the global environment. Technological forecasting led me to the modeling of interactions between energy and the environment, an effort that included one of the earliest long-term forecasts of CO_2 generation and its impact on tropospheric temperature (Smil and Milton 1974).

My interest in modeling was soon supplanted by the study of China's energy. The choice of this topic seemed obvious to me at that time because it was clearly a major gap in our understanding. During the early 1970s information on China's energy was extremely limited in the West. The few extant publications were authored by sinologists whose knowledge of language and culture may have been impeccable but whose writings were not informed by scientific and engineering understanding. Three years after starting this research, I published *China's Energy*, the first comprehensive analysis of the country's energy supplies, uses, and prospects (Smil 1976). Around this same time I became intrigued by the emerging field of energy analysis (energy cost accounting), and eventually I was helped by Tom Long and Paul Nachman in producing the first book-length account of the energy cost of America's most important crop, grain corn (Smil, Nachman, and Long 1982).

My work on China's energy led to comparisons with other major modernizing countries and to broader evaluations of the energy options facing nations seeking economic advancement: *Energy in the Developing World* (1980), a volume I edited with William Knowland, was the outcome of this effort.

Because all poor nations continued to rely heavily on wood and crop residues and because the rising prices of crude oil and other fossil energies spurred a search for alternative supplies in rich nations, there was during this period a rising interest in biomass energies. I came to feel strongly that these energy sources should have a limited role because of the undesirable environmental consequences of their large-scale exploitation. Some of the proposals that were seriously considered during the late 1970s and the early 1980s ranged from uncritically naïve (making gasohol the dominant fuel in Western transportation) to ridiculous (fueling the United States by gasification of kelp cultivated in giant Pacific Ocean plantations). I wrote *Biomass Energies* (Smil 1983) to help sort out such intellectual muddles.

The collapse of crude oil prices in 1985 ended not just the Western world's twelve-year-long preoccupation with energy matters but also many worthy efforts aimed at reducing energy consumption and moderating the impacts of energy use on the environment. At the same time, though, global warming was becoming a major scientific and public concern and there was continuing argument about our capacity to feed a global population that could double by the year 2050. I examined this inherently interdependent triad in a book entitled simply *Energy, Food, Environment: Realities, Myths, Options* (Smil 1987).

None of my books respects disciplinary boundaries, but this volume was by far my most interdisciplinary foray up to that time, and I was gratified that those critics who wanted to see a big picture approved. After completing this project I returned briefly to the study of China and revised and expanded an IDRC report commissioned by Ashok Desai into *Energy in China's Modernization* (Smil 1988). This book examined the rapidly growing and rapidly innovating energy industries created to support Deng Xiaoping's drive to transform an enormously inefficient, ideologically driven state into a modern economy, as well as the inevitable environmental and social consequences of this unprecedented development.

Immediately afterwards I turned to what remains my most comprehensive energy book. In *General Energetics* (Smil 1991) I used several fundamental and unifying principles—most notably power density (W/m^2) and energy intensity (J/g)—to perform comprehensive, comparative quantitative analyses of energy flows in the biosphere and throughout civilization. The biospheric coverage ranged from the Earth's radiation balance to trophic efficiencies in food webs, while the civilization segments covered the energetics of processes ranging from traditional farming to modern chemical syntheses. It took more than two years to complete this book. The greatest challenge was to write it in such a way that the forest of grand energy designs would not be obscured by the necessary inclusion of too many specific trees: if not treated properly, the thousands of numbers in the book would be simply overwhelming. This was a difficult challenge, and my greatest reward was when Philip Morrison, reviewing the book in *Scientific American,* called it "a work of tightly controlled audacity."

A 1987 walk in the Harvard Forest in Petersham with William McNeill led, five years later, to my decision to contribute to his series of books on world history. *Energy in World History* (Smil 1994a) was the first book to look at the world's history explicitly, but not deterministically, from an energy perspective. I tried in this book to draw a more universal picture of energy and history by adding to the histories of energy sources and conversions, and their socioeconomic consequences, appendices on topics such as the energy costs and benefits of coastal whaling, traditional field irrigation, and American draft horses; the contribution of sailing ships to the Dutch Golden Age; charcoal demand by early blast furnaces; and the destructive power of the Hiroshima bomb.

In 1993, I published two environmental books with, inevitably, strong energy components: *Global Ecology* (Smil 1993a) and *China's Environmental Crisis* (Smil 1993b). The first, written in a direct, personal style, concentrated on the links between modern civilization's rising consumption of water, energy, food, and materials and the biosphere's capacity to provide these vital resources. The second revisited a subject whose study I had explored a decade earlier in *The Bad Earth* (Smil 1984). That book, the first comprehensive volume on China's ecosystemic degradation and environmental pollution, met with some attention as well as some disbelief. Some readers were unwilling to accept the fact that China's environmental problems were so widespread, so acute, and in many instances so intractable. A decade later, the world knew more about these problems. Rather than presenting another systematic survey I now inquired deeper into the biophysical limits of national development, contrasting China's population history and prospects with its natural resources and its food and energy requirements.

After publishing two books each on China's environment and China's energy, I felt that I needed to focus my attention elsewhere. During the mid-1990s I decided to revisit another topic I had first approached ten years earlier. In 1985 I had published *Carbon-Nitrogen-Sulfur* (Smil 1985), a wide-ranging appraisal of the three key biogeochemical cycles. A decade later I framed the same challenge even more broadly in a volume on biospheric cycles written for the Scientific American Library. After the cycles came yet another return. I decided to use the comprehensive approach of *General Energetics*—from the Sun to electronic civilization—as a framework for exploring both fundamental and unexpected facets of natural and man-made phenomena ranging from extraterrestrial impacts to the energy cost of pregnancy and from bird flight to nuclear reactors.

Energies: An Illustrated Guide to the Biosphere and Civilization (Smil 1999a) consisted of more than 80 essays accompanied by more than 300 historical and modern images. Judging by its many reviews, this approach worked. *Energies* was my first book published by the MIT Press, a relationship initiated when Phil Morrison suggested that I get in touch with Larry Cohen. My second MIT Press book, *Feeding the World: Challenge for the Twenty-first Century* (Smil 2000c), was written as an extended, broadly founded answer to that often-asked question: will we be able to feed the growing humanity? The third one, *Enriching the Earth: Fritz Haber, Carl Bosch, and the Transformation of World Food Production* (Smil 2001) is an homage to the fact that more than one-third of humanity is alive today thanks to a largely unappreciated technical advance: the synthesis of ammonia from its elements.

If I were asked to recommend just one of my books to read before the present volume, I would suggest my fourth MIT Press book, *The Earth's Biosphere*. The reason is simple: preserving the integrity of the biosphere, which took more than four billion years to evolve, is the greatest challenge of the current century. Once you appreciate this fact, you will understand the critical importance of taking the right steps in securing our energy supply.

There are huge swaths of energy studies in which I have little interest or that I simply cannot handle because of my limited skills, but the matters that I have studied and tried to understand, explain, criticize, evaluate, and promote or decry sweep some broad arcs. These range from the local (e.g., the efficiency of small-scale biogas generation in rural Sichuan) to the global (e.g., combustion-driven interference in grand biogeochemical cycles), from the technical (e.g., detailed analyses of particular production and conversion processes) to the social (e.g., the linkages between energy use and the quality of life), and from historic writings to long-term forecasts. This book reflects these experiences and offers some strongly held opinions, but it does not rigidly forecast or arrogantly prescribe—two activities that rank high on my list of intellectual sins. All it does is explain the realities of our predicament, delineate the many limits and potentials that shape our quest, and point in the direction of pursuits that have perhaps the best chances of creating a more rational future.

Finally, a note on quantification and verification. As I make clear throughout the book, I believe that basic numeracy is all that is needed to do many revealing reality checks, and that back-of-the-envelope calculations (their less-than-confidence-inspiring name notwithstanding) are useful to understand whether or not the results of complex modelling exercises make sense. Heeding a recommendation of a reviewer, I would ask those seeking either an introduction or a deeper understanding of these simple but effective quantitative approaches to consult two books that are unequalled in this respect: John Harte's *Consider a Spherical Cow* (1988), and its more advanced sequel, *Consider a Cylindrical Cow* (2001). The only other recommended item is a good solar-powered scientific calculator, possibly the most cost-effective investment an inquisitive mind can make in a lifetime (I have done nearly all calculations for my books and papers on a TI-35 Galaxy Solar, whose four small PV cells and 42 keys remain as when I bought the gadget nearly 20 years ago).

Energy at the Crossroads

1 Long-term Trends and Achievements

The most fundamental attribute of modern society is simply this: ours is a high-energy civilization based largely on combustion of fossil fuels. Ever since the onset of sedentary farming and the domestication of draft animals all traditional societies secured their requisite mechanical energy by deploying human and animal muscles, and their thermal energy needed for comfort and cooking (and also light) by burning biomass fuels. Even the simplest water- and wind-driven mechanical prime movers (waterwheels and windmills) were completely or nearly absent in some traditional preindustrial societies, but they eventually came to play major roles in a few early modern economies. Wind-driven, and peat-fueled, Dutch Golden Age of the seventeenth century is perhaps the foremost example of such a society (DeZeeuw 1978).

In any case, average per capita availability of all forms of energy in preindustrial societies remained low and also stagnant for long periods of time. This situation was only marginally different in a few regions (most notably in England, today's Belgium, and parts of North China) where coal had been used in limited amounts for centuries both for heating and in local manufacturing plants. And although the nineteenth century saw widespread industrialization in parts of Europe and North America, most of today's affluent countries (including the United States and Japan) remained more dependent on wood than on coal until its closing decades. In addition, in wood-rich countries the absolute gain in total per capita energy use that accompanied the transition from wood to coal was hardly stunning. For example, the U.S. consumption of fossil fuels surpassed that of wood only in the early 1880s; and during the second half of the nineteenth century the average per capita supply of all energy increased by only about 25% as coal consumption rose tenfold but previously extensive wood burning was cut by four-fifths (Schurr and Netschert 1960).

In contrast, human advances during the twentieth century were closely bound with an unprecedented rise of total energy consumption (Smil 2000a). This growth was accompanied by a worldwide change of the dominant energy base as hydrocarbons have relegated coal almost everywhere to only two essential applications, production of metallurgical coke and, above all, generation of electricity. This latter use of coal is a part of another key transformation that took place during the twentieth century, namely the rising share of fossil fuels used indirectly as electricity. Other sources of electricity—hydro and nuclear generation—further expanded the supply of this most convenient kind of commercial energy.

Substantial improvements of all key nineteenth-century energy techniques and introduction of new, and more efficient, prime movers and better extraction and transportation processes resulted in widespread diffusion of labor-saving and comfort-providing conversions available at impressively lower prices. Technical advances also ushered in an unprecedented mobility of people and goods. As a result, widespread ownership of private cars and mass air travel are among the most important social transformations of the second half of the twentieth century. Emergence of extensive global trade in energy commodities opened the paths to affluence even to countries lacking adequate fuel or hydro resources.

The most recent trend characterizing high-energy civilization has been the rising amount and faster delivery of information. Availability of inexpensive and precisely controlled flows of electricity allowed for exponential growth of information storage and diffusion, first by analog devices and after 1945 by harnessing the immense digital potential. For nearly four decades these innovations were increasingly exploited only for military, research, and business applications; a rapid diffusion among general population began in the early 1980s with the marketing of affordable personal computers, and its pace was speeded up with the mass adoption of the Internet during the latter half of the 1990s.

Although modern societies could not exist without large and incessant flows of energy, there are no simple linear relationships between the inputs of fossil fuels and electricity and a nation's economic performance, social accomplishments, and individual quality of life (for many details on these linkages see chapter 2). Predictably, international comparisons show a variety of consumption patterns and a continuing large disparity between affluent and modernizing nations. At the same time, they also show similar socioeconomic achievements energized by substantially different primary energy inputs. Many of the key twentieth-century trends—including higher reliance on natural gas, slow diffusion of renewable energy techniques, efficiency gains in all kinds of energy conversions, and rising per capita use of energy in low-income countries—will continue during the coming generations, but there will have to be also some fundamental changes.

The key reason for these adjustments is the necessity to minimize environmental impacts of energy use in general, and potentially very worrisome consequences of anthropogenic generation of greenhouse gases in particular. Extraction, transportation, and conversion of fossil fuels and generation and transmission of electricity have always had many local and regional environmental impacts ranging from destruction of terrestrial ecosystems to water pollution, and from acidifying emissions to photochemical smog. Carbon dioxide from the combustion of fossil fuels poses a different challenge: it remains the most important anthropogenic greenhouse gas, and its rising emissions will be the main cause of higher tropospheric temperatures.

Consequently, the future use of energy may not be determined just by the availability of resources or by techniques used to extract and convert them and by prices charged for them—but also by the need to ensure that the global energy consumption will not change many other key biospheric parameters beyond the limits compatible with the long-term maintenance of global civilization. Prevention, or at least moderation, of rapid global warming is the foremost, although not the sole, concern in this category, and it may turn out to be one of the most difficult challenges of the twenty-first century. Loss of biodiversity, human interference in the biogeochemical nitrogen cycle, and the health of the world ocean are other leading environmental concerns associated with the rising use of energy.

A Unique Century

Only infrequently is the human history marked by truly decisive departures from long-lasting patterns. That is why the twentieth century was so remarkable as it offered a greater number of such examples, all of them closely connected with the dramatically higher use of energy, than the entire preceding millennium. They range from veritable revolutions in food production (now irrevocably dependent on synthetic nitrogenous fertilizers, pesticides, and mechanization of field tasks) and transportation (private cars, flying) to even more rapid advances in communication (radio, television, satellites, the Internet). Most of the post-1900 advances in basic scientific understanding—from the new Einsteinian physics whose origins date to the century's first years (Einstein 1905) to the deciphering of complete genomes of about twenty microbial species by the late 1990s (TIGR 2000)—would have been also impossible without abundant, inexpensive, and precisely controlled flows of energy.

As far as the evolution of human use of energy is concerned, practically all pre–twentieth-century technical and managerial advances were gradual processes rather than sudden breaks. Any short list of such events would have to include domestication of large draft animals (cattle, horses) whose power greatly surpasses that of humans, construction and slow diffusion of first mechanical prime movers converting indirect flows of solar energy (waterwheels, windmills), and, naturally, the invention of the steam engine, the first machine powered by combustion of a fossil fuel. The epochal transition from renewable to fossil energies proceeded first fairly slowly. Fossil fuels became the dominant source of human energy needs only about two centuries after Newcomen introduced his first inefficient machines during the first decade of the eighteenth century, and more than a century after James Watt patented (1769, renewal in 1775) and mass-produced his greatly improved steam engine (Dickinson 1967; fig. 1.1).

As there are no reliable data on the worldwide use of biomass energies, whose combustion sustained all civilizations preceding ours, we cannot pinpoint the date but we can conclude with a fair degree of certainty that fossil fuels began supplying more than half of the world's total primary energy needs only sometime during the 1890s (UNO 1956; Smil 1994a). The subsequent substitution of biomass energies proceeded rapidly: by the late 1920s wood and crop residues contained no more than one-third of all fuel energy used worldwide. The share sank below 25% by 1950 and during the late 1990s it was most likely no more than 10% (fig. 1.2; for more on biomass energy use see chapter 5). This global mean hides national extremes that range from more than 80% in the poorest African countries to just a few percent in affluent Western nations.

My personal experience spans this entire energy transition in 30 years and it includes all of the four great sources of heat. During the mid-1950s we, as most of our neighbors, still heated our house in the Bohemian Forest on the Czech-German border with wood. My summer duty was to chop small mountains of precut trunks into ready-to-stoke pieces of wood and stack them in sheltered spaces to air-dry, then getting up early in dark winter mornings and using often recalcitrant kindling to start a day's fire. During my studies in Prague and afterward, when living in the North Bohemian Brown Coal Basin, virtually all of my energy services—space heating, cooking, and all electric lights and gadgets—depended on the combustion of

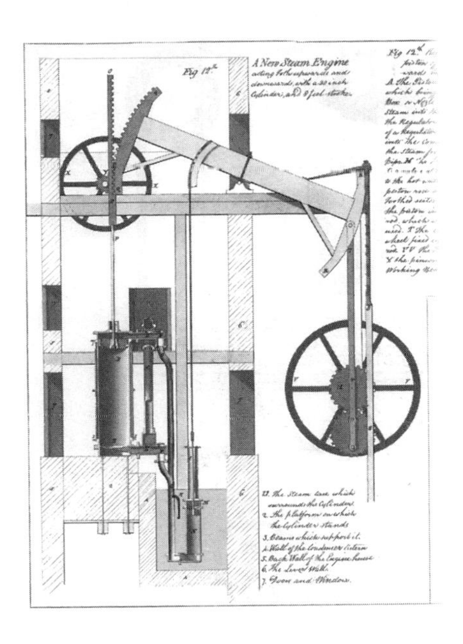

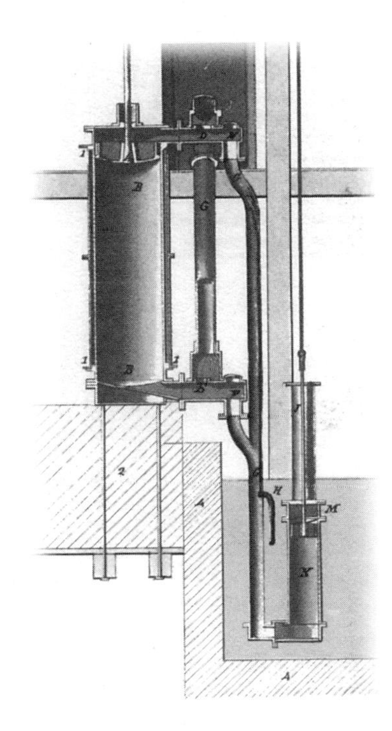

Figure 1.1
Complete drawing of James Watt's improved steam engine built in 1788 and a detail of his key innovation, the separate condenser connected to an air pump. Reproduced from Farey (1827).

lignite. After we moved to the United States the house whose second floor we rented was, as virtually all of its neighbors in that quiet and leafy Pennsylvanian neighborhood, heated by fuel oil. In our first Canadian house, bought in 1973, I had to reset a thermostat to restart a standard natural gas furnace (rated at 60% efficiency), but even that effort has not been necessary for many years. In our new superinsulated passive solar house a programmable thermostat will regulate my superefficient natural gas-fired furnace (rated at 94%) according to a weekly sequence of preset temperatures.

The completed transition means that for the Western nations the entire twentieth century, and for an increasing number of modernizing countries its second half, was the first era energized overwhelmingly by nonrenewable fuels. During the 1990s biomass fuels, burned mostly by households and industries in low-income countries, contained at least 35 EJ/year, roughly 2.5 times as much as during the crossover

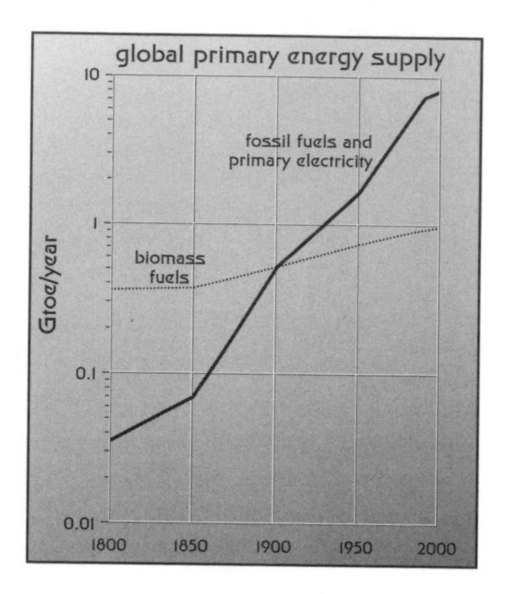

Figure 1.2
Global consumption of biomass and fossil fuels, 1800–2000. Based on Smil (1994a) and on additional data from BP (2001) and UNO (2001).

decade of the 1890s. In contrast, between 1900 and 2000 consumption of fossil fuels rose almost fifteenfold, from about 22 EJ to 320 EJ/year, and primary electricity added about 35 EJ/year (UNO 1956; BP 2001; fig. 1.2). This large expansion of fossil fuel combustion meant that in spite of the near quadrupling of global population—from 1.6 billion in 1900 to 6.1 billion in 2000—average annual per capita supply of commercial energy more than quadrupled from just 14 GJ to roughly 60 GJ, or to about 1.4 toe (Smil 1994a; UNO 2001; BP 2001; fig. 1.3).

But as the global mean hides enormous regional and national inequalities it is more revealing to quote the consumption means for the world's three largest economies (fig. 1.3). Between 1900 and 2000 annual per capita energy supply in the United States, starting from an already relatively high base, more than tripled to about 340 GJ/capita (Schurr and Netschert 1960; EIA 2001a). During the same time the Japanese consumption of commercial energies more than quadrupled to just over 170 GJ/capita (IEE 2000). In 1900 China's per capita fossil fuel use, limited to small quantities of coal in a few provinces, was negligible but between 1950, just after the establishment of the Communist rule, and 2000 it rose thirteenfold from just over 2 to about 30 GJ/capita (Smil 1976; Fridley 2001).

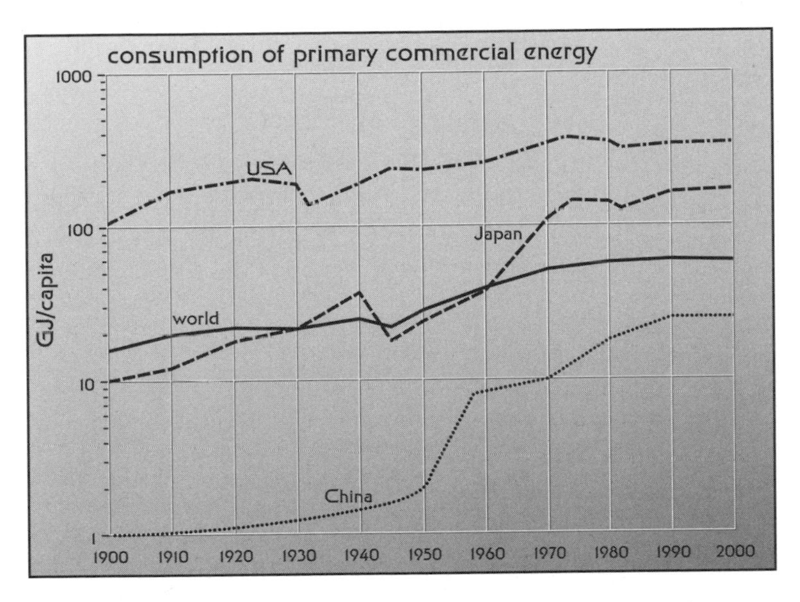

Figure 1.3
Average per capita consumption of primary commercial energy during the twentieth century is shown as the global mean and as national rates for the world's three largest economies, the United States, Japan, and China. Based on Smil (1994a) and on additional data from BP (2001), UNO (2001), and Fridley (2001).

These gains appear even more impressive when they are expressed not by comparing the initial energy content of commercial fuels or energies embodied in generation of primary electricity but in more appropriate terms as actually available energy services. Higher conversion efficiencies will deliver more useful energy and industrialized countries have made these gains by a combination of gradual improvements of such traditional energy converters as household coal stoves, by boosting the performance of three devices introduced during the late nineteenth century—electric lights, internal combustion engines (ICEs), and electric motors—and by introducing new techniques, ranging from natural gas furnaces to gas turbines. As a result, affluent nations now derive twice, or even three times, as much useful energy per unit of primary supply than they did a century ago.

When this is combined with higher energy use they have experienced eightfold to twelvefold increases in per capita supply of energy services as well as welcome improvements in comfort, safety, and reliability, gains that are much harder to quantify. And efficiency gains have taken place much faster and have been even more impressive in some of the most successful industrializing countries. Households replaced their traditional stoves (often no more than 10% efficient) by kerosene heaters and, more recently in cities, by natural gas-fueled appliances (now at least 60% efficient). Returning to my personal experience of exchanging a wood stove for a coal stove, coal stove for an oil-fired furnace, and oil-fired furnace for a standard and later a superefficient natural gas furnace, I am now receiving perhaps as much as six times more useful heat from one Joule of natural gas as I did from a Joule of wood.

Many households in industrializing countries have also exchanged incandescent light bulbs for fluorescent tubes, thus effecting an order of magnitude gain in average efficiency. Industrial gains in efficiency came after importing and diffusing state-of-the-art versions of basic smelting (iron, aluminum), synthesis (ammonia, plastics), and manufacturing (car, appliance assemblies) processes. Consequently, in those modernizing economies when such large efficiency gains have been accompanied by rapid increases in overall energy consumption per capita availabilities of useful energy have risen 20, or even 30 times in just 30–50 years.

Post-1980 China has been perhaps the best example of this rapid modernization as millions of urban families have switched from wasteful and dirty coal stoves to efficient and clean natural gas heating, and as industries abandoned outdated industrial processes with an unmistakably Stalinist pedigree (that is, ultimately, derivations of American designs of the 1930s) and imported the world's most advanced processes from Japan, Europe, and North America. Consequently, per capita supplies of useful energy rose by an order of magnitude in a single generation! And although any global mean can be only approximate and subsumes huge national differences, my conservative calculations indicate that in the year 2000 the world had at its disposal about 25 times more useful commercial energy than it did in 1900. Still, at just short of 40% during the late 1990s, the overall conversion efficiency of the world's primary fuel and electricity consumption to useful energy services remains far below the technical potential.

An even more stunning illustration of the twentieth century advances in energy use is provided by the contrasts between energy flows controlled directly by individuals in the course of their daily activities, and between the circumstances experienced by the users. At the beginning of the twentieth century America's affluent Great Plains farmers, with plentiful land and abundance of good feed, could afford to maintain more draft animals than did any other traditional cultivators in human history. And yet a Nebraska farmer holding the reins of six large horses while plowing his wheat field controlled delivery of no more than 5 kW of steady animate power (Smil 1994a).

This rate of work could be sustained for no more than a few hours before the ploughman and his animals had to take a break from a task that required strenuous exertion by horses and at least a great deal of uncomfortable endurance (as he was perched on a steel seat and often enveloped in dust) by the farmer. Only when the horses had to be prodded to pull as hard as they could, for example when a plow was stuck in a clayey soil, they could deliver briefly as much as 10 kW of power. A century later a great-grandson of that Nebraska farmer plows his fields while sitting in an upholstered seat of air-conditioned and stereo-enlivened comfort of his tractor's insulated and elevated cabin. His physical exertion is equal merely to the task of typing while the machine develops power of more than 300 kW and, when in good repair, can sustain it until running out of fuel.

I cannot resist giving at least two more examples of this centennial contrast. In 1900 an engineer operating a powerful locomotive pulling a transcontinental train at a speed close to 100 km/h commanded about 1 MW of steam power. This was the maximum rating of main-line machines permitted by manual stoking of coal that exposed the engineer and his stoker, sharing a confined space on a small, rattling metal platform, to alternating blast of heat and cold air (Bruce 1952). A century later a pilot of Boeing 747-400 controls four jet engines whose total cruise power is about 45 MW, and retraces the same route 11 km above the Earth's surface at an average speed of 900 km/h (Smil 2000b). He and his copilot can actually resort to the indirect way of human supervision by letting the computer fly the plane: human control is one step removed, exercised electronically through a software code.

Finally, in 1900 a chief engineer of one of hundreds of Europe's or North America's small utility companies supervising a coal-fired, electricity-generating plant that was supplying just a section of a large city controlled the flow of no more than 100,000 W. A century later a duty dispatcher in the main control room of a large interconnected electrical network that binds a number of the U.S. states or allows large-scale transmission among many European countries, can reroute 1,000,000,000 W, or four orders of magnitude more power, to cope with surges in demand or with emergencies. Other quantitative and qualitative jumps are noted throughout this chapter as I describe first the diversification of fuel use and technical innovations that made such advances possible and as I outline consumption trends and their socioeconomic correlates.

Changing Resource Base

In 1900 less than 800 Mt of hard coals and lignites accounted for about 95% of the world's *t*otal *p*rimary *e*nergy *s*upply (UNO 1956; using the acronym TPES, favored by the International Energy Agency). That total was doubled by 1949, to nearly 1.3 Gt of hard coals, and about 350 Mt of lignites, overwhelmingly because of the expansion of traditional manual mining of underground seams (fig. 1.4). Some of these exploited seams were as thin as 25–30 cm, and some thicker seams of high-quality hard coal and anthracite were worked hundreds of meters below the surface. Yet another doubling of the total coal tonnage (in terms of hard coal equivalent) took place by 1988 and the global extraction peaked the next year at nearly 4.9 Gt, with about 3.6 Gt contributed by hard coals and 1.3 Gt by lignites. About 40% of that year's lignite production was coming from the now defunct Communist states of East Germany and the Soviet Union, whose lignites were of particularly low quality averaging, respectively, just 8.8 and 14.7 GJ/t (UNO 2001).

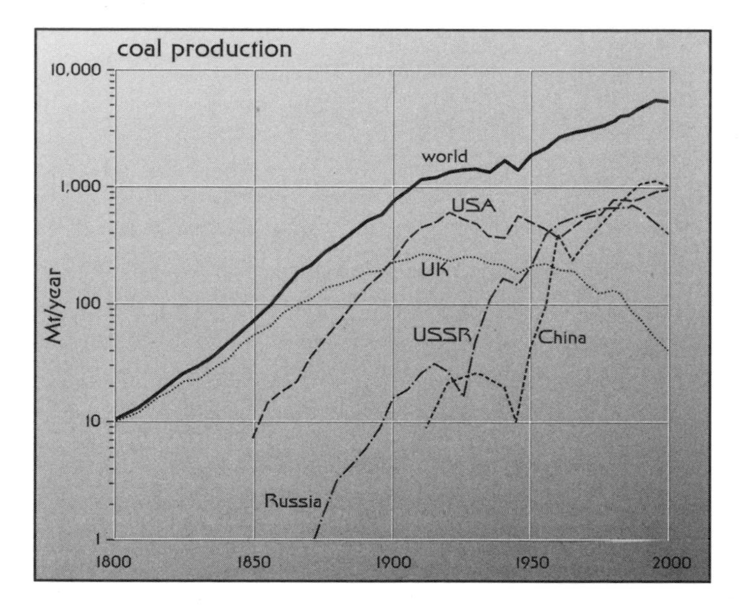

Figure 1.4
Coal production during the nineteenth and twentieth centuries: the ascending global total hides substantial changes in the output of major coal-producing countries, most notably the decline and collapse of the U.K. extraction (see also fig. 1.8) and the rise of China's coal industry. Based on Smil (1994a) and on additional data from BP (2001) and UNO (2001).

Changing makeup of coal extraction was not the only thing that had become different during the second half of the twentieth century. In contrast to the pre-WWII years nearly all of the additional underground production came from highly mechanized workfaces. For example, in 1920 all of the U.S. coal mined underground was manually loaded into mine cars, but by the 1960s nearly 90% was machine-loaded (Gold et al. 1984). Mechanization of underground mining lead to the abandonment of traditional room-and-pillar technique that has left at least half of all coal behind. Where the thickness and layout of seams allowed it, the longwall extraction became the technique of choice. This advancing face of coal cutting protected by moveable steel supports can recover over 90% of coal in place (Barczak 1992; fig. 1.5).

Similar or even higher recoveries are achieved in surface (opencast) mines that have accounted for more than half of America's new coal-producing capacities after 1950. These mines are fundamentally giant earth-moving (more precisely overburden-moving) enterprises aimed at uncovering one or more thick seams from which the

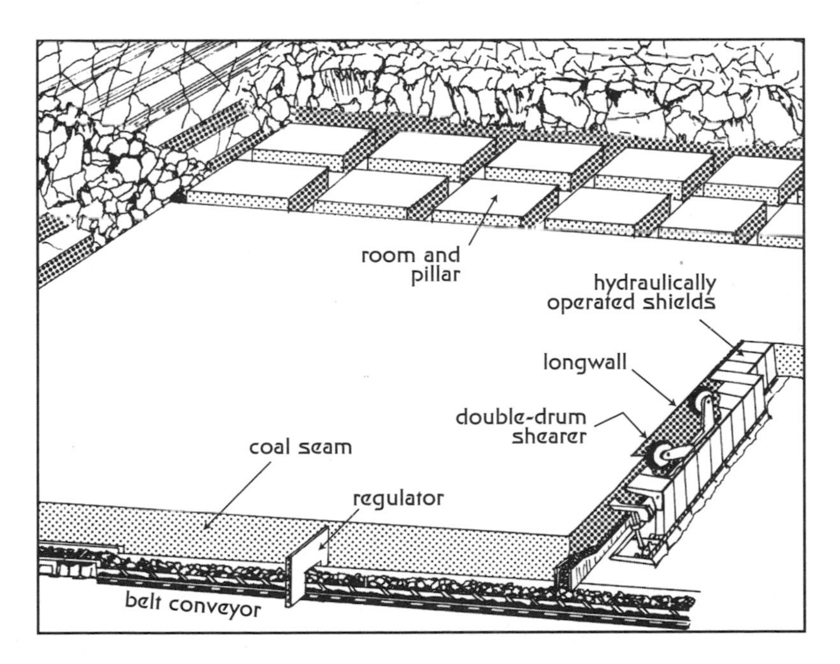

Figure 1.5
Longwall mining—in contrast to the traditional room-and-pillar extraction that leaves at least 50% of coal in place—can recover virtually all coal from level, or slightly inclined, seams. Based on a figure in Smil (1999a).

coal can be mined by not quite as large, but still distinctly oversized, machines. Growth of earth-moving machinery, exemplified by electric shovels and dragline excavators with dippers over 100 m^3, made it possible to exploit seams under as much as 200 m of overburden and to operate mines with annual production in excess of 10 Mt.

Two of the three coal-mining superpowers, the United States and the former Soviet Union, pursued aggressively this method of mining characterized by superior productivity and higher safety. In 1950 only 25% of the U.S. coal originated in surface mines but by the year 2000 the share rose to 65% (Darmstadter 1997; OSM 2001a). About 40% of the Russian coal output now originates in opencast mines, and only the third coal-mining superpower still relies mostly on underground mining. China has always had many small rural surface mines (unmechanized and inefficient operations producing low-quality fuel) but the country's large-scale mining remained an almost exclusively underground affair until the early 1980s, and even now less than 10% of the country's coal originates in modern opencast operations.

Besides impressive gains in fuel recovery these innovations have raised labor productivity and improved occupational safety. Productivity of underground mining rose from less than 1 t/man-shift at the century's beginning to more than 3 t/man-hour in highly mechanized modern pits using longwall or continuous mining systems, while productivities in the largest surface mines in Australia and the United States exceeds 20 t/worker hour (Darmstadter 1997). Fatalities in modern mining have followed the opposite trend. The U.S. statistics show more than a 90% decline of accidental deaths since the early 1930s, and 29 fatalities in 1998 prorated to 0.03 deaths per million tonnes of extracted coal (MSHA 2000). In contrast, death rates in China's coal mines remain extremely high, surpassing five fatalities per million tonnes of extracted coal during the late 1990s (Fridley 2001), and the recent Ukrainian death toll has been higher still.

Completely mechanized surface mining of thick seams raised the annual output of the largest mines to levels approaching, or even matching, annual coal output of smaller coal-producing countries. A number of countries, including the United States, Russia, Germany, and Australia, opened up mines with annual capacities of 15–50 Mt/year. Inevitably, the lower quality subbituminous coals and lignites extracted from shallow seams depressed the average energy content of the fuel. In 1900 a tonne of mined coal was equivalent to about 0.93 tonne of standard fuel (hard coal containing 29 GJ/t); by 1950 the ratio fell to about 0.83, and by the century's end it slipped to just below 0.7 (UNO 1956; UNO 2001).

Energy content of extracted coals thus increased less than 4.5 times between 1900 and 2000 while the world's total fossil fuel consumption rose fifteenfold during the same period. Moreover, the second half of the century saw a notable increase in the generation of primary (hydro and nuclear) electricity. As a result, coal's share in the global supply of primary energy declined during every year of the twentieth century, falling to below 75% just before the beginning of the WWII and to less than 50% by 1962. OPEC's oil price rises of the 1970s engendered widespread hopes of coal's comeback, mainly in the form of gases and liquids derived from the fuel by advanced conversion methods (Wilson 1980), but such hopes were as unrealistic as they were ephemeral (for more on this see chapter 3). Coal's share slipped to just below 30% of the global TPES by 1990 and in 2000 the fuel supplied no more than 23% of all primary commercial energy.

As so many other global means, this one is rather misleading. By the year 2000 only 16 countries extracted annually more than 25 Mt of hard coals and lignites, and six largest producers (in the order of energy content they are the United States, China, Australia, India, Russia, and South Africa) accounted for slightly more than 20% of the world's total coal output in the year 2000. Most notably, in the year 2000 the United Kingdom, the world's second largest coal producer in 1900, extracted less than 20 Mt/year from seventeen remaining private pits, and its peak labor force of 1.25 million miners in the year 1920 was reduced to fewer than 10,000 men (Hicks and Allen 1999; fig. 1.6). Many African and Asian countries use no coal at all, or the fuel supplies only a tiny fraction of their energy consumption, while it still provides nearly 80% of South Africa's, two-thirds of China's, and nearly three-fifths of India's energy demand, but only 25% of the United States and less than 20% of Russia's TPES. In China coal also dominates the household heating and cooking market, as it does in parts of India.

But the fuel has only three major markets left in affluent countries: to generate electricity and to produce metallurgical coke and cement. More efficient iron smelting cut the use of coke by more than half during the twentieth century: today's best blast furnaces need an equivalent of less than 0.5 t of coal per tonne of hot metal, compared to 1.3 t/t in 1900 (Smil 1994a; de Beer, Worrell, and Blok 1998). Extensive steel recycling (some 350 Mt of the scrap metal, an equivalent of about 40% of annual global steel output, is now reused annually) and slowly growing direct iron reduction reduced the role of large blast furnaces, and hence of coking coal. The latest reason for the declining use of coke is the injection of pulverized coal directly into a blast furnace, a practice that became widespread during the

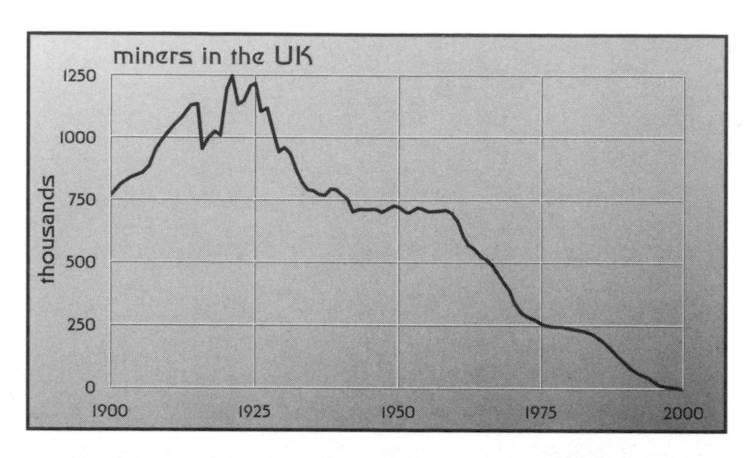

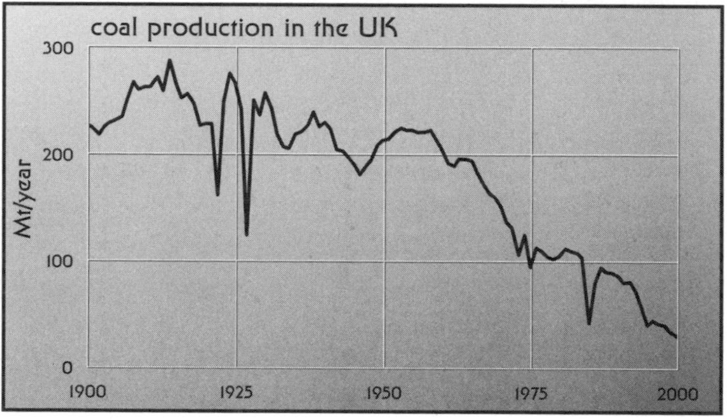

Figure 1.6
The United Kingdom's coal extraction was in decline for most of the twentieth century and the total labor employed in mining peaked during the 1920s. Based on graphs in Hicks and Allen (1999).

1990s: injection of 1 tonne of coal displaces about 1.4 tonnes of coking coal (WCI 2001). Global average of total coal inputs per tonne of crude steel fell from 0.87 in 1980 to 0.73 t by the year 2000 (a 15% decline), and the global demand for metallurgical coke now amounts to only about 17% of extracted hard coal, or just over 600 Mt in the year 2000 (WCI 2001).

Rising demand for electricity has provided the only globally growing market for coal—bituminous and lignite. Almost 40% of the world's electricity is now generated in coal-fired plants (WCI 2001). National shares among major producers are nearly 60% in the United States, close to 70% in India, roughly 80% in China, 85%

in Australia, and 90% in South Africa. Largest coal-fired stations, most of them dating from the 1960s, are either located near huge open-cast or big underground mines, or they are supplied by unit coal trains, permanently coupled assemblies of about 100 wagons with total capacities up to 10,000 t that constantly peddle between a mine and a plant (Glover et al. 1970). But in the long run even the demand for steam coal may weaken substantially, particularly if major coal consumers were to take aggressive steps to reduce the overall level of their carbon emissions (for more on this see the closing section of this chapter; coal's future is assessed in chapter 4).

The only other commercial coal market that has seen a steady growth has been the production of cement. More than 1.5 Gt of cement were produced annually during the late 1990s (MarketPlace Cement 2001) and the processing, requiring mostly between 3–9 GJ/t, has been in many countries increasingly energized by oil or natural gas. Multiplying the global cement output by 0.11, the average conversion factor recommended by the World Coal Institute, results in about 150 Mt of coal used in cement production, mostly in China (now the world's leading producer), with Japan, the United States, and India, each using less than 10 Mt/year (WCI 2001).

Coal was supplying more than 50% of the world's primary commercial energy until 1962, and it remained the single most important commercial fuel until 1966. More importantly, coal that was mined during the twentieth century contained more energy than any other primary resources, about 5,500 EJ. In contrast, the cumulative energy content of all crude oil extracted between 1901 and 2000 was about 5,300 EJ, less than 4% behind the coal aggregate, but during the century's second half, crude oil's total energy surpassed that of coal roughly by one-third. As a whole, the twentieth century can be thus seen as an energetic draw between coal and oil—but coal's rapid post-1960 loss of the global consumption share and its retreat into just the three major markets mark the fuel as a distinct has-been. At the same time, crude oil's recent rise to global prominence (between 1981 and 2000 it supplied nearly 50% more of energy than did coal), its dominance of the transportation market, unpredictable fluctuations of its world price, and concerns about its future supply put it repeatedly into the center of worldwide attention.

The combination of crude oil's high energy density and easy transportability is the fuel's greatest asset. Crude oils vary greatly in terms of their density, pour point, and sulfur content. Differences in density (specific gravity) are due to varying amounts of paraffins and aromatics. Densities are commonly measured by using a reverse °API scale, with heavy Saudi oils rating as low as 28 °API and light Nigerian oils going up to 44 °API (Smil 1991). Pour points extend from −36 °C for the lightest Nigerian crudes to 35 °C for waxy Chinese oil from the Daqing field, and sulfur content ranges between less than 0.5% (sweet oils) to more than 3% (sour crudes). But unlike coals, crude oils have very similar energy content, with nearly all values between 42–44 GJ/t, or about 50% more than the standard hard coal and three to four times as much as poor European lignites (UNO 2001). Unlike the case of coal, the wave of rising demand for crude oil products swept first North America (crude oil has been supplying more than 25% of the country's TPES since 1930), with Europe and Japan converting rapidly to imported liquid fuels only during the 1960s.

Worldwide transition to oil, and particularly its rapid post–World War II phase, was made possible by a combination of rapid technical progress and by discoveries of immensely concentrated resources of the fuel in the Middle East. Every infrastructural element of oil extraction, processing, and transportation had to get bigger, and more efficient, in order to meet the rising demand. Naturally, this growth of ratings and performances captured the often-considerable economies of scale that have made unit costs much lower. The fact that most of these infrastructures had reached size and performance plateaux is not because of the inevitably diminishing returns or insurmountable technical limits but rather because of environmental, social, and political considerations.

Early in the twentieth century, oil extraction began benefiting from the universal adoption of rotary drilling, which was used for the first time at the Spindletop well in Beaumont, Texas in 1901, and from the use of the rolling cutter rock bit introduced by Howard Hughes in 1909 (Brantly 1971). Deepest oil wells surpassed 3,000 m during the 1930s, and production from wells deeper than 5,000 m is now common in several hydrocarbon basins. By far the greatest post-1980 innovation has been a routine use of horizontal and directional drilling (Society of Petroleum Engineers 1991; Cooper 1994). Because horizontal wells can intersect and drain multiple fractures they are more likely to strike oil and to increase productivity (fig. 1.7).

Many horizontal wells can produce 2 to 5 times as much oil as vertical and deviated wells in the same reservoir (Valenti 1991; Al Muhairy and Farid 1993). Progress of horizontal drilling has been remarkable. Initially the drilling and completion costs of horizontal wells were 5–10 times the cost of a vertical bore, but by the late 1980s they declined to as little as 2 times its cost. Horizontal wells are now routinely used for extraction of thin formations and they are particularly rewarding in offshore drilling where a single platform can be used to exploit hydrocarbon-bearing layers

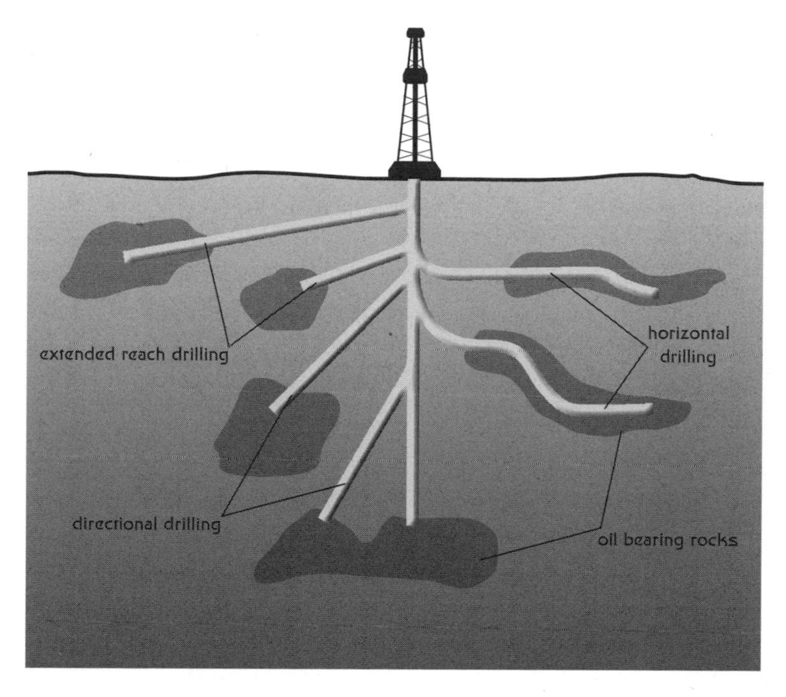

Figure 1.7
Directional drilling (with progressively greater deviation off the vertical), extended-reach drilling (up to 80° off the vertical) and horizontal drilling have made it possible to exploit better several hydrocarbon-bearing structures from a single site and to increase the rate of recovery of oil and gas deposits.

far from the primary hole. The longest horizontal wells are now around 4,000 m, nearly as long as the deepest vertical wells 50 years ago.

In 1947 the first well was completed out of land sight off Louisiana (Brantly 1971). Half a century later offshore extraction was producing about 30% of the global oil output (Alexander's Gas & Oil Connections 1998). This has been made possible by using submersible, semisubmersible, and floating drilling rigs and production platforms that have kept moving to deeper, and also stormier, waters. In 2000 there were 636 mobile offshore drilling units in private and state-owned fleets, about 60% of them being jack-ups and 25% semisubmersibles (World Oil 2000). Some of these rigs are now working in waters up to 2,000 m deep and in 2001 an ultradeepwater drillship, *Discoverer Spirit,* set the record by drilling in 2,900 m of water in the Gulf of Mexico (Transocean Sedco Forex 2001). Offshore production platforms are among the most massive structures ever built. The record-holder in 2000 was the *Ursa* tension leg platform, a joint project of a group of companies lead by the Shell Exploration and Production Company (Shell Exploration and Production Company 1999). The platform has a total displacement of about 88,000 t (more than a Nimitz-class nuclear aircraft carrier), rises 145 m above water and it is anchored 1,140 m below water with 16 steel tendons.

Refining of crude oils—yielding a range of liquid fuels perfectly suited to a variety of specific applications ranging from the supersonic flight to the powering of massive diesel locomotives—was transformed by the introduction of high-pressure cracking after 1913 and of catalytic cracking in 1936. Without these processes it would be impossible to produce inexpensively large volumes of lighter distillates from intermediate and heavy compounds that dominate most of the crude oils. Unlike coals, crude oils are readily pumped on board of large ships and the size of modern crude oil tankers and the cheap diesel fuel they use means that the location of oilfields is virtually of no consequence as far the exports of the fuel are concerned. And crude oil can be sent across countries and continents through the safest, most reliable, and environmentally most benign means of energy transportation, a buried pipeline (for more on tankers and pipelines see the trade section later in this chapter).

Discoveries of the world's largest oilfields (supergiants in oil geology parlance) began during the 1930s and continued for more than two decades. Kuwaiti al-Burgan, now the world's second largest supergiant, was found in 1938. Saudi al-Ghawar, the world's largest oilfield holding almost 7% of the world's oil reserves in the year 2000, was discovered a decade later (Nehring 1978; EIA 2001b). By the time OPEC began increasing its crude oil price in the early 1970s the Middle East was known to contain 70% of all oil reserves, and the region (excluding North Africa) had 50% of the world's oil-producing capacity (fig. 1.8). Global crude oil extraction in 1900 was only about 20 Mt, the mass that is now produced in only about two days. This means that the worldwide crude oil output rose more than 160-fold since 1900 and nearly eightfold since 1950 when the world consumed just over 500 Mt of refined products that provided 25% of all primary commercial energy.

Although it is highly unevenly distributed, today's crude oil extraction is less skewed than the global coal production. Nearly 30 countries now produce annually more than 25 Mt of crude oil, and the top six producers account for 45% (vs. coal's 75%) of the total (BP 2001). In the year 2000, 3.2 Gt of crude oil supplied two-fifths of all commercial primary energy, about 10% below the peak share of about 44% that prevailed during the 1970s (UNO 1976; BP 2001). Crude oil's role in

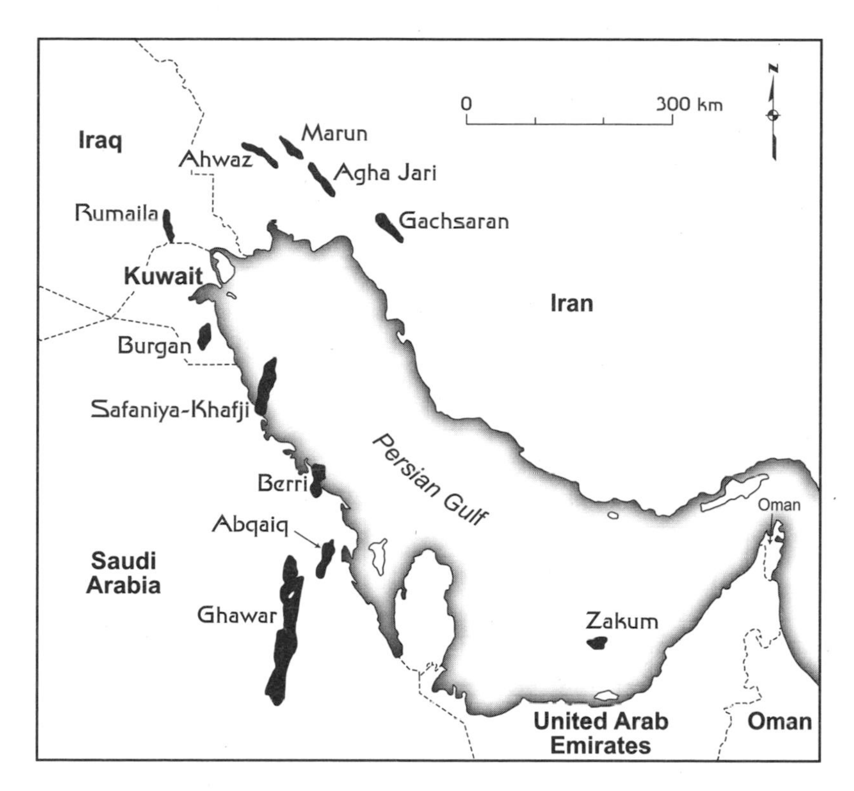

Figure 1.8
Giant Middle Eastern oil fields. Based on oil field maps published in various issues of *Oil & Gas Journal.*

modern societies is even greater than is suggested by its share of the TPES as refined fuels provide more than 90% of energy for the world's transportation. Air transport, one of the twentieth century greatest innovations with enormous economic, military, and social consequences, is unthinkable without refined fuels. So is, of course, the first century of mass public ownership of private cars.

Economic, social, and environmental consequences of automobilization have been even more far-reaching than have been the effects of flying. Land transport was also considerably facilitated by ready availability of inexpensive paving materials derived from crude oil. Crude oil also claims very high shares of the total commercial energy use in many low-income countries that still have only very modest per capita consumption of liquid fuels but rely on them more heavily than most of the affluent world with their more diversified energy supply. Because of these critical supply roles we will go to great lengths in order to secure adequate flows of the fuel that in so many ways defines the modern civilization.

Although crude oil, unlike coal, will never claim more than half of the world's primary commercial energy use I will present, in chapter 4, detailed arguments in order to show that the fuel's future is robust. Spectacular discoveries of supergiant oilfields and expansions that characterized the rise of oil to its global prominence during the twentieth century cannot be replicated in the coming generations, but a strong and globally important oil industry will be with us for generations to come. Its future will be shaped to a large degree by the advances of natural gas industry with which it is either directly commingled or closely associated.

During the first decade of the twentieth century, natural gases contributed only about 1.5% of the world's commercial primary energy consumption, and most of it was due just to the slowly expanding U.S. production. When expressed in energy equivalents the crude oil/natural gas ratio was about 3.1 during the 1910s and the gap between the two hydrocarbon fuels has been narrowing ever since. By the 1950s the ratio was 2.9, by the 1970s, 2.5. Post-1973 slowdown in the growth of oil output contrasted with continuing high increases of natural gas extraction that had doubled during the century's last quarter and lowered the oil/gas ratio to 1.7 during the 1990s. Because of its cleanliness natural gas has been the preferred fuel for space heating, as well as for electricity generation. Unlike many coals and crude oils, its content of sulfur is usually very low, or the gas can be easily stripped of any unwanted pollutants before it is put into a pipeline. Natural gas is now also sought after because it releases the lowest amount of CO_2 per unit of energy (see the last section of this chapter).

Natural gas now supplies 25% of the world's commercial primary energy and all hydrocarbons, ranging from virtually pure CH_4 to heavy crude oils, provide nearly two-thirds of the total. Future growth of the total share of hydrocarbon energies will be almost totally due to the increasing extraction of natural gas (see chapter 4). The two new sources of primary energy supply that could limit the relative share of hydrocarbons—electricity generated by nuclear fission and by converting direct and indirect solar energy flows—are also the ones with very uncertain futures. Coal's dominance of the global commercial primary energy supply lasted about three human generations (70 years), extending from the mid-1890s when it overtook wood to the mid-1960s when it was overtaken by hydrocarbons. Recent years have seen many claims about the imminent peak of global oil output: if true we would be already about halfway through the hydrocarbon era. As I will show in chapter 4 these claims may miss their mark by decades rather than by years. In any case, what is much more difficult to foresee than the timing of the midpoint of global oil extraction is what resource will become dominant after the hydrocarbon extraction begins its inevitable decline.

Technical Innovations

Technical advances that transformed the twentieth-century energy use can be logically divided into three interrelated categories. First are the impressive improvements of several key pre-1900 inventions, most of them originating during the incredibly innovative period between 1880–1895. Second are inventions of new extraction, conversion, and transportation techniques and their subsequent commercialization and refinements. Third are innovations that were introduced for reasons not related to energy production or use but whose later applications to numerous energy-related endeavors have greatly improved their accuracy, reliability, and efficiency. Improved performances of three out of the world's five most important prime movers are the best example in the first category: ICE, electric motor, and steam turbogenerator were all invented during the late nineteenth century. Their inventors would readily recognize the unchanged fundamentals of today's machines but they would marvel at the intervening improvements in performance and at the much higher power ratings of the latest designs.

Two new prime movers, gas turbines and rocket engines, should top a long list of inventions that could be cited in the second category. Both were commercialized only by the middle of the twentieth century but both have subsequently undergone a rapid development. The century's other commercially successful fundamental energy innovations also include two new modes of energy conversion, the now troubled nuclear fission and gradually ascendant photovoltaic generation of electricity. Examples of the last category of technical innovation are hidden everywhere as computerized controls help to operate everything from oil-drilling rigs to power plants, and from room thermostats to car engines. No less fundamentally, new communication, remote sensing, and analytical techniques have greatly transformed operations ranging from the search for deeply buried hydrocarbons to optimized management of interconnected electricity networks.

In spite of this diversity of advances there have been some notable commonalities dictated by the great upheavals of the twentieth century. Diffusion of all kinds of technical advances was set back by World War I, as well as by the economic crisis of the 1930s, but World War II accelerated the introduction of three major innovations: nuclear fission, gas turbines, and rocket propulsion. The two decades following WWII saw a particularly rapid growth of all energy systems, but since the late 1960s most of their individual components—be they coal mines, steam turbines in large thermal stations, transmission voltages, or giant tankers—had reached clear growth plateaux, and in some cases their typical unit sizes or capacities have actually declined.

Mature markets, excessive unit costs, and unacceptable environmental impacts, rather than technical limits to further growth, were the key reasons for this change, as higher efficiency and reliability and a greater environmental compatibility became the dominant design goals of the last two decades of the twentieth century. Next I include only the most important examples in the three principal categories of technical innovations before concentrating in a greater detail on what is perhaps the twentieth century's most far-reaching, long-term energy trend whose course is still far from over, the rising importance of electricity.

Modern life continues to be shaped by several substantially improved late nineteenth-century inventions, above all by electricity generation and transmission systems and by internal combustion engines. Steam engine, the quintessential machine of the early phases of industrialization, continued to be an important prime mover during the first few decades of the twentieth century. By that time its best specimens were nearly ten times more efficient and 100 times more powerful than were the top units at the beginning of the nineteenth century (Smil 1994a). But even these impressive advances could not change the machine's inherently low efficiency and high mass/power ratio. Nascent electricity-generating industry thus rapidly embraced the just-invented steam turbine and once electricity became readily available, electric motors displaced steam engines in countless manufacturing tasks. And, of course, the steam engine could never compete with internal combustion engines as a prime mover in land or airborne transportation.

Every aspect of those two great late nineteenth-century inventions was improved by subsequent innovation, resulting in better performance and reduced environmental impacts. In 1900 efficiencies of thermal electricity generation, with boilers burning lump coal on moving grates, steam pressure at less than 1 MPa and steam temperatures of less than 200 °C, were as low as 5%. Today's best thermal plants, burning pulverized coal and operating at steam pressures in excess of 20 MPa and temperatures above 600 °C, have conversion efficiencies of just over 40% but cogeneration can raise this rate to almost 60% (Weisman 1985; Gorokhov et al. 1999; for more on high-efficiency conversions see chapter 4).

Experiments with milled coal began in England already in 1903 but first large boilers fired with finely pulverized coal were put in operation in 1919 at London's Hamersmith power station. Unit sizes of steam turbines were slow to rise: Parsons' first 1 MW steam turbine was built in 1900 but 100 MW units were widely used only after 1950. But then it took less then two decades to raise the capacity by an order of magnitude as the first 1 GW unit went online in 1967 (fig. 1.9). The largest thermal turbines in coal-fired or nuclear stations now rate about 1.5 GW, but units of 200–800 MW are dominant.

Transmission losses were cut by using better and larger transformers, higher voltages, and direct current links. Peak transformer capacities had grown 500 times during the century. Typical main-line voltages were 23 kV before the WWI, 69 kV during the 1920s, 115 kV during the 1940s, and 345 kV by 1970 (Smil 1994a). Today's top AC links rate 765 kV, with the world's first line of that voltage installed by Hydro-Québec in 1965 to bring electricity 1,100 km south from Churchill Falls in Labrador to Montréal. And the age of long-distance, high-voltage DC transmission began on June 20, 1972 when Manitoba Hydro's 895 km long ±450 kV DC line brought electricity from Kettle Rapids hydro station on the Nelson River to Winnipeg (Smil 1991). Now we have DC links of up to 1,500 kV connecting large plants and major load centers in urban and industrial areas. Creation of regional grids in North America and more extensive international interconnections in Europe (in both latitudinal and longitudinal direction) improved supply security while reducing the requirements for reserve capacities maintained by individual generating systems.

The combination of Daimler's engine, Benz's electrical ignition, and Maybach's float-feed carburetor set a lasting configuration for the expansion of the automobile industry at the very beginning of the automotive era during the mid-1880s, and the subsequent development of Otto-cycle engines has been remarkably conservative (Flink 1988; Newcomb and Spurr 1989; Womack et al. 1991). Still, the industry has seen major technical advances. By far the most important twentieth-century changes included much higher compression ratios (from 4 before WWI to between 8 and 9.5) and declining engine weight. Typical mass/power ratios of ICEs fell from more than 30 g/W during the 1890s to just around 1 g/W a century later (Smil 1994a). Diesel engines have also become both lighter (mass/power ratio is now down to 2 g/W) and much more powerful, particularly in stationary applications.

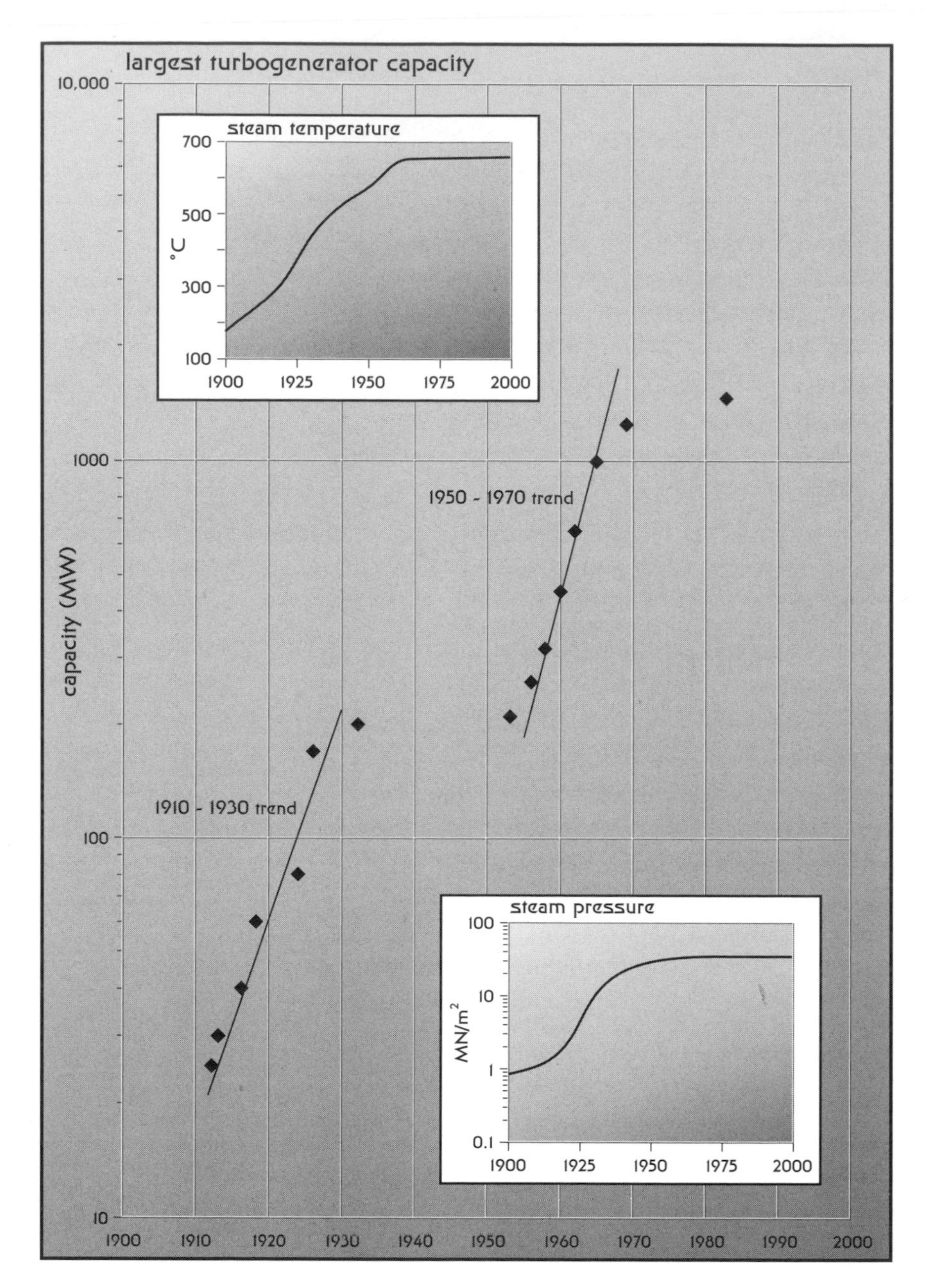

Figure 1.9
Record ratings of the U.S. turbogenerators during the twentieth century. Growth of the highest capacities was interrupted by the economic crisis, WWII, and the postwar recovery; afterward the precrisis growth rate resumed for another two decades. Both the top operating temperatures and the highest pressure used in modern turbogenerators have not increased since the 1960s. Based on data in FPC (1964), various issues of *Power Engineering*, and a figure in Smil (1999a).

But the environmental advantages of better internal engine performance were negated for decades by rising average car power ratings. Ford's celebrated model T, which was sold between 1908 and 1927, rated originally less than 16 kW (21 hp), while even small American cars of the early 1970s had in excess of 50 kW (67 hp). Consequently, the specific fuel consumption of new American passenger cars, which averaged about 14.8 L/100 km (16 mpg) during the early 1930s, kept deteriorating slowly for four decades and it became as high as 17.7 L/100 km (13.4 mpg) by 1973 (EIA 2001a).

This undesirable trend was finally reversed by the OPEC's oil price increases. Between 1973 and 1987 the average fuel demand of new cars on the North American market was about halved as the Corporate Average Fuel Economy (CAFE) standard fell to 8.6 L/100 km (27.5 mpg). Unfortunately, the post-1985 slump in crude oil prices first stopped and then actually reversed this legislative and technical progress. Assorted vans, SUVs and light trucks—with power often in excess of 100 kW and, being exempt from the 27.5 mpg CAFE that applies only to cars, with performance that does not often even reach 20 mpg—have gained more than half of new vehicle market by the late 1990s (Ward's Communications 2000).

Both of the century's new prime movers were adopted so rapidly because of the WWII and the subsequent superpower rivalry. In a remarkable case of a concurrent but entirely independent invention, the first designs of gas turbines took place during the late 1930s when Frank Whittle in England and Hans Pabst von Ohain in Germany built their experimental engines for military planes (Constant 1981). Introduced at the war's very end, jet fighters made no difference to the war's outcome, but their rapid postwar development opened the way for commercial applications as nearly all first passenger jets were modifications of successful military designs. Speed of sound was surpassed on October 14, 1947 with the Bell X-1 plane. The British 106 Comet 1 was the first passenger jet to enter scheduled service in 1952 but structural defects of its fuselage led to its failure. The jet age was ushered in successfully in 1958 by the Boeing 707 and by a redesigned 106 Comet 4.

A decade later came the wide-bodied Boeing 747, the plane that revolutionized transoceanic flight. Pan Am ordered it first in 1966, the prototype plane took off on February 9, 1969, and the first scheduled flight was on January 21, 1970 (Smil 2000b). The first 747s had four Pratt & Whitney turbofan engines, famous JT9D each with a peak thrust of 21,297 kg and with mass/power ratio of 0.2 g/W. Three decades later the Boeing 747-300—the holder of speed and distance (20,044.2 km from Seattle to Kuala Lumpur) records for passenger jets (Boeing 2001)—is powered by twin engines from the same company (PW 4098) whose maximum thrust of 44,452 kg is more than twice as high (Pratt & Whitney 2001). Thrust/weight ratio of these engines is now more than 6, and turbines powering the supersonic military aeroplanes are even better, with thrust/weight ratios as high as 8 (fig. 1.10).

The impact of gas turbines goes far beyond transforming the aerial warfare and worldwide long-distance travel. These prime movers have also found very important stationary applications. They are used to power centrifugal compressors in pumping stations of natural gas pipelines, by many chemical and metallurgical industries, and during the past 15 years they have been increasingly chosen to drive electricity generators (Williams and Larson 1988; Islas 1999). Rising demand for peak electricity generation has lead to steadily higher stationary gas turbine ratings (commonly in excess of 100 MW by the late 1990s) and to efficiency matching the performance of the best steam turbines (fig. 1.10).

The only prime mover that can develop even more power per unit of mass than a gas turbine is the rocket engine. Its large-scale development began only during the WWII with ethanol-powered engines for the infamous German V-1 and V-2 used against England. After a decade of slow development the superpower rocket race started in earnest with the launch of the Earth's first artificial satellite, the Soviet *Sputnik* in 1957. Subsequent advances were driven by the quest for more powerful, but also more accurate, land- and submarine-based intercontinental ballistic missiles. No other prime mover comes close to immense power liberated, necessarily only very briefly, by the largest rocket engines. The U.S. Saturn C5 rocket, which on July 16, 1969 sent Apollo spacecraft on its journey to the Moon, developed about 2.6 GW during its 150-second burn (von Braun and Ordway 1975).

Moon flights were an ephemeral endeavor but satellites launched by relatively inexpensive rockets ushered the age of cheap intercontinental telecommunications, more reliable weather forecasting, and real-time monitoring of extreme weather events that made it possible to issue life-saving warnings. Satellites have also given us unprecedented capacities to monitor the Earth's land use changes, ocean dynamics, and photosynthetic productivity from space (Parkinson 1997; Smil 2002)—and to pinpoint our locations through the global positioning system (Hofmann-Wellenhof et al. 1997).

Discovery of nuclear fission introduced an entirely new form of energy conversion but its rapid commercial adaptation uses the heat released by this novel transformation for a well-tested process of generating steam for electricity generation. The sequence of critical developments was extraordinarily rapid. The first proof of fission

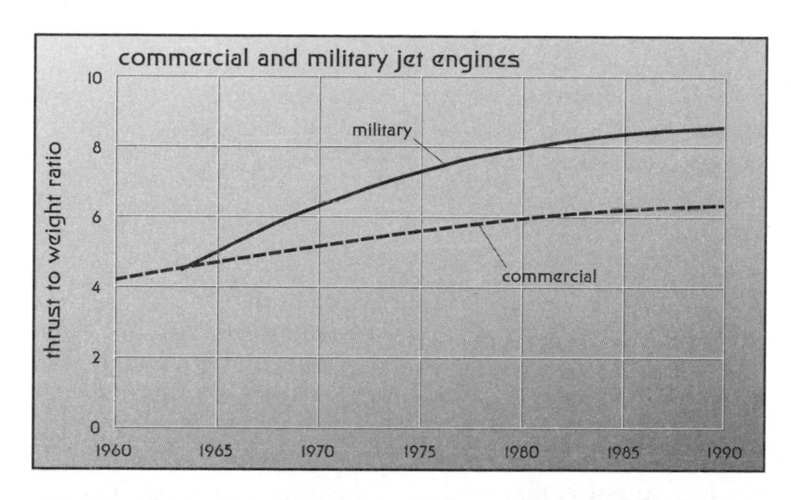

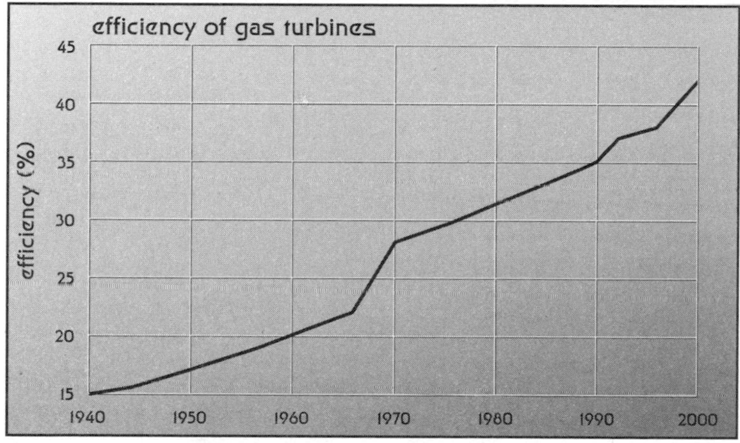

Figure 1.10
Two illustrations of the improving performance of gas turbines. The first graph shows the the increasing thrust ratio of military and commercial jet engines, the other one charts the rising efficiency of stationary gas turbines used for electricity generation. Based on data published in various energy journals.

was published in February 1939 (Meitner and Frisch 1939). The first sustained chain reaction took place at the University of Chicago on December 2, 1942 (Atkins 2000). Hyman Rickover's relentless effort to apply reactor drive to submarines led to the launch of the first nuclear-powered vessel, *Nautilus,* in January 1955 (Rockwell 1991). Rickover was put immediately in charge of, almost literally, beaching the Westinghouse's pressurized water reactor (PWR) used on submarines and building the first U.S. civilian electricity-generating station in Shippingport, Pennsylvania. The station reached initial criticality on December 2, 1957, more than a year after the world's first large-scale nuclear station, British Calder Hall (4x23 MW), was connected to the grid on October 17, 1956 (Atkins 2000; fig. 1.11).

PWR became the dominant choice as this new electricity-generating technique entered the stage of precipitous adoption. Ten years between 1965 and 1975 saw the greatest number of new nuclear power plant orders, and European countries (including the former Soviet Union) eventually ordered about twice as many power reactors

Figure 1.11
Aerial view of Calder Hall on the Cumberland coast, the world's first commercial nuclear electricity-generating station. Photo, taken in May 1962, courtesy of the U.K. Atomic Energy Authority.

as did the United States. As I will detail in the third chapter, the expert consensus of the early 1970s was that by the century's end the world would be shaped by ubiquitous and inexpensive nuclear energy. In retrospect, it is obvious that the commercial development of nuclear generation was far too rushed and that too little weight was given to the public acceptability of commercial fission (Cowan 1990).

Arguments about the economics of fission-produced electricity were always dubious as calculations of generation costs did not take into account either the enormous subsidies sunk by the governments into nuclear R&D (see chapters 2 and 6) or the unknown costs of decommissioning the plants and storing safely the highly radioactive waste for the future millennia. And looking back Weinberg (1994, p. 21) conceded that "had safety been the primary design criterion [rather than compactness and simplicity that guided the design of submarine PWR], I suspect we might have hit upon what we now call inherently safe reactors at the beginning of the first nuclear era. . . ." More fundamentally, promoters of nuclear energy did not take seriously Enrico Fermi's warning (issued even before the end of the WWII at one of the University of Chicago meetings discussing the future of nuclear reactors) that the public may not accept an energy source that generates large amounts of radioactivity as well as fissile materials that might fall into the hands of terrorists (Weinberg 1994).

By the early 1980s a combination of other unexpected factors—declining demand for electricity (see chapter 3 for details), escalating costs in the era of high inflation and slipping construction schedules, and changing safety regulations that had to be accommodated by new designs—helped to turn the fission's prospects from brilliant to dim. Many U.S. nuclear power plants eventually took twice as long to build as originally scheduled, and cost more than twice as much than the initial estimates. Safety concerns and public perceptions of intolerable risks were strengthened by an accident at the Three Mile Island plant in Pennsylvania in 1979 (Denning 1985). By the mid-1980s the shortlived fission era appeared to be over everywhere in the Western world with the exception of France. Accidental core meltdown and the release of radioactivity during the Chernobyl disaster in Ukraine in May 1986 made matter even worse (Hohenemser 1988). Although the Western PWRs with their containment vessels and much tighter operating procedures could have never experienced such a massive release of radiation as did the unshielded Soviet reactor, that accident only reinforced the erroneous but widely shared public perception of all nuclear power being inherently unsafe.

Still, by the century's end nuclear generation was making a substantial contribution to the world's TPES (Beck 1999; IAEA 2001a). By the end of the year 2000 there were 438 nuclear power plants in operation with a total net installed capacity of 351 GW. Fission reactors accounted for about 11% of all installed electricity-generating capacity but because of their high availability factors (global average of about 80% during the late 1990s) they generated about 16% of all electricity (IAEA 2001a). The highest national contributions were in France, where 76% of electricity was generated by PWRs. Lithuania, with its large Soviet-built station in Ingalina, came second with nearly 74% and Belgium third (57%). Japan's share was 33%, the United States' share 20%, Russia's 15%, India's 3%, and China's just over 1% (IAEA 2001a). I will assess the industry's uncertain future in chapter 4.

I would also put photovoltaics (PV), another remarkable nineteenth-century invention, into the category of important new twentieth-century energy conversions. This placement has a logical justification: unlike other conversion techniques that were invented and began to be commercialized before 1900, PV's first practical use took place during the late 1950s. Discovery of the PV phenomenon can be dated precisely to young Edmund Becquerel's 1839 finding that electricity generation in an electrolytic cell made up of two metal electrodes increased when exposed to light (PV Power Resource Site 2001). Little research was done on the PV effect during the subsequent three decades, but the 1873 discovery of the photoconductivity by selenium made it possible for W. G. Adams and R. E. Day to make the first PV cell just four years later. Selenium wafer design was described by Charles Fritts in 1883 but conversion efficiencies of such cells were a mere 1–2%. Einstein's work on the photoelectric effect (Einstein 1905), and not his more famous studies of relativity, earned him a Nobel Prize 16 years later, but had little practical impact on PV development. Nor did Jan Czochralski's fundamental 1918 discovery of how to grow large silicon crystals needed to produce thin semiconductor wafers.

The breakthrough came only in 1954 when a team of Bell Laboratories researchers produced silicon solar cells that were 4.5% efficient, and raised that performance to 6% just a few months later. By March 1958, when Vanguard-I became the first PV-powered satellite (a mere 0.1 W from about 100 cm^2), Hoffman Electronics had cells that were 9% efficient, and began selling 10%-efficient cells just one year later (PV Power Resource Site 2001). In 1962 Telstar, the first commercial telecommunications satellite, had 14 W of PV power, and just two years later Nimbus rated 470 W. PV cells became an indispensable ingredient of the burgeoning satellite industry but land-based applications remained uncommon even after David Carlson and Christopher Wronski at RCA Laboratories fabricated the first amorphous silicon PV cell in 1976. Worldwide PV production surpassed 20 MW of peak capacity (MW_p) in

1983 and 200 MW_p by the year 2000 as solar electricity became one of the fastest growing energy industries (Markvart 2000). Still, the total installed PV capacity was only about 1 GW in 1999, a negligible fraction of more than 2.1 TW available in fossil-fueled generators (EIA 2001c).

The last category of technical, and management, innovations resulting from the diffusion of computers, ubiquitous telecommunications, and common reliance on automatic controls and optimization algorithms has transformed every aspect of energy business, from the search for hydrocarbons to the design of prime movers, and from the allocation of electricity supplies to monitoring of tanker-borne crude oil. An entire book could be devoted to a survey of these diverse innovations that are largely hidden from public view. Its highlights would have to include, among others, a veritable revolution in searching for hydrocarbons, unprecedented accuracy and intensity of monitoring complex dynamic networks, and dematerialized design of prime movers and machines.

Advances in the capabilities of electronic devices used in remote sensing and orders of magnitude higher capacities to store and process field data are behind the revolutionary improvements in the reach and the quality of geophysical prospecting. By the mid-1990s traditional two-dimensional seismic data used in oil exploration were almost completely replaced by three-dimensional images and the latest four-dimensional monitoring (time-lapse three-dimensional) of reservoirs makes it possible to trace and to simulate the actual flow of oil in hydrocarbon-bearing formations and to interpret fluid saturation and pressure changes. This knowledge makes it possible to increase the oil recovery rates from the maxima of 30–35% achievable before 1980 to at least 65% and perhaps even to more than 75% (Morgan 1995; Lamont Doherty Earth Observatory 2001). Global positioning system makes it possible for a company to be instantaneously aware of the exact location of every one of its trucks crisscrossing a continent or every one of its tankers carrying crude oil from the Middle East—and an optimizing algorithm receiving the information about road closures and detours, or about extreme weather events (cyclones, fog) can minimize fuel consumption and time delays by rerouting these carriers.

The Rising Importance of Electricity

There are many reasons to single out electricity for special attention. After millennia of dependence on the three basic energy conversions—burning of fuels, that is fresh or fossilized biomass, use of human and animal muscles, and the capture of indirect solar flows of water and wind—large-scale generation of electricity introduced a new form of energy that has no rival in terms of its convenience and flexibility. No other kind of energy affords such an instant and effortless access. Electricity's advantage, taken utterly for granted by populations that have grown up with its cheap and ubiquitous supply, is evident to anybody who managed a household in the preelectrical era, or who lived in places where expensive electricity was used just for inadequate lighting.

To all those who have never faced daily chores of drawing and hauling water, preparing kindling in morning darkness and cold, washing and wringing clothes by hand, ironing them with heavy wedges of hot metal, grinding feed for animals, milking cows by hand, pitchforking hay up into a loft, or doing scores of other repetitive manual tasks around the house, farmyard, or workshop, it is not easy to convey the liberating power of electricity. I am aware of no better literary attempt to do so than two chapters in an unlikely source, in the first volume of Robert Caro's fascinating biography of Lyndon Johnson (Caro 1982).

Caro's vivid descriptions of the repetitive drudgery, and physical dangers, experienced by a preelectric society are based on recollections of life in the Texas Hill Country during the 1930s. These burdens, falling largely on women, were much greater than the exertions of subsistence farmers in Africa or Latin America because the Hill Country farmers tried to maintain a much higher standard of living and managed much larger farming operations. The word *revolution* is then no exaggeration to describe the day when transmission lines reached the homes of such families.

Electricity's advantages go far beyond instant and effortless access as no other form of energy can rival the flexibility of its final uses. Electricity can be converted to light, heat, motion, and chemical potential and hence it can be used in every principal energy-consuming sector with the exception of commercial flying. Unmanned solar-powered flight is a different matter. AeroVironment's Pathfinder rose to 24 km above the sea level in 1998, and a bigger Helios—a thin, long curved and narrow-flying wing (span of just over 74 m, longer than that of Boeing 747, width of 2.4 m) driven by 14 propellers powered by 1 kW of bifacial solar cells—became the world's highest flying plane in August 2001 as it reached the altitude of almost 29 km (AeroVironment 2001; fig. 1.12).

In addition to its versatility, electricity use is also perfectly clean and silent at the point of consumption and it can be easily adjusted with very high precision to provide desirable speed and accurate control of a particular process (Schurr 1984). And once a requisite wiring is in place it is easy to accommodate higher demand or a

Figure 1.12
The solar-electric *Helios Prototype* flying wing during its record-setting test flight above Hawaiian islands on July 14, 2001. NASA photo ED 01-0209-6 available at <http://www.dfrc.nasa.gov/gallery/photo/HELIOS/HTML/EDO1-0209-6.html>.

greater variety of electricity converters. Finally, electricity can be converted without any losses to useful heat (it can also be used to generate temperatures higher than combustion of any fossil fuel), and it can be turned with very high efficiency (in excess of 90%) into mechanical energy. Among all of its major uses only lighting is still generally less than 20% efficient.

The combination of these desirable attributes brought many profound changes to the twentieth-century use of energy and hence to the functioning of modern economies and the conduct of everyday life. The universal impact of this new form of energy is attested to by the fact that electrification became the embodiment of such disparate political ideals as Lenin's quest for a new state form and Roosevelt's New Deal. Lenin summarized his goal in his famously terse slogan "Communism equals the Soviet power plus electrification." Roosevelt used extensive federal involvement in building dams and electrifying the countryside as a key tool of his New Deal program of economic recovery (Lilienthal 1944).

As with so many other energy-related innovations, the United States pioneered the introduction and mass diffusion of new electric conversions, with Europe and

Japan lagging years to decades behind, and with a large share of today's poor world still undergoing only the earliest stages of these fundamental transformations. The three truly revolutionary shifts—affordable, clean, and flexible lighting, conversion of industrial power from steam to electricity, and the adoption of an increasing variety of household energy converters—proceeded concurrently during the century's early decades, and lighting caused the first large-scale electricity-powered socioeconomic transformation.

Although Edison's incandescent carbon filament lamp, patented in 1879, was about 20 times as efficient as a candle (which converted a mere 0.01% of the burning paraffin into light) it would not have been affordable to mass-produce a device that turned just 0.2% of expensively generated electricity to light. Efficiency comparisons for lighting are done usually in terms of efficacy, the ratio of light (in lumens) and the power used (in W). The earliest light bulbs produced less than 2 lumens/W, and although osmium filaments, introduced in 1898, tripled that rate, they still produced only fairly dim light (no more than a modern 25-W lamp) whose cost was unacceptably high to illuminate properly households or public places. Steady advances during the course of the twentieth century improved the best light efficiencies by an order of magnitude (fig. 1.13; Smithsonian Institution 2001). Light bulb performance was

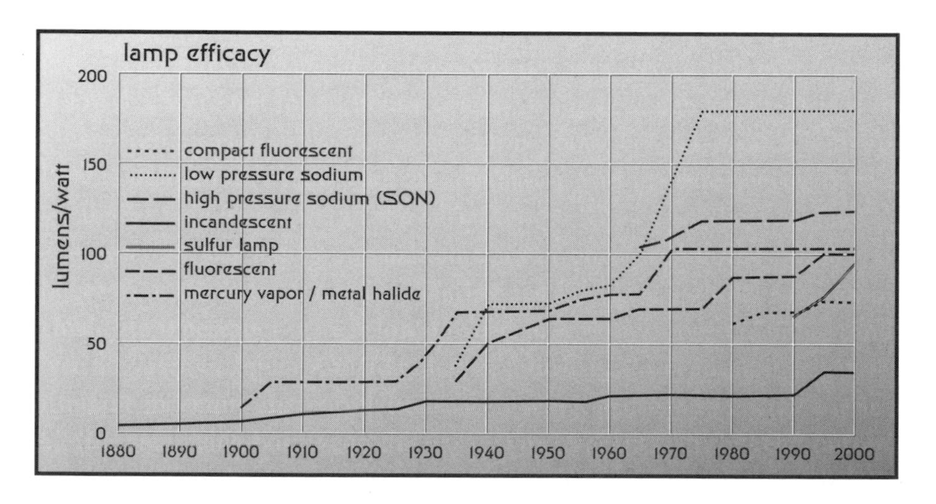

Figure 1.13
Increasing efficacy (lumens/watt) of various kinds of electric lights during the twentieth century. Based on a Smithsonian Institute graph available at <http://americanhistory.si.edu/lighting/chart.htm>.

improved first by squirted tungsten filaments, available after 1905, then by tungsten filaments in vacuum, and by argon-filled lamps with coiled filaments, invented by Irving Langmuir in 1913 (Bowers 1998).

Several breakthroughs came during the 1930s with the introduction of low-pressure sodium lamps (LPS), mercury-vapor lamps (both for the first time in Europe in 1932) and fluorescent lights. LPS, whose stark yellow light dominates street lighting, are the most efficient lights available today. With 1.47 mW/lumen being the mechanical equivalent of light, the efficacy of 175 lumens/W means that LPS lamps convert just over 25% of electric energy into light (fig. 1.13). Mercury-vapor lamps put out initially about 40 lumens/W of blue- and green-tinged white light.

Early fluorescent lights had the same efficacy, and as their efficiency more than doubled and as different types were introduced to resemble more the daylight spectrum they eventually became the norm for institutional illumination and made major inroads in the household market. Today's best fluorescent lights have efficiencies in excess of 100 lumens/W, about equal to metal halide lamps (mercury-vapor lamps with halide compounds) that are now the dominant lighting at sporting events and other mass gatherings that are televised live (fig. 1.13). High-pressure sodium lamps, introduced during the 1960s, produce a more agreeable (golden yellow) light than LPS sources but with about 30% lower efficiency.

For consumers the combination of rising lighting efficacies and falling prices of electricity (see chapter 2) means that a lumen of electric light generated in the United States now costs less than 1/1,000 than it did a century ago. In addition there are obvious, but hard-to-quantify, convenience advantages of electric light compared to candles or whale-oil or kerosene lamps. On a public scale the twentieth century also witnessed spectacular use of light for aims ranging from simple delight to political propaganda. First many American industrialists used concentrated lighting to flood downtowns of large cities with "White Ways" (Nye 1990). Later, Nazis used batteries of floodlights to create immaterial walls to awe the participants at their party rallies of the 1930s (Speer 1970). Now outdoor lighting is a part of advertising and business displays around the world—and in the spring of 2002 two pillars of light were used to evoke the destroyed twin towers of the World Trade Center. The total flux of indoor and outdoor lighting has reached such an intensity that night views of the Earth show that all densely inhabited affluent regions now have more light than darkness, and the only extensive unlighted areas are the polar regions, great deserts, Amazon, and Congo basin—and North Korea (fig. 1.14).

Figure 1.14
Composite satellite image of the Earth at night is a dramatic illustration of electricity's impact on a planetary scale. The image and more information on the Earth at night are available at <http://antwrp.gsfc.nasa.gov/apod/ap001127.html>.

An even more profound, although curiously little appreciated, process was underway as people in industrializing countries were illuminating their homes with better light bulbs: electrification revolutionized manufacturing even more than did the steam engines. This shift was so important not because electric motors were more powerful than steam engines they replaced but because of unprecedented gains in the reliability and localized control of power. These critical gains did not accompany the previous prime-mover shift from waterwheels, or windmills, to steam engines. All of these machines used systems of shafts and toothed wheels and belts to transmit mechanical energy to the point of its final use. This was not a problem with simple one-point uses such as grain milling or water pumping, but it entailed often complex transmission arrangements in order to deliver mechanical power to a multitude of workplaces so it could be used in weaving cloth or machining metals.

Space under factory ceilings had to be filled with mainline shafts that were linked to parallel countershafts in order to transfer the motion by belts to individual machines (fig. 1.15). Accidental outage of the prime mover or a failure anywhere along the chain of transmission shut down the entire arrangement, and even when running flawlessly, such transmission systems lost a great deal of energy to friction: overall mechanical efficiency of belt-driven assemblies was less than 10% (Schurr and Netschert 1960). Continuously running belts were also idling much of the time and made it impossible to control power at individual workplaces. Everything changed only when electric motors dedicated to drive individual machines became the industrial norm. Electrification did away with the overhead clutter (and noise) of transmission shafts and belts, opened up that space for better illumination and ventilation, sharply reduced the risk of accidents, and allowed for flexible floor plans that could

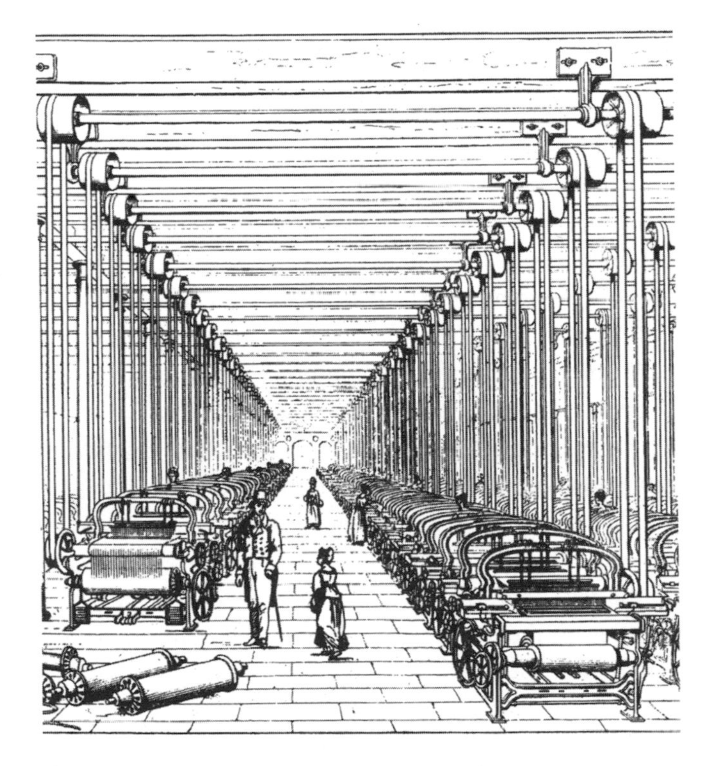

Figure 1.15
Rotating axles and transmission belts were needed to transfer mechanical energy from a central steam engine to individual machines. These cumbersome, dangerous, and inefficient arrangements disappeared with the introduction of electric motors.

easily accommodate new configurations or new machines, and more efficient (at least 70%, often more than 90%) and more reliable power supplies and their accurate control at the unit level raised average labor productivities.

In the United States this great transformation began around 1900 and it took about three decades to complete. At the century's beginning electric motors made up less than 5% of all installed mechanical power in America's industries; by 1929 they added up to over 80% (Devine 1983; Schurr 1984). And the process did not stop with the elimination of steam power as electric motors came to occupy a growing number of new niches to become the most ubiquitous and hence the most indispensable energy converters of modern civilization. In this sense their material analog is steel, an indispensable structural foundation of modern affluence.

The alloy sustains our standard of living in countless ways. A choice list could start with such spectacular applications as supertanker hulls, tension cables suspending graceful bridges, and pressure vessels containing the cores of nuclear reactors. The list could continue with such now mundane machines as semisubmersible oil drilling rigs, electricity-generating turbines or giant metal-stamping presses; and it could close with such hidden uses as large-diameter transcontinental gas pipelines and reinforcing bars in concrete. Steel is indispensable even for traditional materials or for their latest substitutes. All wood and stone are cut and shaped by machines and tools made of steel, all crude oils yielding feedstocks for plastics are extracted, transported, and refined by machines and assemblies made of steel, as are the injection machines and presses moulding countless plastic parts. Not surprisingly, steel output (almost 850 Mt in 2000) is almost 20 times as large as the combined total of five other leading metals, aluminum, copper, zinc, lead, and nickel (IISI 2001).

Ubiquity and indispensability of electric motors is similarly unnoticed. Everything we eat, wear, and use has been made with their help: they mill grain, weave textiles, saw wood, and mould plastics. They are hidden in thousands of different laboratory and medical devices and are being installed every hour by thousands aboard cars, planes, and ships. They turn the fans that distribute the heat from hydrocarbons burned by household furnaces, they lift the increasingly urbanized humanity to high-rise destinations, they move parts and products along assembly lines of factories, whether producing Hondas or Hewlett Packards. And they make it possible to micromachine millions of accurate components for machines ranging from giant turbofan jet engines to endoscopic medical diagnostic devices.

Modern civilization could retain all of its fuels and even have all of its electricity but it could not function without electric motors, new alphas (in baby incubators) and omegas (powering compressors in morgue coolers) of high-tech society. Consequently, it is hardly surprising that electric motors consume just over two-thirds of all electricity produced in the United States, and it is encouraging that they are doing so with increasing efficiencies (Hoshide 1994). In general, their conversion efficiencies increase with rated power; for example, for six-pole open motors full-load efficiencies are 84% at 1.5 hp, 90.2% at 15 hp, and 94.5% at 150 hp. At the same time, it is wasteful to install more powerful motors to perform tasks where they will operate at a fraction of their maximum load. Unfortunately, this has been a common occurrence, with about one-quarter of all U.S. electric motors operating at less than 30% of maximum loads, and only one-quarter working at more than 60% of rated maxima (Hoshide 1994).

The third great electricity-driven transformation, the proliferation of household appliances, has been due, for the most part, to the use of small electric motors but its origins were in simpler heat-producing devices. General Electric began selling its first domestic electrical appliances during the late 1880s but during the 1890s their choice was limited to irons and immersion water heaters and it also included a rather inefficient "rapid cooking apparatus" that took 12 minutes to boil half a liter of water. In 1900 came the first public supply of three-phase current and new electric motor-driven appliances were then introduced in a fairly quick succession. Electric fans were patented in 1902, washing machines went on sale in 1907, vacuum cleaners ("electric suction sweepers") a year later, and first refrigerators in 1912.

Ownership of refrigerators and washing machines is now practically universal throughout the affluent world. A detailed 1997 survey showed that in the United States 99.9% of households had at least one refrigerator and 92% households in single-family houses had a washing machine (EIA 1999a). Ownership of color TVs was also very high: 98.7% of households had at least one set, and 67% had more than two, and the portable vacuum cleaner has metamorphosed in many homes to a central vacuum. Electrical appliances have been also diffusing rapidly in many modernizing countries. In 1999 China's urban households averaged 1.1 color TV sets, 91% of them had a washing machine, and 78% owned a refrigerator (fig. 1.16; NBS 2000).

And there are also many indispensable conversions of electricity where motors are not dominant. Without inexpensive electricity it would be impossible to smelt aluminum, as well as to produce steel in electric arc furnaces. And, of course, there would be neither the ubiquitous feedback controls (from simple thermostats to fly-by-wire wide-bodied jets) nor the omnipresent telecommunications, computers, and

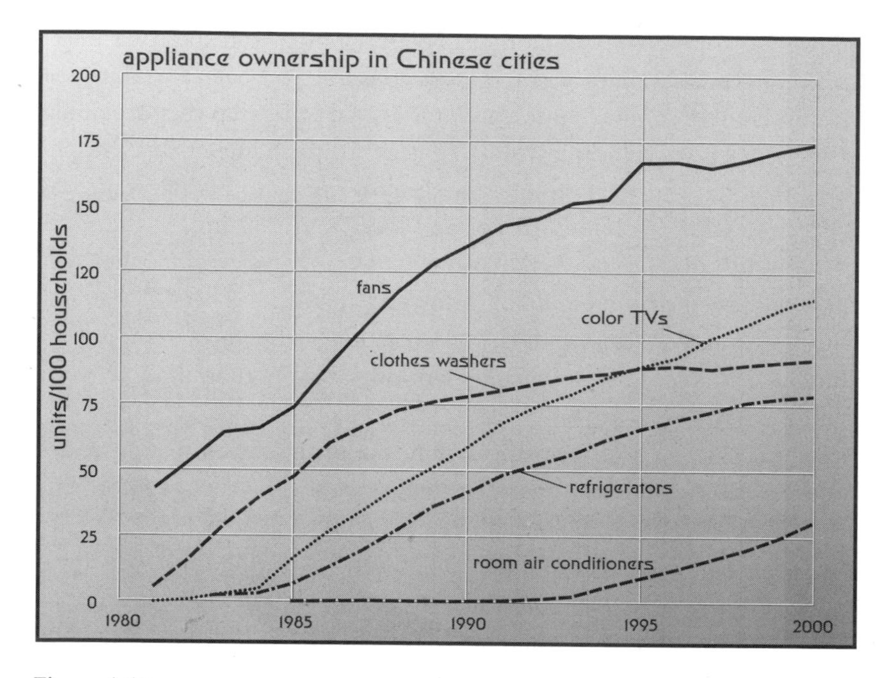

Figure 1.16
Rising ownership of electric appliances in China, 1980–2000. Based on a graph in Fridley (2001) and on data published annually in *China Statistical Yearbook*.

the Internet. Electricity use by the Internet and by a multitude of PC-related devices has been a matter of considerable controversy that began with publications by Mills (1999) and Huber and Mills (1999). They estimated that the Internet-related electricity demand was as much as 8% of the U.S. consumption in 1999. In contrast, Romm (2000) argued that a partial dematerialization of the economy effected by the Internet actually brings net energy savings (Romm 2000), and other analysts found that electricity consumed by computers, their peripheral devices, and the infrastructure of the Internet (servers, routers, repeaters, amplifiers) adds up to a still small but rising share of national demand (Hayes 2001).

A detailed study of the U.S. electricity consumption in computing and other office uses (copiers, faxes, etc.) put the total annual demand at 71 TWh, or about 2% of the nation's total (Koomey et al. 1999). In spite of their large unit capacity (10–20 kW) some 100,000 mainframes used just over 6 TWh compared to more than 14 TWh/year for some 130 million desktop and portable computers whose unit power rates usually only in 100 W. The entire controversy about electricity demand and the Internet—more on this controversy can be found at RMI (1999)—is an excellent example of difficulties in analyzing complex and dynamic systems, yet regardless of what exactly the specific demand may be, the new e-economy will almost certainly increase, rather than reduce, the demand for electricity. I will return to these matters of more efficient energy use in chapter 6.

Inherently high losses of energy during the thermal generation of electricity are the key drawback of the rising dependence on the most flexible form of energy. In 1900 they were astonishingly high: the average U.S. heating rate was 91.25 MJ/kWh, which means that just short of 4% of the available chemical energy in coal got converted to electricity. That efficiency more than tripled by 1925 (13.6%) and then almost doubled by 1950 to 23.9% (Schurr and Netschert 1960). Nationwide means surpassed 30% by 1960 but it has stagnated during the past 40 years, never exceeding 33% (EIA 2001a; fig. 1.17). Only a few best individual stations have efficiencies of 40–42%. Similar averages and peaks prevail elsewhere in the Western world.

The good news is that the average performance of thermal power plants had risen an order of magnitude during the twentieth century. The unwelcome reality is that a typical installation will still lose two-thirds of all chemical energy initially present

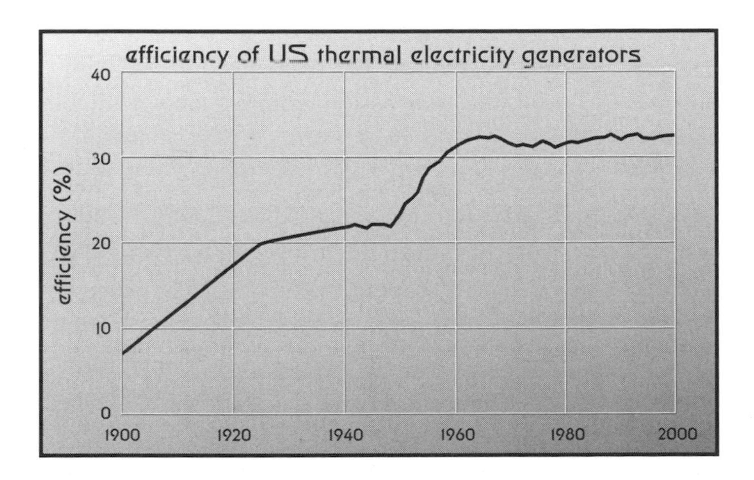

Figure 1.17
By 1960 the average efficiency of the U.S. thermal generation of electricity surpassed 30% and while the best power plants now operate with efficiencies just above 40% the nationwide mean has stagnated for 49 years, clearly an unacceptable waste of resources. Plotted from data in USBC (1975) and EIA (2001a).

in a fossil fuel, or of nuclear energy charged into a reactor in fissile isotopes. This decades-long stagnation of average power plant efficiency is clearly one of the most important performance failures of the modern energy system. For comparison, during the late 1990s, energy wasted annually in U.S. electricity generation surpassed Japan's total energy consumption and it was nearly one-quarter larger than Latin America's total supply of all fossil fuels and primary electricity. I will return to this intolerable inefficiency problem in chapter 4 where I will outline various technical options—some of them already available, others to become commercial soon—to raise this poor average not just above 40% but well over 50%.

But numerous advantages of electricity override the inherent inefficiency of its thermal generation, and the twentieth century saw a relentless rise of the share of the total fossil fuel consumption used to generate electricity. The U.S. share rose from less than 2% in 1900 to just over 10% by 1950 and to 34% in the year 2000 (EIA 2001a). The universal nature of this process is best illustrated by a rapid rate with which China has been catching up. The country converted only about 10% of its coal (at that time virtually the only fossil fuel it used) to electricity in 1950, but by 1980 the share surpassed 20% and by the year 2000 it was about 30%, not far behind the U.S. share (Smil 1976; Fridley 2001). Because of this strong worldwide trend even the global share of fossil fuels converted to electricity is now above 30%, compared to 10% in 1950, and just over 1% in 1900. And, of course, the global supply of electricity has been substantially expanded by hydro generation, whose contribution was negligible in 1900, and by nuclear fission, commercially available since 1956.

Harnessing of hydro energy by larger and more efficient water wheels and, beginning in 1832 with Benoit Fourneyron's invention, by water turbines, was a leading source of mechanical power in early stages of industrialization (Smil 1994a). Two new turbine designs (by Pelton in 1889, and by Kaplan in 1920) and advances in construction of large steel-reinforced concrete dams (pioneered in the Alps, Scandinavia, and the United States before 1900) ensured that water power remained a major source of electricity in the fossil-fueled world. Hydro generation was pushed to a new level before World War II by state-supported projects in the United States and the Soviet Union. Two U.S. projects of that period, Hoover Dam on the Colorado (generating since 1936), and Bonneville on the Columbia, surpassed 1 GW of installed capacity. Giant Grand Coulee on the Columbia (currently 6.18 GW) began generating in 1941 (fig. 1.18), and since 1945 about 150 hydro stations with capacities in excess of 1 GW were put onstream in more than 30 countries (ICOLD 1998).

Figure 1.18
Photograph of the nearly completed Grand Coulee dam taken on June 15, 1941. The station remains the largest U.S. hydro-generating project. U.S. Bureau of Reclamation photograph available at <http://users.owt.com/chubbard/gcdam/highres/build10.jpg>.

Top technical achievements in large dam construction include the height of 335 m of the Rogun dam on the Vakhsh in Tajikistan, reservoir capacity of almost 170 Gm^3 held by the Bratsk dam on the Yenisey, and more than 65 km of embankment dams of the Yacyretâ 3.2 GW project on the Paraná between Paraguay and Argentina (ICOLD 1998). Paraná waters also power the largest hydro project in the Western hemisphere, 12.6 GW Itaipu between Brazil and Paraguay. The world's largest hydro station—the highly controversial Sanxia (Three Gorges) rated at 17.68 GW—is currently under construction across the Chang Jiang in Hubei (Dai 1994). In total about 150 GW of new hydro-generating capacity is scheduled to come online before 2010 (IHA 2000).

Almost every country, with the natural exception of arid subtropics and tiny island nations, generates hydroelectricity. In thirteen countries hydro generation produces virtually all electricity, and its shares in the total national supply are more than 80%

in 32 countries, and more than 50% in 65 countries (IHA 2000). But the six largest producers (Canada, the United States, Brazil, China, Russia, and Norway) account for almost 55% of the global aggregate that, in turn, makes up about 18% of all electricity generation. Combined hydro and nuclear generation, together with minor contributions by wind, geothermal energy, and photovoltaics, now amounts to about 37% of the world's electricity.

Electricity's critical role in modern economies is perhaps best illustrated by comparing the recent differences in intensity trends. As I will explain in some detail in the next chapter, many forecasters were badly mistaken by assuming that a close link between the total primary energy use and GDP growth evident in the U.S. economy after World War II can be used to predict future energy demand. But the two variables became uncoupled after 1970: during the subsequent 30 years the U.S. inflation-adjusted GDP grew by 260% while the primary energy consumption per dollar of GDP (energy intensity of the economy) declined by about 44%. In contrast, the electricity intensity of the U.S. economy rose about 2.4 times between 1950 and 1980, but it has since declined also by about 10%, leaving the late 1990s' intensity almost exactly where it was three decades ago.

Consequently, there has been no decisive uncoupling of economic growth from electricity use in the U.S. case. Is this gentle and modest decline of the electricity intensity of the U.S. economy during the past generation a harbinger of continuing decoupling or just a temporary downturn before a renewed rise of the ratio? Future trends are much clearer for populous countries engaged in rapid economic modernization because during those stages of economic development electricity intensity of an economy tends to rise rapidly.

Trading Energy

Modern mobility of people has been more than matched by the mobility of goods: expanding international trade now accounts for about 15% of the gross world economic product, twice the share in 1900 (Maddison 1995; WTO 2001). Rising trade in higher value-added manufactures (it accounted for 77% of all foreign trade in 1999) makes the multiple much larger in terms of total sales. Total merchandise sales have topped $6 trillion, more than 80 times the 1950 value when expressed in current monies (WTO 2001). Even after adjusting for inflation this would still be about a twelvefold increase.

Global fuel exports were worth almost exactly $400 billion in 1999, equal to just over 7% of the world's merchandise trade. This was nearly 2.6 times the value of all other mining products (about $155 billion) but about 10% behind the international food sales, which added up to $437 billion. Only in the Middle East does the value of exported fuels dominate the total foreign sales, with Saudi Arabia accounting for about 11% of the world's fuel sales. A relatively dispersed pattern of the global fuel trade is illustrated by the fact that the top five fuel exporters (the other four in terms of annual value are Canada, Norway, United Arab Emirates, and Russia) account for less than 30% of total value (WTO 2001).

But the total of annually traded fuels greatly surpasses the aggregate tonnages of the other two extensively traded groups of commodities, metal ores and finished metals, and of food and feed. World iron ore trade totaled about 450 Mt (more than one-third coming from Latin America) and exports of steel reached 280 Mt in 2000, with Japan and Russia each shipping about one-tenth of the total (IISI 2001). Global agricultural trade is dominated by exports of food and feed grains that had grown to about 280 Mt/year by the late 1990s (FAO 2001). In contrast, the tonnage of fuels traded in 2000—just over 500 Mt of coal, about 2 Gt of crude oil and refined products, and only about 95 Mt of natural gas (converting 125 Gm^3 by using average density 0.76 kg/m^3) added up to roughly 2.6 Gt.

Although only about 15% of the global coal production are traded, with some two-thirds of the total sold for steam generation and one-third to produce metallurgical coke, the fuel has surpassed iron ore to become the world's most important seaborne dry-bulk commodity and hence its exports set the freight market trends (WCI 2001). Australia, with more than 150 Mt of coal shipped annually during the late 1990s (roughly split between steam and coking fuel), has become the world's largest exporter, followed by South Africa, the United States, Indonesia, China, and Canada. Japan has been the largest importer of both steam and coking coal (total of over 130 Mt during the late 1990s), followed now by the South Korea, Taiwan, and, in a shift unforeseeable a generation ago, by two former coal-mining superpowers, Germany and the United Kingdom, which import cheaper foreign coal mainly for electricity generation.

Crude oil leads the global commodity trade both in terms of mass (close to 1.6 Gt/year during the late 1990s) and value (just over $200 billion in 1999). Almost 60% of the world's crude oil extraction is now exported from about 45 producing countries and more than 130 countries import crude oil and refined oil products.

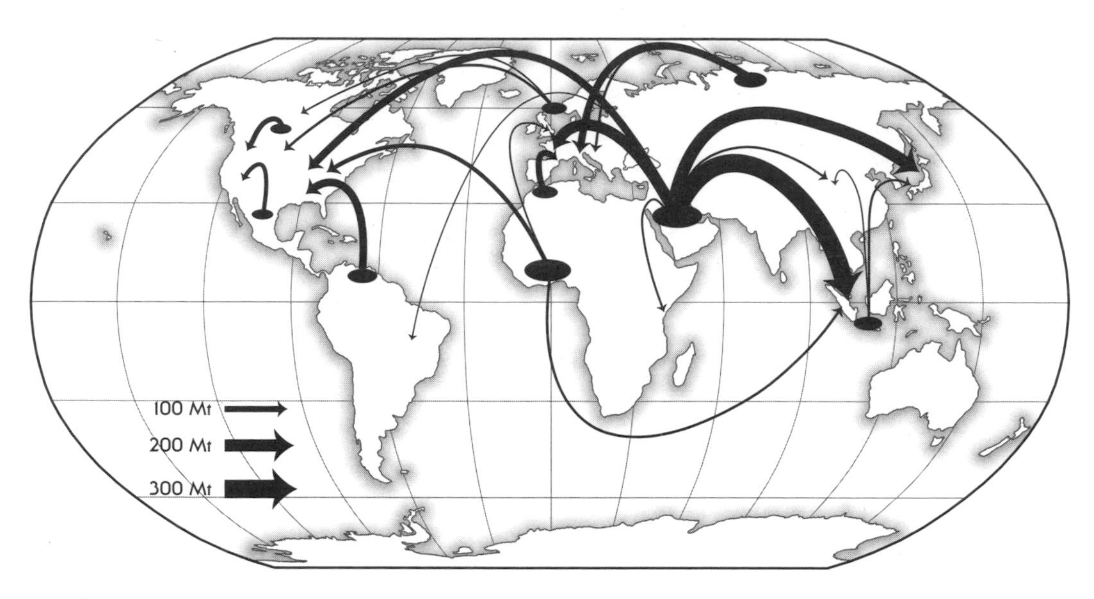

Figure 1.19
Crude oil exports are dominated by flows from the Middle East. Venezuela, Western Siberia, Nigeria, Indonesia, Canada, Mexico, and the North Sea are the other major sources of exports. Based on a figure in BP (2001).

Global dominance of the Middle Eastern exports is obvious (fig. 1.19). The six largest exporters (Saudi Arabia, Iran, Russia, Norway, Kuwait, and the UAE) sell just over 50% of the traded total, and six largest importers (the United States, Japan, Germany, South Korea, Italy, and France) buy 70% of all shipments (BP 2001; UNO 2001). Rapidly rising post-WWII demand for crude oil in Europe and Japan stimulated the development of larger oil tankers. After going up from just over 2,000 dead-weight tons (dwt) to over 20,000 dwt between the early 1880s and the early 1920s, capacities of largest tankers stagnated for over a generation. They took off again only after WWII when the size of largest tankers began doubling in less than 10 years and it reached a plateau in excess of 500,000 dwt by the early 1980s (Ratcliffe 1985; Smil 1994a).

As already noted earlier in this chapter, pipelines are superior to any other form of land transportation. Their compactness (1-m-diameter line can carry 50 Mt of crude oil a year), reliability and safety (and hence the minimal environmental impact) also translate into relatively low cost of operation: only large riverboats and ocean tankers are cheaper carriers of energy. The United States had long-distance pipelines for domestic distribution of crude oil since the 1870s but the construction of pipelines for the export of oil and gas began only after WWII. Large American lines from the Gulf to the East Coast were eclipsed by the world's longest crude oil pipelines laid during the 1970s to move the Western Siberian crude oil to Europe. The Ust'–Balik–Kurgan–Almetievsk line, 2,120 km long and with a diameter of 120 cm, can carry annually up to 90 Mt of crude oil from a supergiant Samotlor oilfield to European Russia and then almost 2,500 km of branching large-diameter lines are needed to move this oil to Western European markets.

Natural gas is not as easily transported as crude oil (Poten and Partners 1993; OECD 1994). Pumping gas through a pipeline takes about three times as much energy as pumping crude oil and undersea links are practical only where the distance is relatively short and the sea is not too deep. Both conditions apply in the case of the North Sea gas (distributed to Scotland and to the European continent) and Algerian gas brought across the Sicilian Channel and the Messina Strait to Italy. Transoceanic movements would be utterly uneconomical without resorting first to expensive liquefaction. This process, introduced commercially during the 1960s, entails cooling the gas to −162 °C and then volatilizing the liquefied natural gas (LNG) at the destination.

Just over 20% of the world's natural gas production was exported during the late 1990s, about three-quarters of it through pipelines, the rest of it as LNG. Russia, Canada, Norway, the Netherlands, and Algeria are the largest exporters of piped gas, accounting for just over 90% of the world total. Shipments from Indonesia, Algeria, and Malaysia dominate the LNG trade. The longest (6,500 km), and widest (up to 142 cm in diameter) natural gas pipelines carry the fuel from the supergiant fields of Medvezh'ye, Urengoy, Yamburg, and Zapolyarnyi in the Nadym–Pur–Taz gas production complex in the northernmost Western Siberia (fig. 1.20) to European Russia and then all the way to Western Europe, with the southern branch going to northern Italy and the northern link to Germany and France.

The largest importers of piped gas are the United States (from Canadian fields in Alberta and British Columbia), Germany (from Siberian Russia, the Netherlands, and Norway), and Italy (mostly from Algeria and Russia). Japan buys more than half of the world's LNG, mainly from Indonesia and Malaysia. Other major LNG importers are South Korea and Taiwan (both from Indonesia and Malaysia), France and Spain (from Algeria). The U.S. imports used to come mostly from Algeria and Trinidad but recent spot sales bring them in from other suppliers.

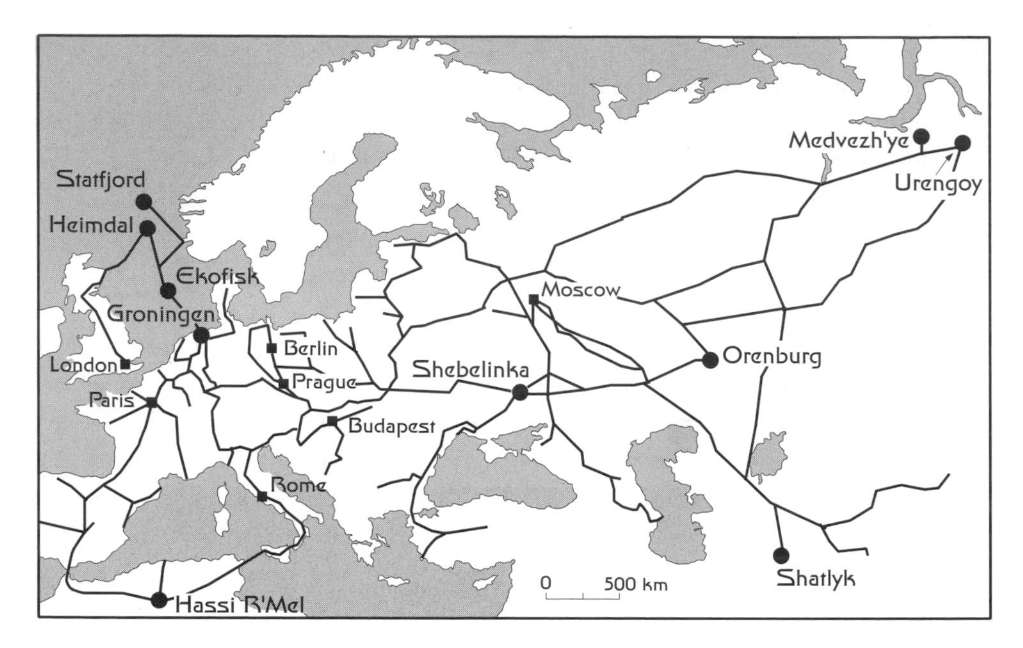

Figure 1.20
The world's longest natural gas pipelines carry the fuel from giant fields in Western Siberia all the way to Western Europe over the distance of more than 4,000 km. Reproduced from Smil (1999a).

In comparison to large-scale flows of fossil fuels the international trade in electricity is significant in only a limited number of one-way sales or multinational exchanges. The most notable one-way transmission schemes are those connecting large hydrogenerating stations with distant load centers. Canada is the world's leader in these exports: during the late 1990s it transmitted annually about 12% of its hydroelectricity from the British Columbia to the Pacific Northwest, from Manitoba to Minnesota, the Dakotas, and Nebraska, and from Québec to New York and the New England states. Other notable international sales of hydroelectricity take place between Venezuela and Brazil, Paraguay, and Brazil, and Mozambique and South Africa. Most of the European countries participate in complex trade in electricity that takes advantage of seasonally high hydro-generating capacities in Scandinavian and Alpine nations as well as of the different timing of daily peak demands.

Combination of continuing abandonment of expensive coal extraction in many old mining regions, stagnation and decline of crude oil and natural gas production in many long-exploited hydrocarbon reservoirs, and the rising demand for cleaner fuels to energize growing cities and industries means that the large-scale trade in fossil fuels and electricity that has contributed so significantly to the transformation of energy use in the twentieth century is yet another trend in the evolution of global energy system that is bound to continue during the twenty-first century.

Consumption Trends

As noted at the beginning of this chapter, the twentieth century saw large increases of not only of aggregate but also of average per capita energy consumption, and these gains look even more impressive when the comparison is done, as it should be, in terms of actual useful energy services. Although the proverbial rising tide (in this case of total energy consumption) did indeed lift all boats (every country now has a higher average per capita TPES than it did a century ago), the most important fact resulting from long-term comparisons of national energy use is the persistence of a large energy gap between the affluent nations and industrializing countries.

High-energy civilization exemplified by jet travel and the Internet is now truly global but individual and group access to its benefits remains highly uneven. Although the huge international disparities in the use of commercial energy had narrowed considerably since the 1960s, an order-of-magnitude difference in per capita consumption of fuels still separates most poor countries from affluent nations, and the gap in the use of electricity remains even wider. There are also large disparities among different socioeconomic groups within both affluent and low-income nations.

At the beginning of the twentieth century, industrializing countries of Europe and North America consumed about 98% of the world's commercial energy. At that time most of the world's inhabitants were subsistence farmers in Asia, Africa, and Latin America and they did not use directly any modern energies. In contrast, the United States per capita consumption of fossil fuels and hydro electricity was already in excess of 100 GJ/year (Schurr and Netschert 1960). This was actually higher than were most of the national European means two or three generations later, but because of much lower conversions efficiencies delivered energy services were a fraction of today's supply. Very little had changed during the first half of the twentieth century: by 1950 industrialized countries still consumed about 93% of the world's commercial energy (UNO 1976). Subsequent economic development in Asia and Latin America finally began reducing this share, but by the century's end affluent countries, containing just one-fifth of the global population, claimed no less than about 70% of all primary energy.

Highly skewed distribution of the TPES is shown even more starkly by the following comparisons. The United States alone, with less than 5% of humanity, consumed about 27% of the world's TPES in 2000, and the seven largest economies of the rich world (commonly known as G7: the United States, Japan, Germany, France, the United Kingdom, Italy, and Canada) whose population adds up to just about one-tenth of the world's total claimed about 45% of the global TPES (BP 2001; fig. 1.21). In contrast, the poorest quarter of mankind—some 15 sub-Saharan African countries, Nepal, Bangladesh, the nations of Indochina, and most of rural India—consumed a mere 2.5% of the global TPES. Moreover, the poorest people in the poorest countries—adding up globally to several hundred million adults and children including subsistence farmers, landless rural workers, and destitute and homeless people in expanding megacities—still do not consume directly any commercial fuels or electricity at all.

National averages for the late 1990s show that annual consumption rates of commercial energy ranged from less than 0.5 GJ/capita, or below 20 kgoe, in the poorest countries of sub-Saharan Africa (Chad, Niger) to more than 300 GJ/capita, or in excess of 7 toe, in the US and Canada (BP 2001; EIA 2001a). Global mean was just over 1.4 toe (or about 60 GJ/capita)—but the previously noted huge and persistent consumption disparities result in the distribution of average national rates that is closest to the hyperbolic pattern rather than to a bimodal or normal curve. This means that the mode, amounting to one-third of the world's countries, is in the lowest consumption category (less than 10 GJ/capita) and that there is little variation in the low frequency for all rates above 30 GJ/capita (fig. 1.21). The global mean consumption rate is actually one of the rarest values with only three countries, Argentina, Croatia, and Portugal, having national averages close to 60 GJ/capita.

Continental averages for the late 1990s were as follows (all in GJ/capita): Africa below 15; Asia about 30; South America close to 35; Europe 150; Oceania 160; and North and Central America 220. Affluent countries outside North America averaged almost 150 GJ/capita (close to 3.5 toe), while the average for low-income economies

Figure 1.21 ▶
Two ways to illustrate the highly skewed global distribution of commercial energy consumption at the end of the twentieth century. The first one is a Lorenz curve plotting the national shares of total energy use and showing the disproportionately large claim on the world's energy resources made by the United States and the G7 countries. The second is the frequency plot of per capita energy consumption displaying a hyperbolic shape. Plotted from data in UNO (2001) and BP (2001).

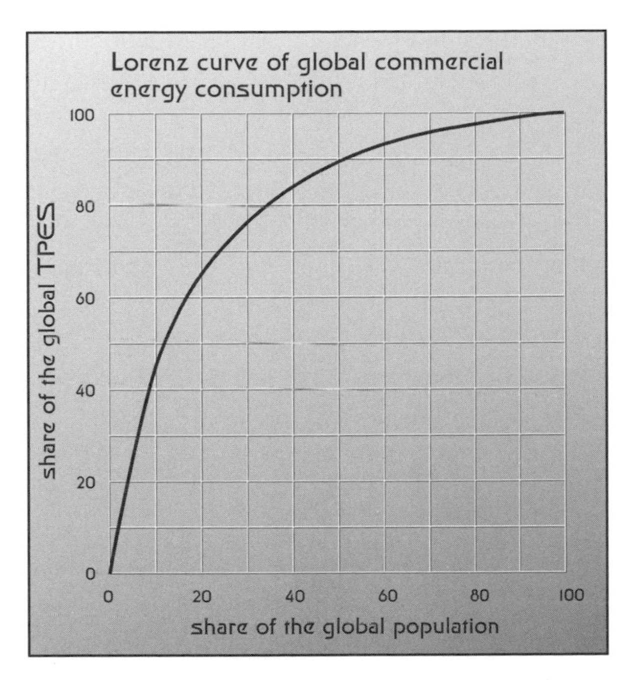
Lorenz curve of global commercial energy consumption
share of the global TPES
share of the global population
0
20
40
60
80
100

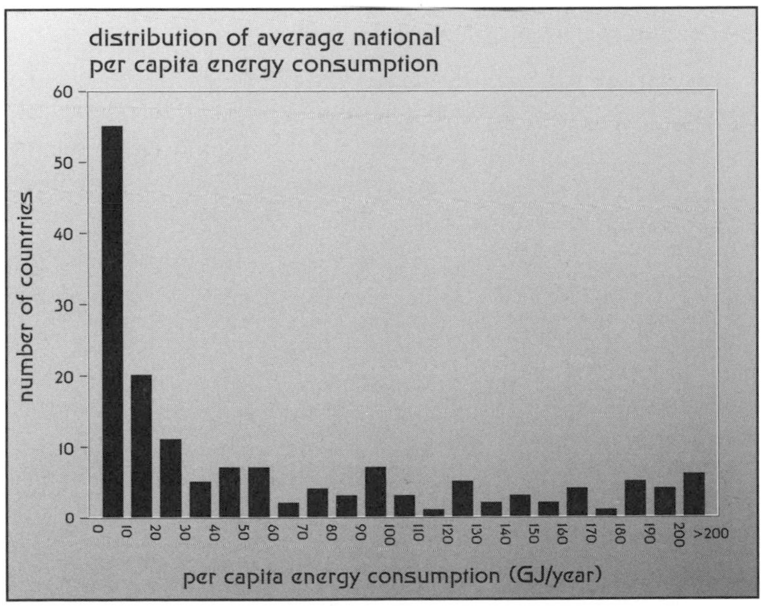
distribution of average national per capita energy consumption
number of countries
per capita energy consumption (GJ/year)
0
10
20
30
40
50
60
70
80
90
100
110
120
130
140
150
160
170
180
190
200
>200

was just 25 GJ/capita (0.6 toe). Leaving the small, oil-rich Middle Eastern states (Kuwait, United Arab Emirates) aside, the highest per capita gains during the second half of the twentieth century were made, in spite of some very rapid population growth rates, in Asia (UNO 1976; UNO 2001). The most notable individual examples are those of Japan (an almost elevenfold gain since 1950) and South Korea (about a 110-fold increase). More than a score of sub-Saharan countries at the opposite end of the consumption spectrum had either the world's lowest improvements or even declines in average per capita use of fuels and electricity.

Formerly large intranational disparities have been greatly reduced in all affluent countries, but appreciable socioeconomic differences remain even in the richest societies. For example, during the late 1990s the U.S. households earning more than U.S. $50,000 (1997)/year consumed 65% more energy than those with annual incomes below U.S. $10,000 (1997) did (U.S. Census Bureau 2002). Large regional difference in household consumption are primarily the function of climate: during the late 1990s the average family in the cold U.S. Midwest consumed almost 80% more energy than those in the warmer Western region (EIA 2001a).

Analogical differences are even larger in low-income economies. During the same period China's annual national consumption mean was about 30 GJ/capita but the rates in coal-rich Shanxi and in the capital Shanghai, the country's richest city of some 15 million people, were nearly 3 times as high and the TPES of the capital's 13 million people averaged about 2.5 times the national mean (Fridley 2001). In contrast, the mean for more than 60 million people in Anhui province, Shanghai's northern neighbor, was only about 20 GJ/capita and for more than 45 million people in landlocked and impoverished Guangxi it was as low as 16 GJ/capita (fig. 1.22). And the differences were even wider for per capita electricity consumption, with the annual national mean of about 0.9 MWh/capita and the respective extremes 3.4 times higher in the country's most dynamic megacity (Shanghai) and 50% lower in its southernmost island province (Hainan).

Household surveys also show that during the late 1990s urban families in China's four richest coastal provinces spent about 2.5 times as much on energy as did their counterparts in four interior provinces in the Northwest [National Bureau of Statistics (NBS) 2000]. Similar, or even larger, differences in per capita energy consumption and expenditures emerge when comparing India's relatively modernized Punjab with impoverished Orissa, Mexico's *maquilladora*-rich Tamaulipas with conflict-riven peasant Chiapas, or Brazil's prosperous Rio Grande do Sul with arid and historically famine-prone Ceará.

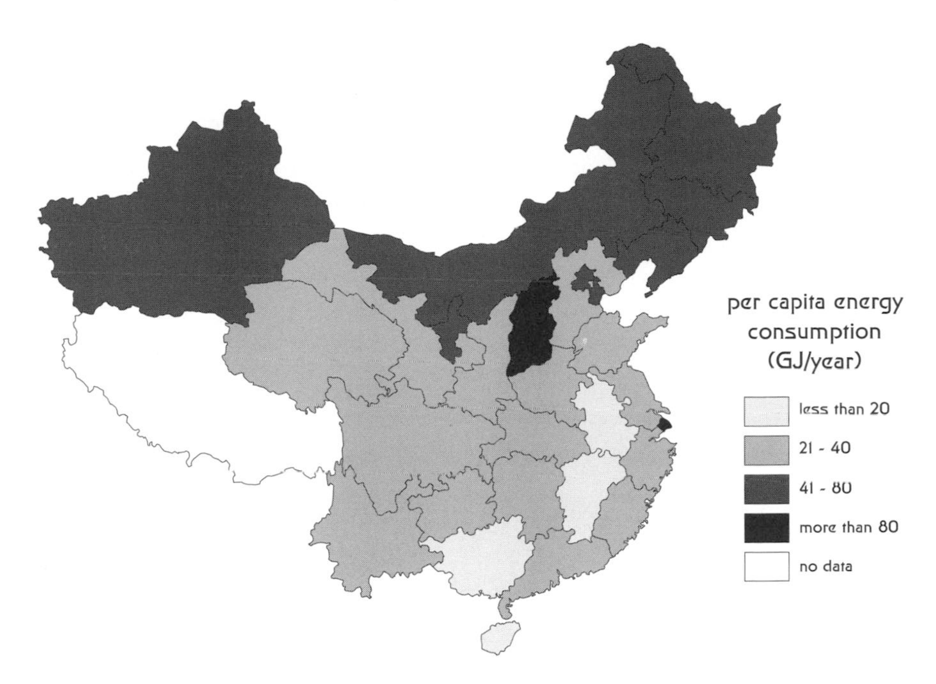

Figure 1.22
In China provincial averages of per capita energy consumption span more than a sevenfold range, from just over 10 GJ/year in Hainan in the tropical south to more than 80 GJ/year in the coal-rich Shanxi in the north. Nationwide annual mean is about 30 GJ/capita. Plotted from 1996 data in Fridley (2001).

Finally, I have to address the changing pattern of final energy uses. Structural transformation of modern economies has brought several major shifts in the sectoral demand for commercial energy. Although universal in nature, these changes have proceeded at a highly country-specific pace. Their most prominent features are the initial rise, and later decline, of the energy share used in industrial production; gradual rise of energy demand by the service sector; steady growth of energy used directly by households, first for essential needs, later for a widening array of discretionary uses; and, a trend closely connected to rising affluence and higher disposable income, an increasing share of energy use claimed by transportation. And although agriculture uses only a small share of the TPES, its overall energy claims, dominated by energies embodied in nitrogenous fertilizers and in field machinery, had grown enormously during the twentieth century, and high energy use in farming now underpins the very existence of modern civilization.

In affluent nations, agriculture, the dominant economic activity of all preindustrial societies, consumes only a few percent of the TPES, ranking far behind industry, households, transportation, and commerce. Agriculture's share in final energy consumption rises when the total amount of fuels and electricity used directly by field, irrigation, and processing machinery is enlarged by indirect energy inputs used to produce machinery and agricultural chemicals, above all to synthesize nitrogen fertilizers (Stout 1990; Fluck 1992). National studies show that in affluent countries during the last quarter of the twentieth century the share of total energy use claimed by agriculture was as low as 3% (in the United States) and as high as 11% in the Netherlands (Smil 1992a). In contrast, direct and indirect agricultural energy uses claimed about 15% of China's TPES, making it one of the major final energy uses in those countries. This is understandable given the fact that the country is now the world's largest producer of nitrogen fertilizers (one-fourth of the world's output) and that it irrigates nearly half of its arable land (Smil 2001; FAO 2001).

The global share of energy used in agriculture is less than 5% of all primary inputs, but this relatively small input is a large multiple of energies used in farming a century ago and it is of immense existential importance as it has transformed virtually all agricultural practices and boosted typical productivities in all but the poorest sub-Saharan countries. In 1900 the aggregate power of the world's farm machinery added up to less than 10 MW, and nitrogen applied in inorganic fertilizers (mainly in Chilean $NaNO_3$) amounted to just 360,000 t. In the year 2000 total capacity of tractors and harvesters was about 500 GW, Haber–Bosch synthesis of ammonia fixed almost 85 Mt of fertilizer nitrogen, fuels and electricity were used to extract, process and synthesize more than 14 Mt P in phosphate fertilizers and 11 Mt K in potash, pumped irrigation served more than 100 Mha of farmland, and cropping was also highly dependent on energy-intensive pesticides (FAO 2001).

I calculated that these inputs required at least 15 EJ in the year 2000 (about half of it for fertilizers), or roughly 10 GJ/ha of cropland. Between 1900 and 2000 the world's cultivated area grew by one-third—but higher yields raised the harvest of edible crops nearly sixfold, a result of more than a fourfold rise of average productivity made possible by roughly a 150-fold increase of fossil fuels and electricity used in global cropping (fig. 1.23). Global harvests now support, on the average, four people per hectare of cropland, compared to about 1.5 persons in 1900. Best performances are much higher: 20 people/ha in the Netherlands, 17 in China's most populous provinces, 12 in the United States on a rich diet and with enough surplus for large-scale food exports (Smil 2000c).

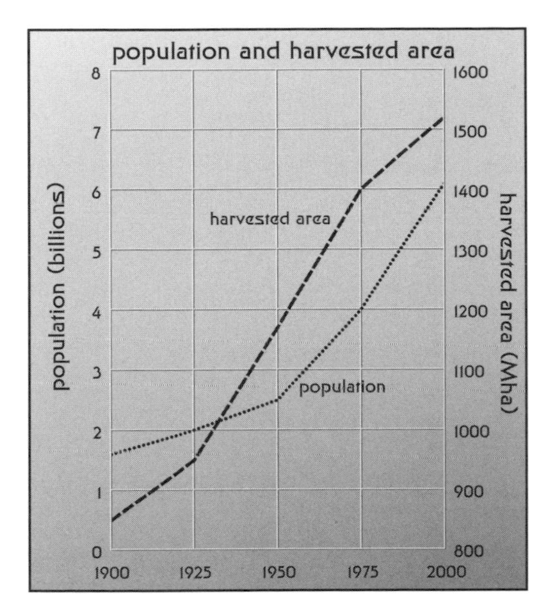

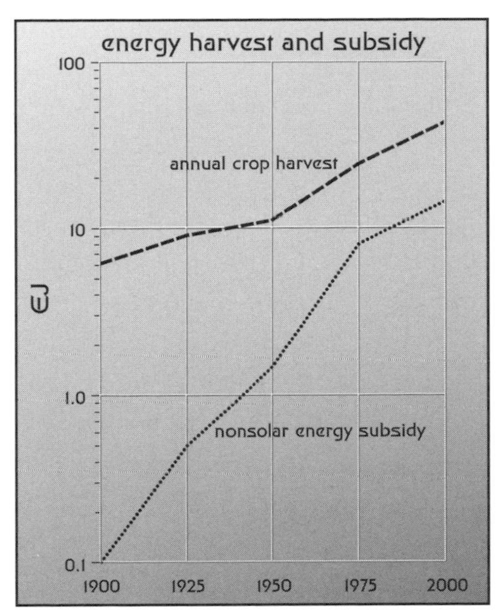

Figure 1.23
One hundred years of agricultural advances are summed up by the trends in harvested area and energy contents of global harvests and nonsolar energy subsidies. Based on Smil (1994a) and unpublished calculations.

In 1900 global crop harvest prorated to just 10 MJ/capita a day, providing, on the average, only a slim safety margin above the minimum daily food needs, and greatly limiting the extent of animal feeding. Recent global harvests have averaged 20 MJ/capita, enough to use a significant part of this biomass (more than 40% globally, up to 70% in the richest countries) for feed. As a result, all affluent nations have surfeit of food (average daily per capita availability in excess of 3,000 kcal) and their diets are extraordinarily rich in animal proteins and lipids. Availability of animal foods is much lower in low-income countries but, on the average and with the exception of chronically war-torn countries, the overall food supply would be sufficient to supply basically adequate diets in highly egalitarian societies (Smil 2000c).

Unfortunately, unequal access to food is common and hence the latest FAO estimate is that between 1996 and 1998 there were 826 million undernourished people, or about 14% of the world's population at that time (FAO 2000). As expected, the total is highly unevenly split, with 34 million undernourished people in the high-income economies and 792 million people in the poor world. The highest shares of undernourished population (about 70% of the total) are now in Afghanistan and

Somalia, and the highest totals of malnourished, stunted, and hungry people are in India and China where dietary deficits affect, respectively, about 20% (total of some 200 million) and just above 10% (nearly 130 million) of all people.

In early stages of economic modernization primary (extractive) and secondary (processing and manufacturing) industries commonly claim more than a half of a nation's energy supply. Gradually, higher energy efficiencies of mineral extraction and less energy-intensive industrial processes greatly reduce, or even eliminate, the growth of energy demand in key industries. As already noted, these improvements have been particularly impressive in ferrous metallurgy and in chemical syntheses. Synthesis of ammonia (the world's most important chemical in terms of synthesized moles; in terms of total synthesized mass ammonia shares the primary position with sulfuric acid) based on the hydrogenation of coal required more than 100 GJ/t when it was commercially introduced in 1913 by the BASF. In contrast, today's state-of-the-art Kellogg Brown & Root or Haldor Topsøe plants using natural gas both as their feedstock and the source of energy need as little as 26 GJ/t NH_3 (Smil 2001; fig. 1.24).

Increasing importance of commercial, household, and transportation uses in maturing economies can be seen in secular trends in those few cases where requisite national statistics are available, or by making international comparisons of countries at different stages of modernization. In the United States the share of industrial energy use declined from 47% in 1950 to 39% in 2000 (EIA 2001a), while in Japan a rise to the peak of 67% in 1970 was followed by a decline to just below 50% by 1995 (IEE 2000). In contrast, industrial production in rapidly modernizing China continues to dominate the country's energy demand: it has been using 65–69% of primary energy ever since the beginning of economic reforms in the early 1980s (Fridley 2001).

Rising share of energy use by households—a trend attributable largely to remarkable declines in average energy prices (see the next chapter for examples of secular trends)—is an excellent indicator of growing affluence. U.S. households now use on-site about 20% of the TPES, compared to 15% in Japan and to only just over 10% in China. Moreover, there has been an important recent shift within this rising demand as nonessential, indeed outright frivolous, uses of energy by households are on the rise. For most of North America's middle-class families these luxury uses began only after World War II, in Europe and Japan only during the 1960s. These trends slowed down, or were temporarily arrested, after 1973, but during the 1990s

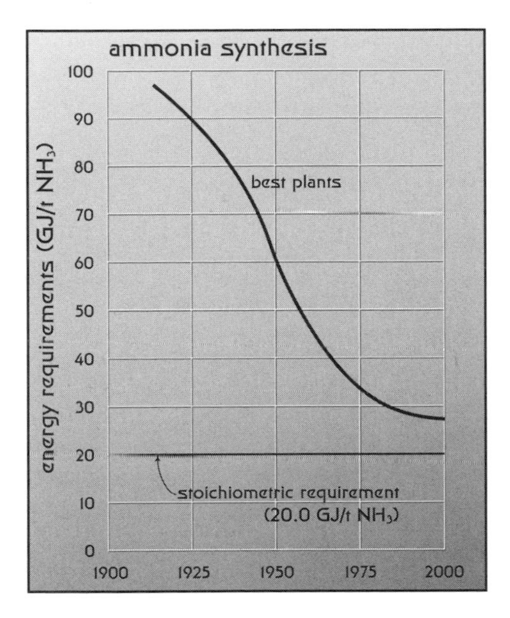

Figure 1.24
Declining energy intensity of ammonia synthesis using the Haber–Bosch process first commercialized in 1913, gradually improved afterward, and made much more efficient by the introduction of single-train plants using centrifugal compressors during the 1960s. From Smil (2001).

they were once again in full flow, energizing the increasingly common displays of ostentatious overconsumption.

Comparisons of electricity use illustrate well this transformation. In 1900 installed capacity of electricity converters in a typical urban U.S. household was limited to a few low-power light bulbs adding up to less than 500 W. Fifty years later at least a dozen lights, a refrigerator, a small electric range with an oven, a washing machine, a television, and a radio in a middle-class house added up to about 5 kW. In contrast, in the year 2000 an all-electric, air-conditioned exurban (i.e., more than 50 km from a downtown) house with some 400 m^2 of living area and with more than 80 switches and outlets ready to power every imaginable household appliance (from a large-capacity freezer to an electric fireplace) can draw upward of 30 kW.

But much more power commanded by that affluent American household is installed in the family's vehicles. Every one of its three cars or SUVs will rate in excess of 100 kW, and a boat or a recreation vehicles (or both, with some of the latter ones equalling the size of a small house), will boost the total power under the household's control close to half of 1 MW! This total is also being enlarged by a proliferation of outdoor energy converters, ranging from noisy gasoline-fueled leaf blowers to massive natural gas-fired pool heaters. Equivalent power—though nothing like the convenience, versatility, flexibility, and reliability of delivered energy services—would have been available only to a Roman *latifundia* owner of about 6,000 strong slaves, or to a nineteenth-century landlord employing 3,000 workers and 400 big draft horses. A detailed survey of the U.S. residential energy use shows that in 1997 about half of all on-site consumption was for heating, and just over one-fifth for powering the appliances (EIA 1999a). But as almost half of all transportation energy was used by private cars the U.S. households purchased about one-third of the country's TPES.

Energy use in transportation amounted to by far the largest sectoral gain and most of it is obviously attributable to private cars. In 1999 the worldwide total of passenger cars surpassed 500 million, compared to less than 50,000 vehicles in 1900, and the grand total of passenger and commercial vehicles (trucks and buses) reached nearly 700 million (Ward's Communications 2000). U.S. dominance of the automotive era had extended almost across the entire century. In 1900 the country had only 8,000 registered vehicles but 20 years later the total was approaching 10 million; in 1951 it surpassed 50 million (USBC 1975). By the century's end it reached 215 million, or 30% of the world total, but the European total was slightly ahead (fig. 1.25).

Passenger travel now accounts for more than 20% of the TPES in many affluent countries, compared to just around 5% in low-income countries. Although the U.S. ownership of passenger cars (2.1 persons per vehicle in 2000) is not that much higher than in Japan (2.4), it is the same as in Italy and is actually lower than in Germany (2.0), the United States remains the paragon of car culture. This is because the mean distance driven annually per American vehicle is considerably longer than in other countries and, incredibly, it is still increasing: the 1990s saw a 16% gain to an average of about 19,000 km/vehicle (EIA 2001a). Average power of U.S. cars is also higher and the annual gasoline consumption per vehicle (about 2,400 L in 2000 compared to 580 L in 1936, the first year for which the rate can be calculated) is commonly 2 to 4 times as high as in other affluent nations. As a result, the country uses a highly disproportionate share of the world's automotive fuel consumption. In 1999 energy content of its liquid transportation fuels (almost 650 Mtoe) was 25%

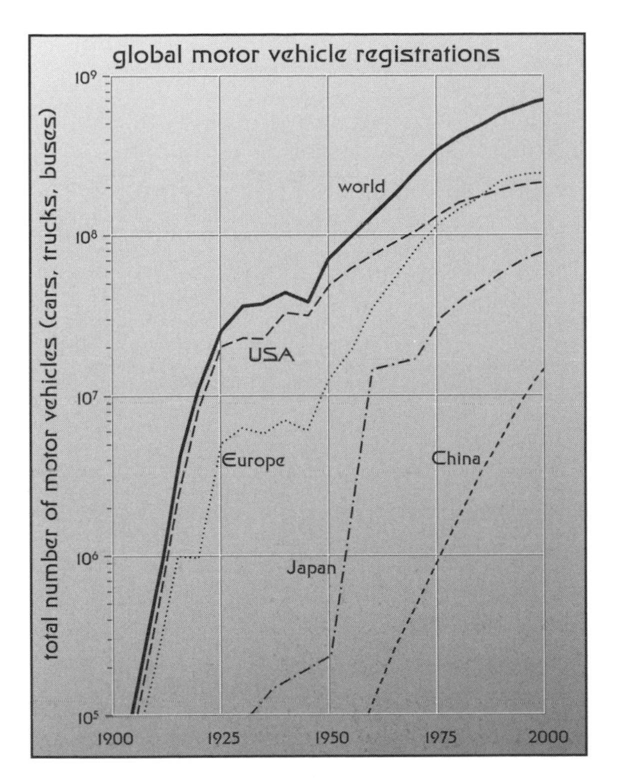

Figure 1.25
Global, United States, European, and Japanese vehicle fleets, 1900–2000. European totals of passenger cars, trucks, and buses surpassed the U.S. vehicle registrations during the late 1980s. Based on a figure in Smil (1999a) with additional data from Ward's Communications (2000).

higher than Japan's *total* primary energy consumption, and it amounted to more than 7% of the global TPES (EIA 2001a; BP 2001).

In contrast to the automobile traffic that has shown signs of saturation in many affluent countries during the 1990s, air travel continued to grow rapidly during the twenty-first century's last decades. Passenger-kilometers flown globally by scheduled airlines multiplied about 75 times between 1950 and 2000 (ICAO 2000)—but in the United States a combination of better engine and airplane design nearly doubled the average amount of seat-kilometers per liter of jet fuel between 1970–1990 (Greene 1992). As with so many other forecasts, any predictions of long-term growth rates of the global aviation will depend on September 11, 2001 being either a tragic singularity or the first in a series of horrific terrorist attacks.

Looking Back and Looking Ahead

Beginnings of new centuries, and in this case also the start of a new millennium, offer irresistible opportunities to look back at the accomplishments, and failures, of the past 100 years and to speculate about the pace and form of coming changes. The third chapter of this book is an extended argument against any long-range particular quantitative forecasting, and even a perfect understanding of past developments is an insufficient guide for such tasks. At the same time, recurrent patterns and general trends transcending particular eras cannot be ignored when outlining the most likely grand trends and constructing desirable normative scenarios. Many energy lessons of the twentieth century are thus worth remembering.

Slow substitutions of both primary energies and prime movers should temper any bold visions of new sources and new techniques taking over in the course of a few decades. The first half of the century was dominated by coal, the quintessential fuel of the previous century, and three nineteenth-century inventions—ICE, steam turbine, and electric motor—were critical in defining and molding the entire fossil fuel era, which began during the 1890s. In spite of currently fashionable sentiments about the end of the oil era (for details see chapter 4), or an early demise of the internal combustion engine, dominant energy systems during first decades of the twenty-first century will not be radically different from those of the last generation.

Because of hasty commercialization, safety concerns, and unresolved long-term storage of its wastes, the first nuclear era has been a peculiarly successful failure, not a firm foundation for further expansion of the industry. And in spite of being heavily promoted and supported by public and private funding, contributions of new nonfossil energy sources ranging from geothermal and central solar to corn-derived ethanol and biogas remain minuscule on the global scale (see chapter 5 for details). Among new converters only gas turbines have become an admirable success in both airborne and stationary applications, and wind turbines have been improved enough to be seriously considered for large-scale commercial generation. Photovoltaics have proved greatly useful in space and in specialized terrestrial applications but not yet in any large-scale generation of electricity.

But the twentieth-century notable lessons go beyond advances in conversions. After all, even a more efficient energy use always guarantees only one thing: higher environmental burdens. Consequently, there remains enormous room for the inverted emphasis in dealing with energy needs—for focusing on deliveries of particular energy services rather than indiscriminately increasing the supply (Socolow

1977). A realistic goal for rationally managed affluent societies is not only to go on lowering energy intensities of their economies but also eventually to uncouple economic growth from the rising supply of primary energy.

And the challenge goes even further. Evolution tends to increase the efficiency of energy throughputs in the biosphere (Smil 1991) and impressive technical improvements achieved during the twentieth century would seem to indicate that high-energy civilization is moving in the same direction. But in affluent countries these more efficient conversions are often deployed in dubious ways. As David Rose (1974, p. 359) noted a generation ago, "so far, increasingly large amounts of energy have been used to turn resources into junk, from which activity we derive ephemeral benefit and pleasure; the track record is not too good." Addressing this kind of inefficiency embedded in consumer societies will be much more challenging than raising the performance of energy converters.

The task is different in modernizing countries where higher energy supply is a matter of existential necessity. In that respect the twentieth century was also a successful failure: record numbers of people were lifted from outright misery or bare subsistence to a decent standard of living—but relative disparities between their lives and those of inhabitants of affluent nations have not diminished enough to guarantee social and political stability on the global scale. Even when stressing innovation and rational use of energy, modernizing economies of Asia, Africa, and Latin America will need massive increases of primary energy consumption merely in order to accommodate the additional 2–3 billion people they will contain by the year 2050—but expectations based on advances achieved by affluent countries will tend to push the demand even higher.

This new demand will only sharpen the concerns arising from the twentieth-century's most worrisome consequence of harnessing and converting fossil fuels and primary electricity—from the extent to which our actions have changed the Earth's environment. We have managed to control, or even to eliminate, some of the worst local and regional effects of air and water pollution, but we are now faced with environmental change on a continental and global scale (Turner et al. 1990; Smil 1997). Our poor understanding of many intricacies involved in this unprecedented anthropogenic impact requires us to base our actions on imperfect information and to deal with some uncomfortably large uncertainties.

Perhaps the best way to proceed is to act as prudent risk minimizers by reducing the burden of modern civilization on the global environment. As long as we depend heavily on the combustion of fossil fuels this would be best accomplished by striving for the lowest practicable energy flows through our societies. There is no shortage of effective technical means and socioeconomic adjustments suited for the pursuit of this strategy but the diffusion of many engineering and operational innovations will not proceed rapidly and broad public acceptance of new policies will not come easily. Yet without notable success in these efforts the century's most rewarding commitment—to preserve integrity of the biosphere—will not succeed.

These lessons of the twentieth century make it easy to organize this book. After explaining a broad range of linkages between energy and the economy, and environment and the quality of life in chapter 2, I will gather arguments against specific quantitative forecasting and in favor of normative scenarios in chapter 3. Then I will discuss in some detail uncertainties regarding the world's future reliance on fossil fuels (in chapter 4) and opportunities and complications present in the development and diffusion of new nonfossil energies ranging from traditional biomass fuels to the latest advances in photovoltaics (in chapter 5). In the book's closing chapter I will first appraise the savings that can be realistically achieved by a combination of technical advances, better pricing, and management and social changes, then show that by themselves they would not be enough to moderate future energy use and in closing I will describe some plausible and desirable goals. Even their partial achievement would go far toward reconciling the need for increased global flow of useful energy with effective safeguarding of the biosphere's integrity.

2

Energy Linkages

All energy conversions are merely means to attain a multitude of desired goals. These range from personal affluence to national economic advancement and from technical excellence to strategic superiority. The energy-economy nexus is particularly strong, much studied, and yet much misunderstood. When seen from a physical (thermodynamic) perspective, modern economies are best described as complex systems acquiring and transforming enormous and incessant flows of fossil fuels and electricity. Some very high correlations between the rate of energy use and the level of economic performance make it possible to conclude that the latter is a direct function of the former. In reality, the relationship between energy use and economic performance is neither linear nor does it follow any other easily quantifiable function.

A closer examination of the relationship between TPES and GDP reveals that the linkage has been complex as well as dynamic. The link changes with developmental stages, and although it displays some predictable regularities there are also many national specificities that preclude any normative conclusions about desirable rates of energy consumption. As a result, some of the seemingly most obvious conclusions based on these correlations are wrong. Most importantly, identical rates of economic expansion in different countries do not have to be supported by virtually identical increases in their TPES, and countries do not have to attain specific levels of energy use in order to enjoy a comparably high quality of life. Perhaps the best way to explore the complexities of the critical link between energy and the economy is to examine national energy intensities.

I will do so by looking both at the long-term changes of these revealing ratios as well as by making detailed international comparisons of energy intensities during a particular year. I will be also deconstructing the measure in order to point out its weaknesses and to find the key reasons of its substantial secular and international variation. Afterwards I will complete the analysis of links between energy and the economy by taking a brief look at long-term trends of energy prices and then making a more detailed examination of real costs of fossil fuels and electricity. They should be appreciably higher than the existing prices that either entirely exclude many externalities associated with the provision and conversion of various forms of energy or account for them only in incomplete fashion.

Relationship between the quality of life and energy use is perhaps even more complex than is the energy–economy link. Good economic performance translated into substantial public investment in health, education, and environmental protection and into relatively high disposable incomes is a clear prerequisite of a high standard of living but there are no preordained or fixed levels of energy use to produce such effects. Irrational and ostentatious overconsumption will waste a great deal of energy without enhancing the quality of life while purposeful and determined public policies may bring fairly large rewards at surprisingly low energy cost. As with the energy–economy link I will use many international comparisons to demonstrate both of these discouraging and encouraging realities.

While higher use of energy may or may not bring desired economic and personal benefits it will always have one worrisome consequence: greater impact on the environment. Extraction of fossil fuels and generation of hydroelectricity are major causes of land use changes caused by surface mines, hydrocarbon fields and large water reservoirs. Transportation of fuels and transmission of electricity contribute to this problem due to extensive rights-of-way for railways, roads, pipelines, and high-voltage lines. Seaborne transport of crude oil is a leading source of polluting the ocean waters, particularly of the coastal areas. Combustion of fossil fuels is the largest source of anthropogenic greenhouse gases as well as of several air pollutants damaging plants, materials, and human and animal health. And while the risk of accidental releases of radioactivity from nuclear power plants may be minimized by careful design, the need for millennia-long storage of nuclear wastes poses unprecedented security and vigilance demands, a challenge that has yet to be solved by any modern society.

Environmental history of the twentieth century shows that concerns shift as our prevention, control, and management capabilities improve and as our scientific understanding deepens. Suffocating episodes of London-type smog, with high concentrations of particulates and SO_2 causing many premature deaths, or streams carrying so much waste that they could ignite have become textbook illustrations of extreme situations encountered before the introduction of effective laws to control basic forms of air and water pollution. But photochemical smog has proved a much more difficult phenomenon to control, as has the biosphere's enrichment by anthropogenic nitrogen; these concerns will remain with us for decades to come. And rapid global warming would have effects that will persevere for centuries. There is thus no doubt that links between energy use and the environment, on spatial scales ranging from local to global and on temporal scales extending from ephemeral events to millennia, will play a critical role in determining our energy futures. That is why I will return to examine this fundamental linkage once more in this book's last chapter.

Energy and the Economy

To begin on the global scale, a very close secular lockstep is revealed by comparing the worldwide consumption of commercial energy with the best available reconstruction of the gross world economic product (GWP) during the course of the twentieth century. In order to eliminate considerable inflationary fluctuations and the effect of biased official exchange rates the GWP should be expressed in constant monies, and purchasing power parities (PPP) should be used in converting national GDPs to a common monetary unit. When this is done the growth rates of the global commercial TPES coincide almost perfectly with those of the GWP, indicating a highly stable elasticity near 1.0. Each variable shows approximately a sixteenfold increase in 100 years, with energy consumption rising from about 22 to 355 EJ, and the economic product [in constant US$(1990)] going from about $2 to $32 trillion (Maddison 1995; World Bank 2001).

Closeness of the relationship is also revealed by a very high correlation between national per capita averages of GDP and TPES. For the year 2000 the correlation between the two variables for 63 countries listed in BP's annual energy consumption statistics was 0.96, explaining 92% of the variance, clearly as close a link as one may find in the often unruly realm of economic and social affairs. The line of the best fit in the scattergram of these variables runs diagonally from Bangladesh in the lower left corner to the United States in the upper right quadrant (fig. 2.1). A closer examination of the scattergram shows that no affluent economy with average per capita GDP (PPP adjusted) in excess of $20,000 consumed annually less than 100 GJ of primary commercial energy per person during the late 1990s while none of the least developed countries, with average per capita PPP GDPs below $1,000 used more than 20 GJ/person.

High correlations can be also found for the secular link between the two variables for a single country. For example, it is enough to glance at a graph plotting the

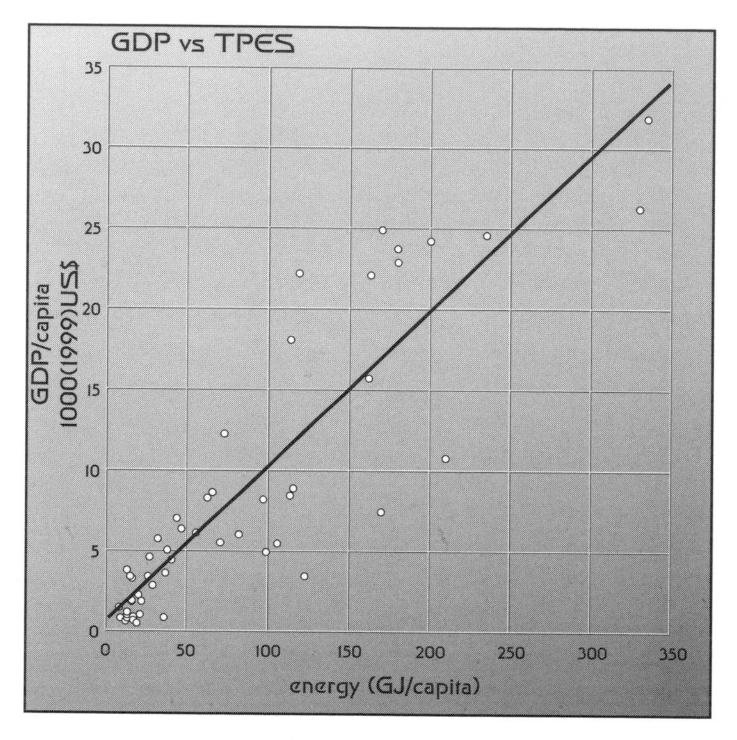

Figure 2.1
A high correlation between TPES and GDP, with both values expressed as per capita averages for the year 2000. Values calculated from data in BP (2001) and UNDP (2001).

course of Japan's energy use and GNP growth to see a very high correlation between the two variables (fig. 2.2). These graphic images and extraordinarily high correlations reinforce a common perception of the economic performance being a direct function of the total energy supply. They also seem to support the conclusion that a given level of economic well-being requires a particular level of energy consumption. This impression begins to weaken with a closer examination of national energy intensities. Energy intensity (EI) of a nation's economy is simply a ratio of the annual TPES (in common energy units) and the GDP. Specific energy intensities for individual fuels or for total electricity consumption provide additional insights.

Energy intensities can be easily calculated from readily available statistical compendia issued annually by the UNO (2001), IEA (2001), EIA (2001a), and the BP (2001). U.N. statistics now offer energy aggregates in terms of both coal and oil equivalents as well as in GJ, while the other data sources use oil equivalent as the

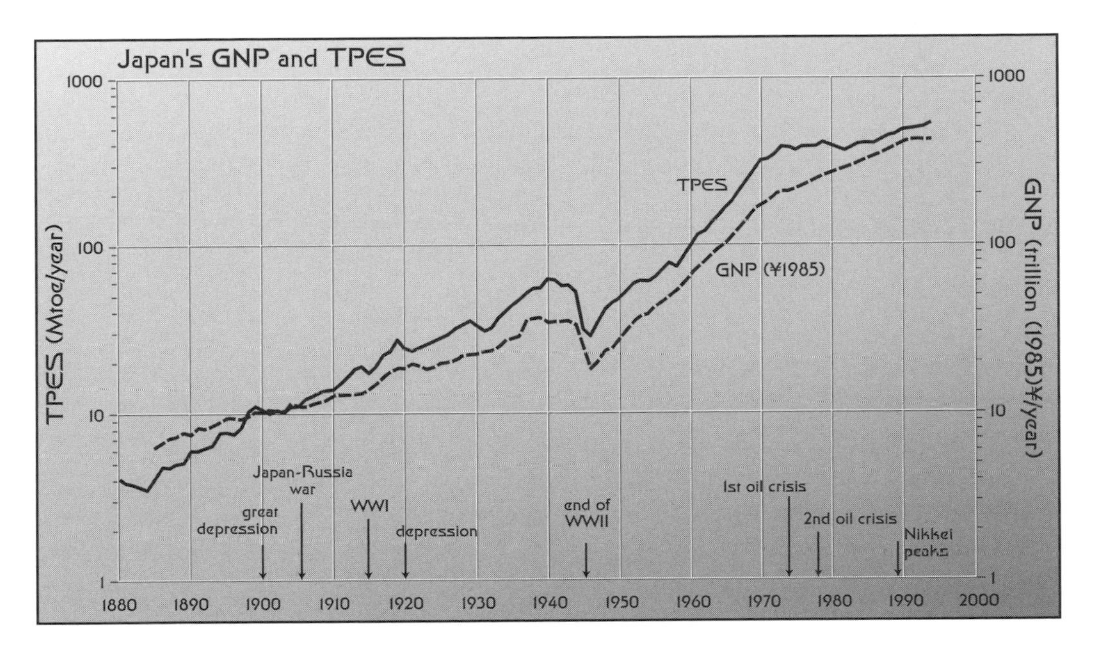

Figure 2.2
Very close relationship between Japan's GNP (expressed in constant ¥) and the country's TPES during the twentieth century. Based on a graph in IEE (2000).

common denominator. Similarly, national GDP figures are readily accessible in a number of statistical compilations, including those published by the UNDP (2001) and the World Bank (2001). In order to find undistorted long-term trends all GDP values must be expressed in constant monies, while international comparisons require conversion to a common currency, that is almost invariably to the U.S.$.

As the concept of EI has become more commonly used, several international data sources carry its annual updates and retrospectives. OECD's *Energy Balances* include energy intensities for all member states calculated with the TPES expressed in Mtoe and with GDPs in constant U.S.$ (OECD 2001). *OPEC Review* has been publishing historic reviews (data starting in 1950) and annual updates of energy intensities—in barrels of oil equivalent per U.S. $1,000 (1985)—for the whole world and for all major economic groups. And the U.S. EIA publishes annually the latest values, in Btu/U.S. $1,000 (1990), as well as historical retrospect for most of the world's countries (EIA 2001d).

Energy intensities can be seen as revealing long-term indicators of national economic development, and their international comparisons are often used as no less revealing indicators illustrating energy efficiencies of individual economies and their relative success, or failure, to use fuels and electricity less wastefully. Heuristic appeal of this aggregate measure is obvious as nations with relatively low consumption of fuels and electricity per unit of GDP should enjoy several important economic, social, and environmental advantages. Logical expectations are that relatively low energy inputs will help to minimize total production costs and that they will make exports more competitive in the global market. If a country is a major fuel importer, then a relatively low EI will help to minimize its costly fuel imports and improve its foreign trade balance. Low EI is also a good indicator of the prevalence of advanced extraction, processing, and manufacturing techniques.

Efficient use of raw materials, extensive recycling of materials with high embedded energy content, flexible manufacturing processes, inventories minimized by just-in-time deliveries, and efficient distribution networks are among the desirable infrastructural attributes contributing to low EI. And—as I will detail in the closing segment of this chapter—because production, processing, transportation, and final uses of energy constitute the world's largest source of atmospheric emissions and a major cause of water pollution and ecosystemic degradation, low EI is highly desirable in order to minimize many inevitable environmental impacts and to maintain good quality of life.

Comparisons of secular trends of national energy intensities show some notable similarities. EI almost invariably rises during the early stages of industrialization and its peak is often fairly sharp and hence relatively short. In most countries the peak has been followed by appreciable decline of EI as mature modern economies use their inputs (be they electricity, steel, or water) more efficiently. This pattern is well illustrated by the British, American, and Canadian examples (fig. 2.3). Energy intensity of the U.S. economy peaked rather sharply around 1920, and by 2000 it was down by nearly 60%. But, as the Japanese trend shows, instead of a sharp peak there may be an extended plateau. National energy intensities plotted in fig. 2.3 also show the specific features of long-term trends. Different timing of EI peaks and different rates of ascent and decline of national lines reflect the differences in the onset of intensive industrialization and the country-specific tempo of economic development and technical innovation. As I will show in the next chapter, this spatial and temporal variability complicates the task of forecasting the trends of EI.

Historical trend of the global EI cannot be quantified with a satisfactory accuracy but the best approximations indicate that the world as a whole is becoming a more rational user of commercial energies. Energy intensity of the global economy was

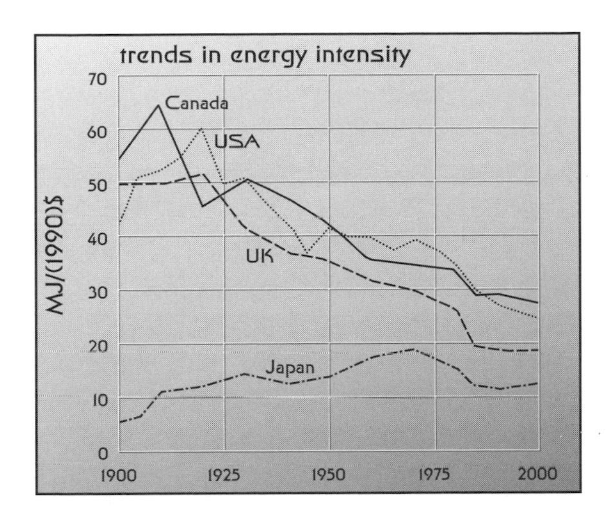

Figure 2.3
A century of energy intensities of national economies of the United States, Canada, the United Kingdom, and Japan. Based on a figure in Smil (2000a).

approximately the same in 1900 and in 2000, about 11 MJ/U.S.$ (1990)—but it had not been stable in between: it peaked around 1970, and it had declined since that time by about one-fifth. These shifts were a product of a complex combination of both concordant and countervailing national trends that follow a broadly universal pattern of transitions but do so at country-specific times and rates.

Recent energy intensities of the world's most important economies span a considerable range of values. When the national GDPs are converted by using purchasing power parities and expressed in constant U.S.$ (1990) the 1999 EI numbers for the G7 economies ranged from less than 7 MJ/$ for both Italy and Japan to less than 8 MJ for Germany and France, close to 9 MJ for the United Kingdom, more than 11 MJ for the United States, and over 13 MJ/$ for Canada. I should also point out that the IEA and EIA values of national EI differ rather substantially even for these major Western economies, and I will explain the major reasons for such discrepancies in the next section of this chapter. In contrast, the 1999 national EI values were around 10 MJ/$ for India and China, about 25 MJ for Russia, and, incredibly, 35 MJ/$ for the Ukraine (fig. 2.4).

These values confirm and reinforce some of the widely held snapshot notions about the economic fortunes of nations: efficient Japan, solidly performing Germany, the relatively wasteful United States, and Canada. China, in spite of its recent

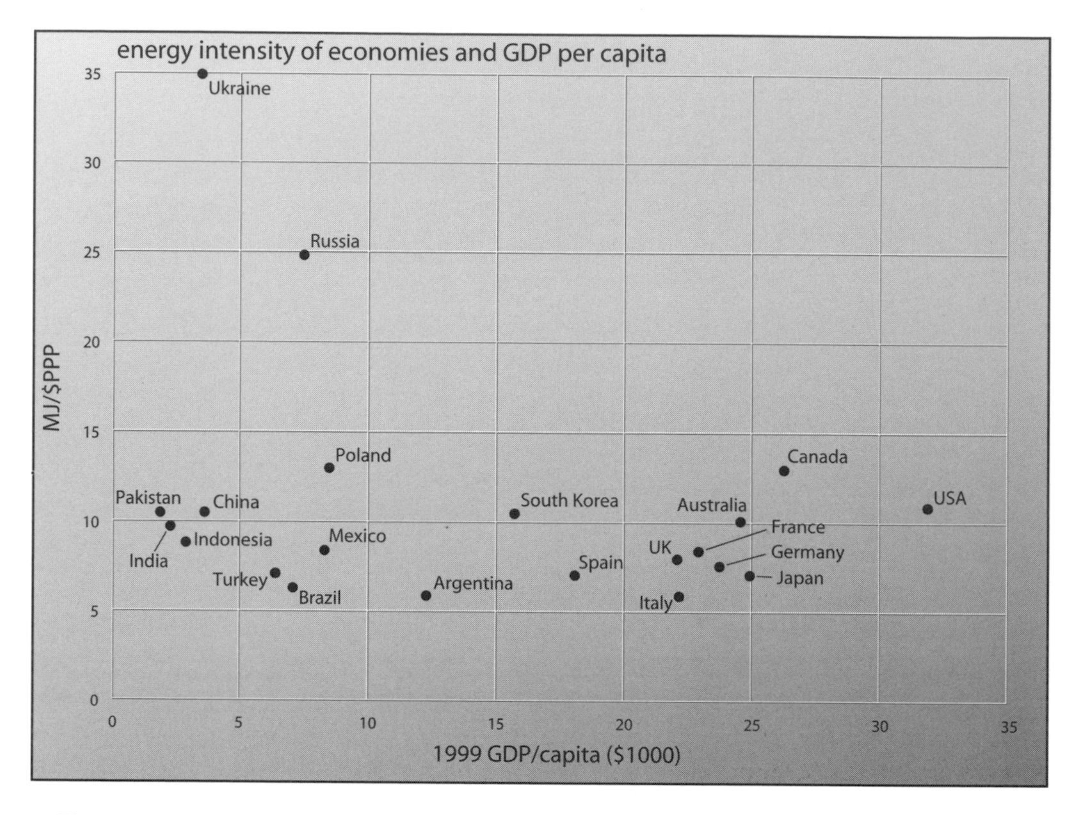

Figure 2.4
Comparison of national energy intensities and average GDP per capita, with both values calculated by using purchasing power parities rather than standard exchange rates. Plotted from data in UNDP (2001).

advances, still has a long way to go to become a truly modern economy; Russia is in an economic swamp and the Ukraine is beyond salvation. EI has thus become a potent indicator of a nation's economic and technical efficiency as well as of its environmental impact. As any aggregate measure, it offers a simple, readily comprehensible, and memorable distillation of complex realities that lends itself to praise or to condemnation of a nation's achievements and policies.

But while its heuristic or exhortatory value is undeniable, the EI ratio must be approached with great caution. If the measure is interpreted in a naïve, ahistorical, and abstract fashion—that is in the way I have done intentionally in the preceding paragraph—its use only reinforces some inaccurate notions, and it misleads more than it enlightens. Deconstruction of the measure offers a deeper understanding of underlying realities, uncovers a number of serious data limitations, leads to a careful interpretation of differences in levels and trends, and helps to avoid simplistic, and hence potentially counterproductive, conclusions.

Deconstructing Energy Intensities

First, both the TPES values converted to a common denominator and the GDP values are far from accurate. Specific national conversion factors for coal and hydrocarbon production and imports are used in order to minimize the possibility of major errors and often may result in relatively minor over- or underestimates. Perhaps the most notable exception is the conversion of China's huge coal extraction, now the world's largest in raw terms. In official Chinese statistics raw coal output is multiplied by 0.71 to get the hard coal equivalent. But during the 1990s more than 33% of China's coal extraction originated from small local pits, and less than 20% of all raw coal output was washed and sorted (Fridley 2001), and hence the real conversion factor to standard fuel could be as much as 10% lower than the official value.

A more important, and indeed an intractable conversion problem is the treatment of primary electricity, that is mostly hydro and nuclear electricity. When the primary electricity is converted by using the straight thermal equivalent (1 kWh = 3.6 MJ) countries relying heavily on hydro or nuclear generation will appear to be much more energy efficient than the nations generating electricity from fossil fuels. The contrast between Sweden and Denmark, two countries with very similar standard of living makes this problem clear. The two neighbors generate, respectively, less than 10% and more than 90% of all electricity by burning fossil fuels. Denmark's 36 TWh generated in 1999 needed about 9.5 Mtoe of fossil fuels—while Sweden's almost 145 TWh produced from water and fission have the thermal equivalent of just 12.3 Mtoe.

Equating primary electricity with the heat content of fossil fuels used to generate a nation's thermal electricity solves that undercount, but it introduces two new difficulties. Conversion ratios must be adjusted annually with changing efficiency of thermal generating stations. This is a logical procedure and it can be justified where thermal generation makes up most of a country's electricity supply—but it makes much less sense where hydro generation is the dominant mode of production, and it is inapplicable where the water power is the only source of electricity. Moreover, although the generation of hydroelectricity may remain basically stable over a long period of time (a common situation in countries where virtually all of the suitable sites have been dammed) its fuel equivalent would be declining with improved efficiency of thermal generation. Widespread diffusion of cogeneration and other high-efficiency techniques (see chapter 4 for details) could raise the average performance of thermal generation by 30–50% during the next 20 years—and hence cut the equivalent contribution of primary electricity by the same amount.

In any case, use of the fuel equivalent will obviously inflate TPES of all large-scale producers of hydroelectricity. If Sweden, where nonfossil electricity accounted for 94% of total generation in 2000, were to generate all of its electricity from coal or hydrocarbons it is certain that it would not be producing so much of it. The contrast between Sweden and Denmark is, once again, illuminating. While the resource-poor Denmark generates less than 7 MWh/capita in thermal stations burning imported fuels, Sweden's nonfossil electricity generation prorates to more than 16 MWh/capita. When this generation gets converted by the prevailing efficiency of fossil-fueled plants it would be equal to about 37 Mtoe.

Both the UN and BP resort to a compromise solution when converting primary electricity to a common denominator: they use thermal equivalent for hydroelectricity and the prevailing efficiency rate (about 33%) of fossil-fueled generation for nuclear electricity. In this case the fuel equivalent comes to 24.3 Mtoe and logical arguments can be then made for using values of 12.3, 24.3, and 37 Mtoe to convert Sweden's nearly 145 TWh of primary electricity. Accounting choices would thus inflate or lower the country's overall TPES and hence its EI. Problems of this kind suggest a need for a universal and transparent system of appropriate energy indexing that would compare in a fair manner flows of very different quality (Smith 1988).

Historical comparisons of EI for affluent countries and international comparisons of the measure across the entire spectrum of economic development mislead if they account only for modern commercial energies and exclude biomass fuels from TPES. This omission greatly underrates the actual fuel combustion during the early stages of industrialization when most of the GDP in farming, services, and manufacturing is generated without any modern energies. As already noted in chapter 1, the U.S. fuelwood combustion was surpassed by coal and crude oil only during the early 1880s (Schurr and Netschert 1960). In China biomass accounted for more than half of all primary energy until the mid-1950s (Smil 1988) and it still provided about 15% of all TPES during the late 1990 (Fridley 2001). During the 1980s biomass fuels supplied more than one-third of all fuel in at least 60 countries, and a decade later that count decreased to fewer than 50 countries (OTA 1992; UNDP 2001).

National GDP figures are much more questionable than aggregate TPES values. Even when leaving aside the fundamental problem of what is really measured—incongruously, GDP grows even as the quality of life may be declining, it increases even as irreplaceable natural resources and services are being destroyed (Daly and Cobb 1989)—specific weaknesses abound (Maier 1991). The first universal problem, arising from the inadequacy of economic accounts, makes it difficult to offer revealing long-term retrospectives as well as meaningful comparisons of affluent and low-income economies. Reconstruction of GDPs going back more than a century are now available for most European countries, the United States, and Canada (Feinstein 1972; USBC 1975) as well as for Japan (Ohkawa and Rosovsky 1973), India (Maddison 1985), and Indonesia (van der Eng 1992). The most comprehensive global data set was published by Maddison (1995). Numerous differences in the scope and quality of primary data mean that even the best of these data sets are only good (or acceptable) approximations of real trends rather than their accurate reconstructions.

Even more importantly, standard national accounts do not include the value of essential subsistence production and barter that still contribute a major part of economic activity in scores of African, Latin American, and Asian countries. Unfortunately, no extensive, uniformly prepared adjustments are available to correct this fundamental omission even for the largest modernizing economies of China or India. And the problem is not limited to low-income economies: black market transactions that are not captured by standard national account have been estimated to raise the real GDPs of the richest Western nations by anywhere between 5–25% (Mattera 1985; Thomas 1999).

The second universal difficulty arises from the need to convert national GDP values to a common denominator. Without such conversion it would be impossible to make revealing international comparisons of energy intensities, be they for a current year when comparing a group of nations, or for a past date when calculating GDPs in constant monies for trend comparisons. Almost always the U.S.$ is chosen to be the common denominator, and most of the past comparisons were done by using nominal exchange rates. This is an unsatisfactory approach as these rates reflect primarily the prices of widely traded commodities and may have few links with those parts of a country's economy that are not involved in foreign exchange. Use of official exchange rates almost always increases the real disparity between the rich and poor nations, and it distorts real GDPs even when comparing the achievements of the world's richest economies whose GDPs are prepared according to a uniform System of National Accounts.

Before 1971 the Bretton Woods system of fixed exchange rates provided a stable basis for currency conversions. Since that time sometimes quite rapid and often rather substantial fluctuations of exchange rates have repeatedly elevated or depressed a country's GDP expressed in U.S.$ by more than 10% within a single year. Using the PPPs circumvents these problems. Their use will tend to raise, often substantially, the real GDPs of all poor countries, and to lower, sometimes considerably, the real GDPs of affluent countries in Europe, Asia, and Oceania. But, in turn, these corrections have their own problems, mainly because it is impossible to specify a universal consumption basket containing typical food purchases, average spending on housing and the make-up and frequency of discretionary expenditures.

OECD's *Main Economic Indicators* carry average annual PPP multipliers for all of its member countries, and PPP-adjusted GDPs for all of the world's countries are used by the UNDP in calculating national indices in the widely quoted Human Development Index (UNDP 2001). Using the PPP-adjusted GDPs rather than simply converting national economic performance by exchange rates results in a substantially different set of EI. PPP adjustments make an enormous difference especially when comparing the performance of industrializing countries with that of postindustrial economies (fig. 2.5). Comparisons of some recent values show India's PPP-corrected GDP five times as large as the total obtained by using the official rate of conversion, while Brazil's rate should be at least 70% higher. PPP-adjusted rates express better the real economic circumstances but they are not necessarily the best choice for calculating more meaningful energy intensities.

Using PPP-converted GDPs could result in exaggerations whose errors would be relatively even larger—but in the opposite direction—than those generated by unrealistically low exchange-rate converted values. The Chinese example is particularly noteworthy. While the country's 1999 exchange-rate converted GDP was less than $800/capita, the UNDP's PPP value was just over $3,600 (UNDP 2001). The first rate would yield, with China's TPES at about 750 Mtoe (without biomass), energy intensity of about 760 kgoe/$1,000. The other one would result in energy intensity of only about 160 kgoe/$1,000. Both results are clearly wrong. In the first case China's energy intensity would be nearly 25% higher than India's roughly 610 kgoe/$1,000 (without biomass)—in the other case it would be actually equal to the Japanese performance.

But no matter how well or poorly they are constructed, national energy intensities do not tell us *why* the countries rank as they do. Six key variables that explain

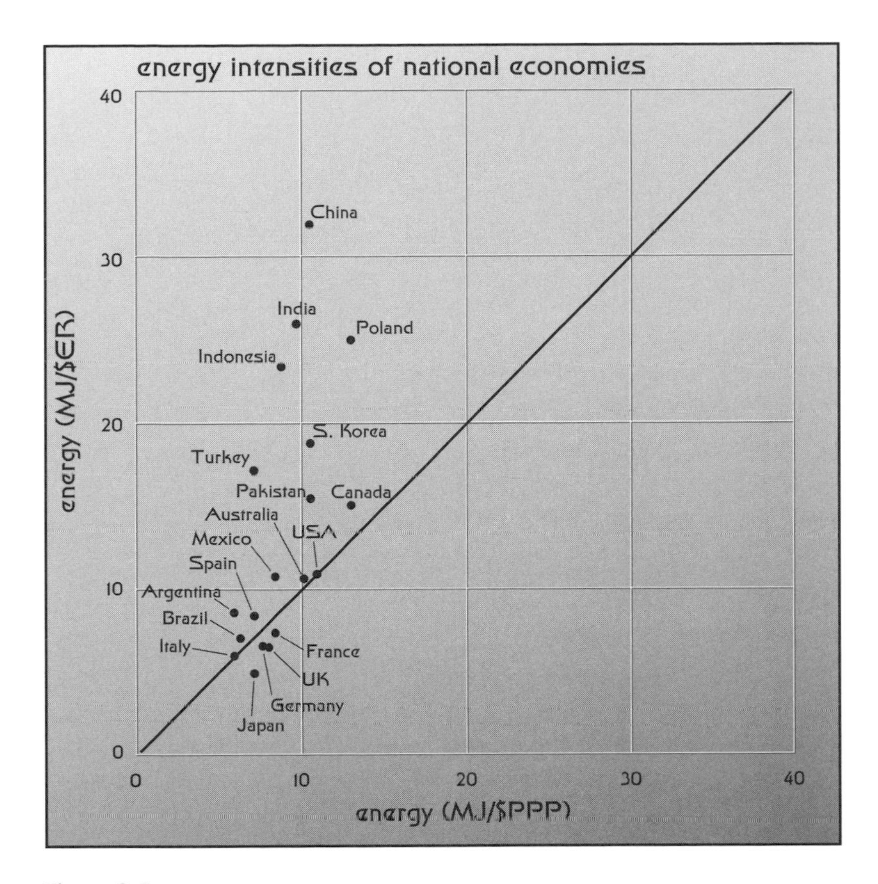

Figure 2.5
Comparison of energy intensities of national economies calculated with GDP values converted by the prevailing rate of exchange and by using purchasing power parities. Plotted from GDP data in UNDP (2001).

most of these differences are the degree of energy self-sufficiency; composition of the primary energy supply; differences in industrial structure and in discretionary personal consumption of energy; country size; and climate. Only closer looks at these factors will provide the ingredients for revealing answers. In virtually every case some of these variables will tend to enhance overall energy intensity while others will be depressing it. As a result, no single variable can act alone as a reliable predictor of national energy intensity: only a combination of at least half a dozen major attributes adds up to a specific—and usually unique—explanation of a nation's energy intensity.

High level of dependence on relatively expensive (and often also highly taxed) imports will generally promote frugality in energy consumption. In contrast, extraction, processing, and distribution of coals and hydrocarbons are all energy-intensive activities, and abundant fuel and electricity supply has historically attracted numerous energy-intensive industries and given rise to dense transportation networks. Consequently, there can be no doubt that the often much envied high degree of self-sufficiency in energy supply is commonly connected with higher overall energy intensities. The most obvious contrast is between the two largest Western economies: energy-intensive United States, the world's largest producer of fossil fuels, which imports about a quarter of its huge demand, and Japan, which has been recently buying just over 80% of its TPES.

Energy cost of underground coal extraction may be more than 10% of the fuel's gross heat content, and long-distance railway transportation is also relatively energy-intensive. Overall energy cost of coal delivery may be thus anywhere between less than 1% (from a large surface mines to an adjoining electricity-generating plant) to more than 15% (from a deep mine to a distant point of consumption) of the fuel's energy content. In contrast, crude oils, easier to extract and cheaper to transport, are delivered to refineries with efficiencies as high 99.5% and rarely lower than 97%. Refining will consume an equivalent of 4–10% of the initial oil input with common crudes, and substantially more with heavy oils requiring special treatment. Overall energy efficiency for different refined products at retail level may be as low as 80% and as high as about 93%. Losses of natural gas during field development, flaring and pipeline distribution may surpass 5% of extracted fuel (see also chapter 4). Energy for pipeline compressors can easily double that share where long-distance transportation is needed. Actually delivered energy may be thus anywhere between 85–98% of the extracted gas.

Composition of the primary energy supply has a number of inescapable consequences for national EI. Differences in conversion efficiencies always favor hydrocarbons over solid fuels. Although modern coal-fired power plants may generate electricity as efficiently as their hydrocarbon-fueled counterparts (that is by converting between 33–42% of the combusted fuel), they have higher internal consumption owing to their now universal controls of fly ash and often also of sulfur oxides. Combustion of coal in mine-mouth plants—as opposed to the burning of cleaner hydrocarbons which can be done closer to load centers—also increases the level of transmission losses. Efficiency disparities are much larger when the fuels are used directly for space and water heating. Even good coal stoves are rarely more than 40%

efficient, and efficiencies of small coal boilers are usually below 70%. In contrast, the lowest mandated efficiency of the U.S. household gas furnaces is now 78%, and the best units can convert more than 95% of the gas into heat (EPA 2001).

Given the intensive global competition and the relatively rapid diffusion of new techniques it is not surprising that total fuel and electricity requirements for principal energy-intensive products, be it aluminum, steel, or nitrogen fertilizers, are very similar in all affluent countries. Existing differences will be only of secondary importance, and the relative scale of production of energy-intensive goods will have much greater influence on overall industrial energy intensities. These specific patterns of industrial production are due to often large differences in natural endowment and economic history.

For example, Canada, the world's largest producer of hydro electricity is also the world's fourth largest producer of energy-intensive aluminum, smelting about 11% of the world's total supply, while Japan, one of the largest consumers of this versatile metal, imports annually more than 2 Mt of it but it smelted mere 7,000 t in the year 2000 (Mitsui 2001). Canada is also a relatively large producer of other energy-intensive nonferrous metals (above all of copper and nickel), steel and nitrogen fertilizers, while low energy intensities of Italian and French industrial production are due to low shares of mining in general, moderate size of both ferrous and nonferrous metallurgy, and relatively high shares of hydrocarbons and electricity in TPES.

Importance of structural differences can be also persuasively demonstrated by comparing national averages of industrial energy intensity with the estimated levels of technical efficiency derived by frontier analysis (Fecher and Perelman 1992). High overall industrial energy intensities may conjure the images of inefficient production, of outdated factories and processes, and of technical inferiority. But when a production frontier is defined as the maximum attainable output level for a given combination of inputs, sectoral estimates do not show any meaningful correlation with overall energy intensities of leading industrial sectors of G7 countries. Canada and the United States, the two countries with the highest EI, actually have the most technically efficient machinery and equipment sector, and U.S. performance also equals or outranks most G7 countries in the chemical sector (Smil 1994b).

Numerators of national energy intensities are not made up solely of fuel and primary electricity inputs used in productive processes: they also include all energies consumed in households and for personal mobility. Because these two categories of final energy uses have been rising steadily with modernization, urbanization and higher disposable incomes, their increasing shares of total consumption are now among the most important determinants of overall national energy intensities in all affluent countries. These demands are strongly influenced by climate and by the size of national territory. Households consume fuels and electricity for final services ranging from cooking and water heating to refrigeration and lighting, but in all colder climates the average number of heating degree days is the single most important predictor of total energy uses in households (Smil 1994b).

More than a 20-fold difference in their annual frequency between southern Florida and northern Maine (fig. 2.6), replicated within the EU (between southern Italy and central Sweden), illustrates the range of heating requirements. Mass ownership of air conditioners means that close attention must be also paid to differences in cooling degree days: in the coterminous United States they range from less than 500 in northern Maine to more than 3,000 in southern Florida (Owenby et al. 2001). Combination of heating (and with the spread of air conditioning also cooling) degree days and average dwelling sizes is thus the dominant reason for a nearly fourfold disparity in per capita use of residential energy in G7 countries.

Higher than predicted U.S. values are explained by relatively large consumption for space cooling and by higher ownership of household appliances, while a higher preferred indoor winter temperature is a major reason for the elevated Italian value. Similarly, national differences in the use of transportation energies cannot be ascribed primarily to different conversion efficiencies, but rather to a combination of spatial, historical, and cultural imperatives. Cars are now the largest consumers of transportation energy in every industrialized country, and the differences in average automobile energy intensities have been lowered to a fraction of the gap prevailing a generation ago. As a result, new American passenger cars (excluding SUVs) now have average fuel needs only marginally higher than the Japanese mean.

Typical travel distance, rather than average fuel efficiency, is a better predictor of energy needs. More pastime driving and traditionally high suburbanization in the United States—and its gradual extension into exurbanization with one-way commuting distances frequently surpassing 80 km—have resulted in steadily increasing distance traveled annually by an average U.S. car. During the late 1990s the total was approaching 19,000 km, or nearly 25% above the 1960 mean (Ward's Communications 2000). Although Japan and the United States have now a very similar rate of car ownership (2.1 versus 2.5 people per car) the average distance traveled annually by a passenger car is now almost twice as long in the United States as it is in Japan.

As far as the flying is concerned, all major airlines use similar mixtures of aircraft (mostly Boeing and Airbus planes), a reality eliminating any possibilities of major

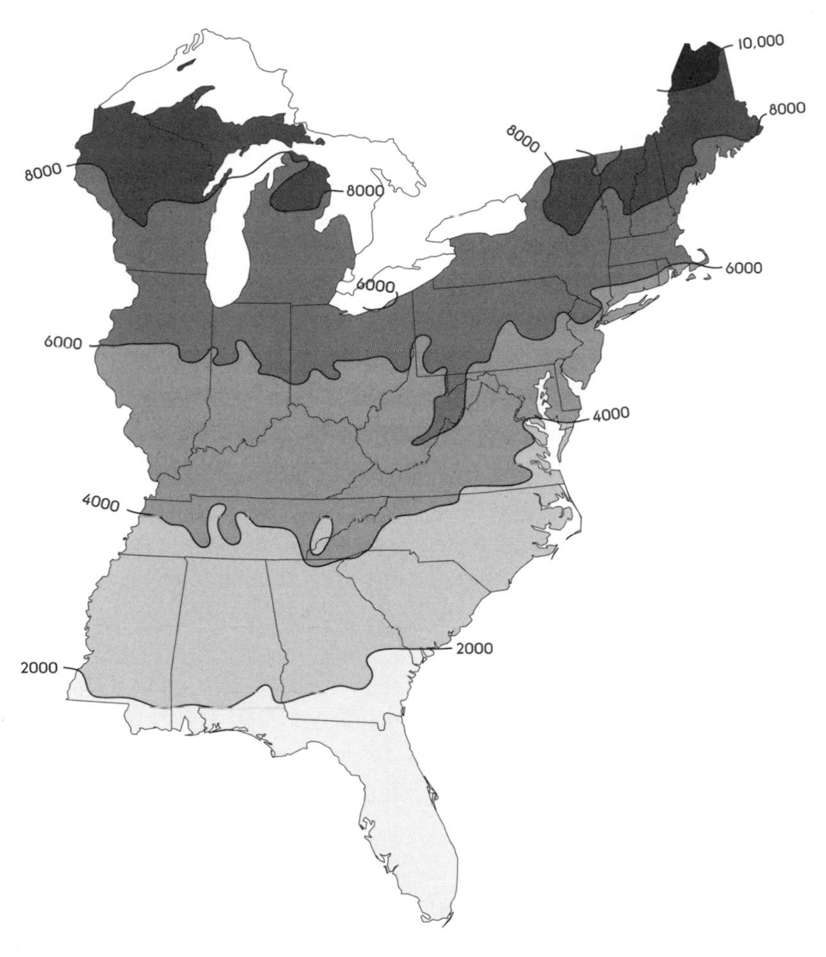

Figure 2.6
Heating degree day map of the Eastern United States (based on 1961–1990 temperatures) indicates large differences arising from a roughly 20°-span in latitude as well as from the contrasts between maritime and continental climates. Simplified from a detailed map available at the online *Climatography of the U.S. No. 81—Supplement # 3*, <http://lwf.ncdc.noaa.gov/oa/documentlibrary/clim81supp3/clim81.html>.

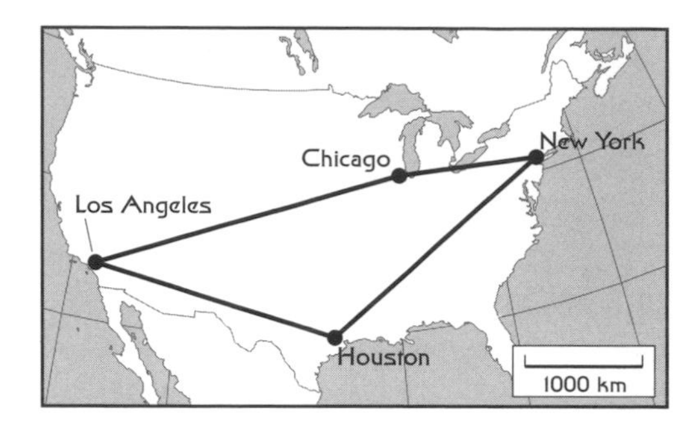

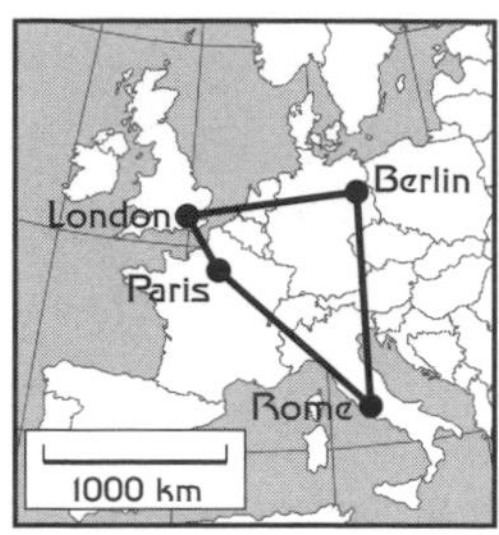

Figure 2.7
Same-scale maps of the United States and Europe showing the distance among the four largest cities in the United States (New York, Los Angeles, Chicago, and Houston) and in the European Union (London, Paris, Berlin, and Rome) illustrate North America's higher need for energy in transportation.

differences in operation efficiency. But total distance connecting the three most populous cities is less than 800 km in Japan as well as in all Western European countries—but it is more than 3,000 km in the United States (New York–Los Angeles–Chicago) and Canada (Toronto–Montreal–Vancouver), and Canadians and Americans also fly more often (fig. 2.7). In the year 1999 the total passenger air travel prorated to about 3,800 km for every person in the United States and almost 2,200 km in Canada—but only about 1,300 km in Japan and Germany and less than 700 km in Italy (ICAO 2000).

Clearly, North America's large private energy consumption, which adds so much to the two countries' high overall energy intensity, is only secondarily a matter of less efficient cars, planes, furnaces, and appliances. The continent's unparalleled private energy use in households is determined largely by climate and dwelling size, and in travel it is dictated mostly by distances and fostered by high personal mobility. In contrast, Japan's low energy intensity is largely the result of repressed private energy consumption in households and in transportation—and not a matter of extraordinary conversion efficiency (Smil 1992b). European countries fit in between these two extremes: their industries are generally fairly energy-efficient, as are their household converters and cars, but their private energy consumption is appreciably higher in Japan because of better housing and higher personal mobility.

Finally, I should note the far from negligible role of energy use by the military in boosting the overall EI. The United States is paying a particularly high energy cost for being the only remaining military superpower. Excluding the additional fuel demands of the Gulf War (1991) and of the Serbia–Kosovo bombing (1998), the three U.S. armed services consumed annually about 25 Mtoe during the 1990s (EIA 2001a). This total is more than the total commercial energy consumption of nearly two-thirds of the world's countries, and it is roughly equal to the annual TPES of Switzerland or Austria. Naturally, such operations as the Gulf War or the war in Afghanistan boost the energy use not only because of the visible air and land actions but also because of attendant long-distance logistics needs.

What is then the verdict on energy intensities? Few national aspirations are as desirable as a quest for lower energy intensity of the economic output. Lower EI evokes the images of economic and energetic efficiency, of doing more with less, of minimizing fuel and electricity inputs, of higher international competitiveness and minimized environmental impacts. Consequently, it would be valuable to know how individual economies score in this critical respect—but the commonly used energy intensities are not such simple indicators of efficiency. They do not tell us how efficiently individual nations use their fuels and electricity in their specific productive activities, how advanced are their converting techniques, how effective is their management, or how large is their potential for energy conservation. Rather, those simple quotients are products of complex natural, structural, technical, historical, and cultural peculiarities. The lower the better may be a fine general proposition, but not one to be applied to national energy intensities in absolute terms and on a worldwide scale.

State-of-the-art processes will not suffice to keep overall EI low if they are a part of inherently energy-intensive metallurgical industries or chemical syntheses in a country where such economic activities account for a relatively large share of GDP (clearly a Canadian case). On the other hand, cramped living conditions may help to make the final quotient look better as people inhabiting such houses will use less energy than would be expected given the country's economic wealth and climate (obviously Japan's case). International comparisons of energy intensities are revealing—but only if they are used as starting points of a deeper search for the reasons behind the differences. Used simply as absolute measures for categorical comparisons of economic and energetic efficiencies they are grossly misleading.

Energy Prices

Simple generalizations are also misplaced as far as the energy prices are concerned. Their secular decline has been essential for turning fuels and electricity into hidden, mundane commodities whose reliable, abundant, and inexpensive supply is now taken for granted. But long-term trends appear in a different light once we consider that energy prices have been repeatedly manipulated, and that they largely exclude many health and environmental costs associated with high rates of energy use. Where inflation-adjusted series can be reconstructed for the entire century there are impressive secular declines, or at least remarkable constancy, of energy prices. These trends illustrate the combined power of technical innovation, economies of scale, and competitive markets.

North America's electricity became a particularly great bargain during the course of the twentieth century. The earliest available U.S. national average price, 15.6 cents/kWh in 1902, would be about $2.50 in 1990 monies; a decade later the real price was down to $1.12 (1990), by 1950 it was about 15 cents, and it continued to fall during the 1960s. The long-term trend was reversed with the oil price jump in 1974 and by 1982 the average price was nearly 60% above the early 1970s level. Subsequent gradual decline brought the average U.S. price back to just around 6 cents by the year 2000 (fig. 2.8). The huge real price drop of nearly 98% between 1900 and 2000 still vastly underestimates the real price decline. With average per capita disposable incomes about 5 times as large and with average conversion efficiencies of lights and electric appliances being 2 or 3 times as high as in 1900, a unit of useful service provided by electricity in the United States was at least 200, and up to 600 times, more affordable in the year 2000 than it was 100 years earlier (fig. 2.8).

Not surprisingly, this U.S. price trend does not represent a universal pattern. For example, post–World War II prices of Japanese electricity have not changed dramatically: average inflation-adjusted price in the late 1990s prices was almost exactly the same as it was in 1970, and it was only about 30% lower than in 1950 (IEE 2000). But as the average per capita income in the year 2000 more than doubled since 1970 and it had increased about tenfold since 1950, Japan's electricity consumption has been also getting steadily more affordable. And although there was no sharp drop in electricity prices during the 1990s electricity demand for lighting went up by an astonishing 50% as did the total residential demand for electricity. Also better-lit Japanese homes are also cooler in hot, humid summers and now contain more electric appliances: by the decade's end, air conditioning was installed in more than twice as many households as in 1989, and four times as many families had a PC. After decades of frugal household management Japanese consumers are finally splurging on home comforts.

In contrast to impressively declining electricity prices the inflation-adjusted U.S. prices of coal and oil have shown some rapid fluctuations but, over the long run,

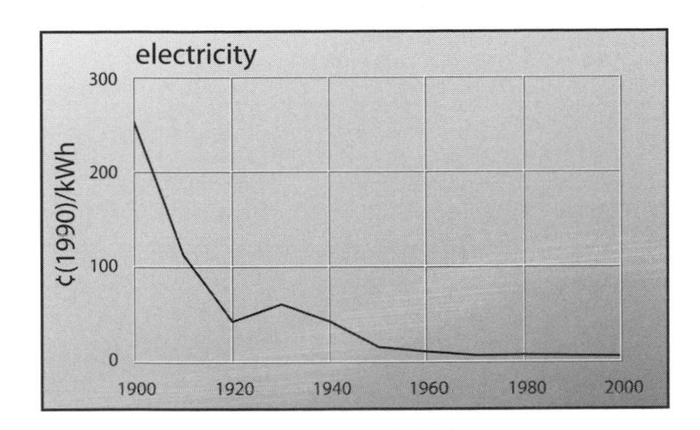

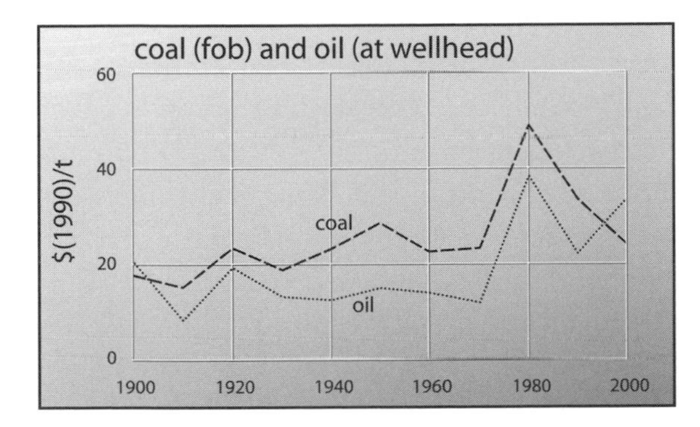

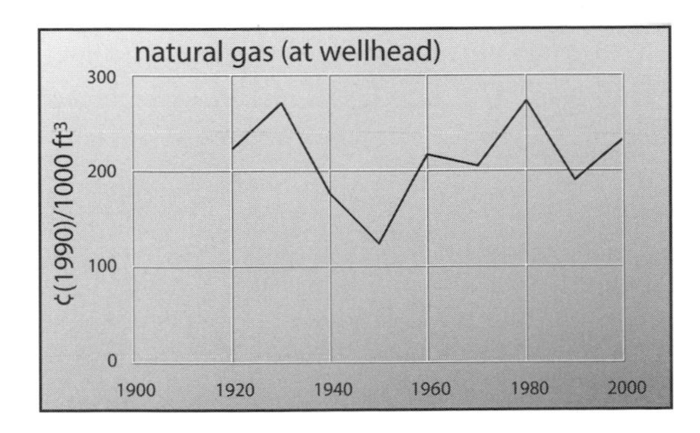

Figure 2.8
One hundred years of average annual prices paid by the U.S. consumers for electricity, coal, oil, and natural gas. All values are in inflation-adjusted monies. Plotted from data in USBC (1975) and EIA (2001a).

they have remained remarkably constant. When expressed in constant monies, the average f.o.b. price of U.S. bituminous coal during the late-1990s was almost exactly the same as in 1950—or in 1920. Similarly, the price of imported light Middle Eastern crude was almost exactly as cheap as was the average price of the U.S.-produced crude during the first years of the twentieth century (fig. 2.8). This relative price constancy means that in terms of the work–time cost gasoline was only about 20% as expensive during the 1990s as it was in the early 1920s, and that the useful energy services derived from oil, or coal, combustion were about ten times more affordable by the century's end than in 1900. Natural gas is the only major form of commercial energy showing a gradual increase in inflation-adjusted prices, going up more than ninefold between 1950 and 1984, then falling by about 60% by 1995 before rebounding to the levels prevailing during the early 1980s by the year 2000 (fig. 2.8).

But it would be naive to see these and virtually any other twentieth-century energy prices either as outcomes of free market competition—or as values closely reflecting the real cost of energy. Long history of governments manipulating energy prices has brought not only unnecessarily higher but also patently lower prices due to subsidies and special regulations affecting distribution, processing, and conversion of fossil fuels, and generation of electricity. During the twentieth century governments provided direct financing, research funds, tax credits, and guarantees to advance and subsidize particular development and production activities, thus favoring one form of supply over other forms of energy.

There are many examples of these price-distorting practices (Kalt and Stillman 1980; Gordon 1991; Hubbard 1991; NIRS 1999; Martin 1998). In the United States one of the most visible governmental interventions has been the state-financed building of large dams initiated during the Roosevelt's New Deal years 1930s by the Bureau of Reclamation and later also undertaken by the U.S. Army Corps of Engineers (Lilienthal 1944). In many nations nuclear electricity generation has benefited from decades of generous public funding of basic research and operational subsidies. For example, federal subsidies to the Atomic Energy of Canada, a crown corporation, amounted to CD$15.8 billion (1998) between 1952 and 1998, representing an opportunity cost (at a 15% rate of return) of just over CD$(1998) 200 billion (Martin 1998). The U.S. nuclear industry received more than 96% of U.S.$145 billion (1998) disbursed by the Congress between 1947 and 1998 (NIRS 1999). Moreover, nuclear industries in the United States, the United Kingdom, France, India, and China have also benefited from experiences and innovations spilling over from an even more intensive support of military nuclear R&D.

Special tax treatment has been enjoyed for an even longer period of time by the U.S. oil companies that can immediately write off so-called intangible drilling costs. And all independent oil and gas producers are allowed to deduct 15% of their gross revenue and this depletion deduction can greatly exceed actual costs (McIntyre 2001). More recently generous tax benefits have been given to the developers of some renewable energies. The most expensive subsidy is given the producers of corn-based alcohol to make gasohol competitive with gasoline. But as the process actually needs more energy than is released by the combustion of the produced ethanol, large-scale production of this biofuel is highly questionable (Giampietro et al. 1997). Nor has this subsidy lowered the U.S. dependence on imported oil: all it has done is to boost the profits of the country's largest agribusiness, Archer–Daniels–Midland. I will have more on this counterproductive effort, and on other subsidies for renewable energy, in chapter 5.

As for the nuclear electricity generation, in 1954 the Price–Anderson Act, section 170 of the Atomic Energy Act, reduced private liability by guaranteeing public compensation in the event of a catastrophic accident in commercial nuclear generation (DOE 2001a). No other energy supplier enjoys such sweeping governmental protection. At the same time, power producers have been heavily regulated in many countries. Gordon (1994) singled out the general underpricing of electricity caused by this regulation as a key market failure and argued that deregulation is preferable to conservation programs that were imposed on the utilities.

A critical look at the price history of today's most important fossil fuel reveals a particularly extensive, and in Gordon's (1991) opinion universally depressing, record of government intervention in the oil industry. This is true even when leaving notoriously meaningless pre-1991 (when the country was the world's largest crude oil producer) Soviet prices aside. Until 1971 the U.S. oil prices were controlled through prorated production quotas by a very effective cartel, the Texas Railroad Commission (Adelman 1997). OPEC was set up in 1960 when the foreign producing companies reduced posted crude oil prices. In order to protect their revenue the OPEC members would not tolerate any further reductions of posted prices, an income tax became an excise tax—and the posted price lost all relation to any market price (Adelman 1997).

A decade later OPEC countries began raising their excise taxes, and in March 1971 the Texas Railroad Commission removed the limits on production: short-term power to control prices thus shifted from Texas, Oklahoma, and Louisiana to the newly cohesive OPEC. Oil production cuts by the OPEC's Arab members in October

1973 led to the quadrupling of crude oil prices within just six months and, after a short spell of stability, between 1979 and 1980 prices rose about 3.5 times above their 1974 level, to an average of about $38 per barrel of Texas oil (BP 2001; EIA 2001a). Consumers in Japan, Europe, and throughout the poor world were immediately affected by these large price increases, but because of the crude oil price controls imposed in 1971 during Nixon's first term, the average inflation-adjusted price of the U.S. gasoline in 1978 was still no higher than it was a decade earlier (GAO 1993). Only when the controls ended in 1981 did the gasoline prices rise to levels unseen since World War II.

As I illustrate in some detail in the next chapter, expectations of further OPEC-driven large oil price hikes became the norm. This error was due largely to three entrenched misconceptions, namely the beliefs in OPEC's price-setting power, in the world running out of oil and in the oil demand that is fundamentally price inelastic. In reality, there was never a time when OPEC's power was such as to entirely overrule the market (Mabro 1992). Mabro also makes the point that since 1984, after the market forces appeared to be dominant, OPEC's influence on global oil pricing has not been abolished. Moreover, the duality of difficult-to-reconcile objectives that is at the root of OPEC's discords—the group being concerned with the price of oil but individual member countries being concerned about maximizing their own revenues—puts as effective a downward pressure on fuel's price now as it did a generation ago.

Recently fashionable verdicts about the imminent peak of oil extraction are addressed in some detail in chapter 4, and the notion of inelastic oil demand was disproved by the early 1980s as a combination of economic recession, conservation efforts, and the inroads by natural gas cut the need for oil and reduced OPEC's importance. In August 1985, with prices at about $27/barrel, Saudi Arabia stopped acting as a swing producer, and one year later prices fell to less than $10 per barrel. The Gulf War caused only a temporary price blip, and for most of the 1990s oil prices fluctuated in the zone between a monopoly ceiling above U.S.$30(1990) and a long-run competitive equilibrium floor which Adelman put at less than U.S.$7 per barrel (Adelman 1990).

Once again, low oil prices were taken for granted and, once again, energy demand began rising even in those rich nations that were already by far the largest users, and importers, of oil. Between 1989 and 1999 energy consumption rose by almost 15% in the United States, 17% in France, 19% in Australia, and, in spite of stagnating and even declining economic performance, by 24% in Japan (IEA 2001). This trend, driven predominantly by private consumption (larger homes with more energy converters, larger cars, more frequent travel), pushed energy use to record levels. By 1999 the U.S. domestic oil extraction was 17% lower than a decade earlier just as the SUVs gained more than half of the new car market. The rapid rise of China's oil imports further strained the export market. Not surprisingly, OPEC's share of oil output rose once more above 40% and oil prices rose from a low of about $10 per barrel in March 1999 to more than $25 per barrel by the year's end, and they briefly surpassed $30 per barrel in September 2000 only to fall once again with the long-overdue burst of the U.S. bubble economy and with the onset of global economic recession in the wake of the terrorist attack on the United States on September 11, 2001 (fig. 2.8; see also chapter 3 for more details on recent price swings).

One thing has remained constant throughout these eventful shifts: the world price of crude oil bears very little relation to the actual cost of extracting the fuel. The average cost of oil extraction from most of the large oilfields is low in absolute terms and even lower in comparison to the cost of other commercial energies. Oil prices are not thus primarily determined by production costs but by a variety of supply and demand factors whose specific combinations provide unique outcomes at particular times. And given the extreme relationship between fixed and variable costs in oil production there is a repeated disconnect between prices and the extraction rate. Falling prices may not bring a rapid decline in supply because once the investment has been made a marginal barrel cost is very low; conversely, the need for new capital investment causes a lag between rising prices and expanded supply.

Periodic spikes of the world oil price and mismanagement of national or regional deliveries of other forms of energy also led to renewed calls for energy price controls. The latest U.S. recurrence of this persistent phenomenon took place during the spring of 2001 as gasoline prices rose to levels unseen for a generation, and as California lived through a shocking (and also largely self-inflicted) spell of electricity shortages and escalating prices. In June 2001 a poll found that 56% of Americans favored price caps for energy and many supporters of price controls argued that energy markets are not genuinely competitive to begin with, that collusion leads to overpricing and that legislating a reasonable level of profits would be much preferable (ABC News 2001). Opponents of price caps replied with standard and well-tested arguments that controls in tight markets only worsen the situation by discouraging new investment and conservation. As gasoline prices fell and California's electricity shortages eased, the talk about nationwide energy price controls evaporated—but the well-worn arguments will surely be back with the next price spike.

Given certain assumptions an excellent argument can be made for even lower prices of energy than those that have prevailed during the second half of the twentieth century. At the same time it must also be acknowledged that most of the existing prices do not fully reflect the real costs of fossil fuels and electricity, and that the inclusion of numerous environmental, health, safety, and other externalities would push them appreciably and substantially higher. The quest for more inclusive pricing is not a new phenomenon but it has been invigorated by attempts at environmentally more responsible valuations in general and by an increasing interest in carbon taxes, designed to limit the emissions of CO_2, in particular.

Real Costs of Energy

Unaccounted, or inadequately captured externalities include all immediate and short-term negative impacts related to the discovery, extraction, distribution, and conversion of energy resources as well as the costs related to decommissioning and dismantling of facilities, long-term storage of wastes, and long-lasting effects on ecosystems and human health (Hubbard 1991). There is no doubt that most fuel and electricity prices still either ignore, or greatly undervalue, these externalities that range from largely quantifiable local environmental impacts to global consequences that are impossible to monetize in a satisfactory manner, and from health effects on general population to substantial economic and social burdens of large-scale military interventions.

As I will detail later in this chapter, extraction, distribution, and conversion of various energies generates numerous environmental externalities, and hence also many health impacts. Only a few of these matters have been satisfactorily, or largely, internalized in some countries during the latter half of the twentieth century. Perhaps the most widely known case of an almost immediate payment for a large-scale environmental damage is the grounding of the *Exxon Valdez* on March 24, 1989. The company spent directly about U.S.$2 billion (1990) on oil cleanup and paid half that amount to the state of Alaska (Keeble 1999). Costs of restoring waters and (at least superficially) rocky shores and beaches of Prince William Sound were thus internalized to a degree unprecedented in previous oil spill accidents. There have been other instances of notable internalization. Large increases in insurance premiums for giant (more than 200,000 dwt) tankers reflecting the risks of accidental collisions resulting in extensive crude oil spills have been a major factor in stopping a further growth of these huge vessels.

Environmental impact assessments of proposed energy megaprojects and sustained protests by groups most negatively affected by such developments have led to postponement, design changes, or even cancellation of construction. Several of these noteworthy cases include opposition to hydroelectric dams whose reservoirs would flood the lands of indigenous populations or displace large numbers of people. Passionate campaign waged by Québec Cree against damming the Grande Baleine River was among the major reasons for the New York State Power Authority to cancel a twenty-year $12.6 billion contract to buy 1 GW of electricity that Hydro-Québec would have generated by this megaproject, and a widespread domestic and international opposition forced the World Bank to withdraw from financing the highly controversial Sardar Sarovar dam, the largest of the planned group of 30 dams in India's Narmada river basin (Friends of the River Narmada 2001).

But the most notable systemic progress in internalizing long-term environmental and health consequences of energy-related activities has been made by the U.S. coal industry and coal-fired electricity generation (Cullen 1993). In order to prevent deadly mine explosions, regulations now prescribe maximum allowable levels of dust and methane and the maintenance of these levels requires appropriate ventilation of shafts and corridors and suppression of dust at workfaces and in tunnels. Effectiveness of these and other safety measures is obvious: as already noted in chapter 1, annual U.S. fatalities in coal mining are less than 1% of China's dismal record (MSHA 2000; Fridley 2001). Mining companies must also contribute to the disability and compensation funds put in place to ease the suffering of miners afflicted with black-lung disease (Derickson 1998).

All mining operations must reduce acid drainage and waste runoff containing heavy metals, and operators of surface mines must try to restore the original landscape contours as much as is practicable and either seed the reclaimed land with grasses and eventually establish shrub and tree cover or develop new water surfaces (OSM 2001b). External costs associated with coal mining are relatively small compared to the measures that have to be taken to limit the effect of fuel's combustion on the atmosphere. Every coal-fired power plant operating in the United States must have electrostatic precipitators that capture fly ash leaving the boilers. These devices now have efficiencies in excess of 99% and besides the electricity needed for their operation there are further costs incurred in transporting the captured fly ash and ponding it, burying it in landfills or using it in construction or paving (for more on fly-ash utilization see chapter 4).

Installation of flue gas desulfurization (FGD) processes began during the late 1960s and by 1999 almost 30% of the U.S. coal-fired electricity-generating capacity was equipped with some kind of sulfur emission controls (Hudson and Rochelle 1982). Cost of FGD facilities depends on their type and the amount of sulfur present in coal. Recent costs per installed kW have ranged from less than $50 to nearly $400, and the average cost of $125/kW adds at least 10–15% to the original capital cost and operating them increases the cost of electricity production by a similar amount (DOE 2000a). The latter costs include up to 8% of electricity generated by a plant used in powering the FGD process and the disposal of the captured waste. Most common commercial processes use reactions with ground limestone or lime to convert the gas into calcium sulfate that must be then landfilled. FGD is mandated in order to combat acid deposition, a complex environmental degradation that affects vulnerable biota and entire ecosystems, damages materials and human health and limits visibility (for more, see the last section of this chapter).

There is no doubt that widespread use of FGD, together with the use of cleaner, and costlier, fuels has reduced acid deposition over the Eastern North America (fig. 2.9). Moreover, the findings of the expensive [in excess of $500 million (1985)], decade-long (1981–1990) investigations undertaken by the U.S. National Acid Precipitation Assessment Program (NAPAP) showed only a limited damage to lakes and forests (NAPAP 1991). Consequently, it may be argued that the 1990 tightening of SO_x and NO_x emissions goals for coal-fired power plants already tipped the balance in the direction of excessive controls (Gordon 1991). In contrast, Alewell et al. (2000) found that even after deep European cuts in SO_2 emissions many continental sites have experienced significant delays in returning to their preacidification status, or even no recovery at all.

Such contrasting conclusions are the norm, rather than an exception in debates about the real costs of energy. An ORNL study found that actual damages from the entire coal fuel cycle amount to a mere 0.1 c/kWh (Lee et al. 1995), a negligible fraction of the recent average cost of just over 6 c/kWh and a total many magnitudes apart from Cullen's (1993) estimate of a tenfold increase of actual fuel costs. Similarly, generalized damage values that were actually adopted by some public utility commissions in the United States during the 1990s were not only substantially different but they were typically much higher than damages whose calculations took into account the characteristics of the reference site and incorporated results of the dose–response functions (Martin 1995).

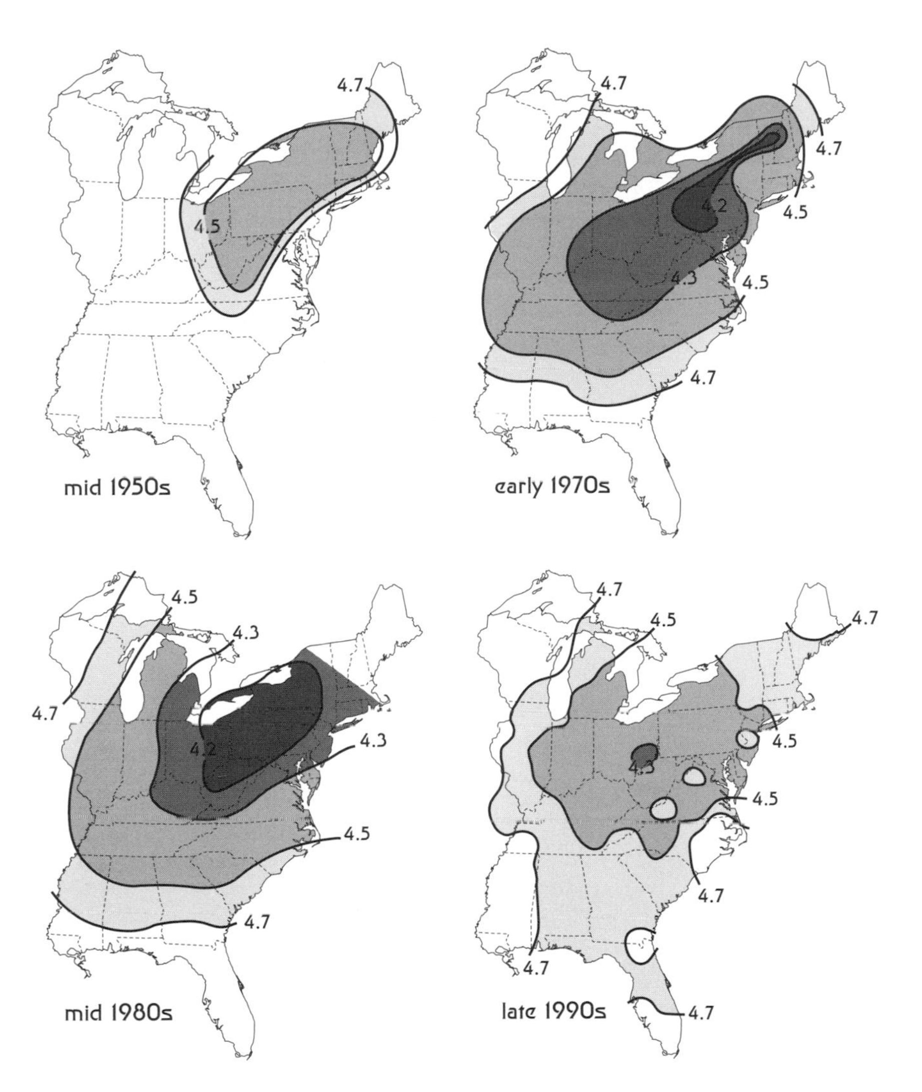

Figure 2.9
Expanding (mid-1950s to early 1970s) and declining (mid-1980s to late 1990s) area of high acid precipitation in the Eastern United States. Reproduced from Smil (2002).

For example, several U.S. utility commissions (in California, Nevada, New York, New Jersey, and Massachusetts) used rates between $884 (1990) and 28,524 (that is more than a 30-fold difference) for environmental damages due to the emissions of one tonne of NO_x and $832 (1990) to $4,226 (1990) for a tonne of SO_x. In contrast, the studies that modelled actual dispersion and the impact of pollutants for specific sites showed that the combined damage for all different generation processes was between 0.3–0.6 c/kWh, and an order of magnitude lower for the natural gas-fired generation (Lee et al. 1995). And while Hohmeyer (1989) calculated the external costs of German electricity production to be of the same order of magnitude as all of the internalized costs, Friedrich and Voss (1993) judged his methodologies unsuitable and the estimates derived from them too high.

The most recent set of detailed studies of external costs of electricity has resulted from a ten-year project funded by the European Commission (EC 2001a). Its conclusions are definitely on the high side, with claims that the cost of producing electricity from coal or oil would double and that of natural gas-generated electricity would go up by 30% if damages to the environment and to health (but not including those attributable to global warming) were taken into account. As expected, hard coal- and lignite-fired generation was found to have the highest externalities (on the average nearly € 6c/kWh, with the range between 2–15 c), hydro generation the lowest (0.03–1 c/kWh). In aggregate, these externalities are equivalent to 1–2% of the EU's gross economic product.

Although the research used the same methodology national damage rate per tonne of pollutant, dominated by effects on human health, differ widely, reflecting the total populations affected. Consequently, a tonne of SO_2 causes, at worst, damage worth € 5,300 in Ireland but as much as € 15,300 in France (ExternE 2001). The difference is even greater for particulates, largely because the waste incineration plants are often located near or in large cities where their emissions can affect more people: a tonne of emitted particulates was calculated to produce damage worth up to € 57,000 in Paris but as little as € 1,300 in Finland.

But even assessments done at the same time for the same populations can end up with very different values because the health impacts of energy uses are so tricky to monetize. Attempts to estimate costs of air pollution, or to value the benefits of clean air, are perhaps the best examples of these challenges. In spite of decades of intensifying research there are no fundamentally new conclusions offering a firmer guidance in assessing long-term effects of many principal air pollutants on human health. Numerous statistical analyses seeking to unravel the relationship between air pollution and mortality have been handicapped by dubious design, irremovable interferences of unaccounted (and often unaccountable) variables, and difficulties of separating the effects of individual pollutants (for example sulfates from general particulate matter) when people are exposed to complex, and varying, mixtures of harmful substances. What Ferris (1969) wrote more than a generation ago is still true: general mortality statistics appear to be too insensitive to estimate effects of chronic low-level air pollution with accuracy.

Consequently, major differences in calculations of health costs of pollution (or health benefits of abatement) are the norm. An excellent example was provided by two studies quantifying the annual benefits of reduced mortality and morbidity benefits accruing from meeting the air quality standards in the Los Angeles Basin during the 1980s. A study sponsored by the California Air Resources Board estimated the benefit at between $2.4–6.4 billion (Rowe et al. 1987), while a 1989 assessment prepared for the South Coast authorities offered a best conservative estimate of $9.4 billion a year, with a plausible range between $5 and $20 billion (Hall et al. 1992). Continued large differences between extreme estimates—a more than eightfold discrepancy in the two California cases—makes cost–benefit studies of air pollution control a matter of unceasing scientific controversies (Krupnick and Portney 1991; Lipfert and Morris 1991; Voorhees et al. 2001).

And then there is, of course, a large class of externalities associated with the production and conversion of fossil fuels that cannot be internalized in any meaningful fashion because the existing state of our understanding makes it impossible to give any reliable quantitative estimates of their long-term impacts. Global warming brought by anthropogenic emissions of greenhouse gases (a closer look at this process ends this chapter) is the prime example of these profound uncertainties. Depending on the rate and extent of eventual global climate change the costs that should be imputed to a kWh of electricity generated by burning a fossil fuel might be range from relatively minor to very large. Because of a number of other fundamental uncertainties regarding the valuation of external costs of electricity it can be argued that political and legislative action should not wait until all the remaining doubts are resolved (Hohmeyer and Ottinger 1991).

And the accounting for military expenditures attributable to securing energy supplies offers a perfect example of scale problems, both spatial and temporal, arising from the choice of boundaries. Unequivocal selection of a proper analytical set may be impossible even when the requisite monetary accounts are readily available, but those, too, may be quite fuzzy. Perhaps the best illustration of this challenge is to try to estimate the real cost of the Persian Gulf oil to American consumers, a concern made so obvious by the Iraqi invasion of Kuwait and the subsequent American response. In order to keep an uninterrupted supply from the world's most important oil-producing region the country led a large military alliance first in a mounting armed deployment in the area (Operation Desert Shield, August 1991–January 1992), and then in a bombing campaign and the ground war (Operation Desert Storm, January–March 1992).

Costs of this commitment differ substantially when expressed as funding requirements, incremental expenditures, and total outlays. The U.S. Department of Defense estimated the total funding requirement for these operations at $47.1 billion, and total incremental cost at $61.1 billion. But the General Accounting Office pointed out that DOD's financial systems are not capable of determining these costs reliably (GAO 1991a). The services accounted for the total costs at the unit level without any adjustment for the costs they would have to bear even in the absence of the Persian Gulf conflict. Such corrections were estimated later at higher reporting levels, but the use of broad and undefined categories makes it difficult to verify if specific expenses were properly charged. Sums involved in such accounts are not trivial. One command subdivided its operation expenses of $1.5 billion for October–December 1990 into 19 categories, but one entitled "Special Activities" took about $1 billion—and the command officials could not specify the subcategories of expenditures under this hazy label.

Consequently, GAO (1991a) believed that the DOD's estimate of incremental costs is too high—but it estimates that the total cost of the operation was actually over $100 billion. About half of this huge total was for direct and indirect costs of raising, equipping, operating, and supporting 540,000 Americans in the Gulf, as well as about $10 billion of related costs, above all forgiving Egypt's $7 billion of debt. Even if we would assume that the available accounts are basically sound and that the whole cost of the Desert Shield/Storm should be charged against the price of the Persian Gulf oil, which sum are we to choose? But the second assumption is hardly any more valid than the first one. Oil was certainly the key reason for the U.S. involvement but the Iraqi quest for nuclear and other nonconventional weapons with which the country could dominate and destabilize the region, and implications of this shift for the security of U.S. allies, risks of another Iraqi–Iranian or Arab–Israeli war and the spread of militant fundamentalist Islam certainly mattered as well.

There is simply no objective procedure to separate the costs of such a multiobjective operation as the Desert Shield/Storm according to its varied but interconnected aims. Contributions of the Desert Shield/Storm to the real cost of oil are even more unclear because of substantial foreign burden sharing. As the total pledges of $48.3 billion of foreign help were almost $800 million above the OMB's estimate of U.S. funding requirements, one could stay on these narrow accounting grounds and argue that the country actually made a profit! And whatever the actual cost of the operation to the U.S., against what amount of oil is this to be prorated? Merely against the U.S. imports of the Persian Gulf crude (which at that time amounted to less than one-fifth of all U.S. purchases) or against all shipments from the region (they go mostly to Japan and Europe) or, because a stable Gulf means a stable OPEC and promotes low world oil prices, against the global production?

And what would be the time scale for prorating the costs? For the Desert Shield/Storm certainly not just the months of the operation—but how long had the war's stabilizing effect lasted? With Iraq defeated but unconquered some would argue that the war did not achieve its intended effect. And it has been followed by substantial annual military expenditures required to enforce the two no-flight zones above Iraq, to do constant reconnaissance, to maintain aircraft carriers on station in the Persian Gulf, to resupply land bases with prepositioned material, and to engage in exercises and contingency planning. And how to account for the entire span of the U.S. naval presence in the Gulf going back to 1949? What kinds of costs should be counted: totals needed to create and to operate various task forces, or only the incremental outlays? Should we count only the ships and the planes in the Gulf, or also those in the Indian Ocean and the Mediterranean fleet? And what share of this naval power operated for decades in the area as a part of the superpower conflict rather than as a guarantee of unimpeded oil flow?

And is it correct to leave out from the real price of the Persian Gulf oil the costs of U.S. economic assistance to the countries of the region? Between 1962 and 2000 the net U.S. disbursements to Israel, Egypt, and Jordan added up to nearly $150 billion, with about three-fifths of this total going to Israel. Clearly a sizeable portion of these expenses—be it for purchase of military hardware, food imports, or industrial development—was spent in order to prop up the two friendly Arab regimes and to strengthen the country's principal strategic ally in the region, and such investments could be logically seen as yet another cost of Middle Eastern oil.

Obviously, there can be no nonjudgmental agreement about what share of that aid should be charged against the cost of crude oil exported from the region. And the increased U.S. military presence in the region following the September 11, 2001 attack on New York and Washington only complicates any attempts at assessing the real cost of the Middle Eastern oil. Fight against terrorist networks is undoubtedly paramount in this new effort, but larger geopolitical considerations concerning Egypt, Israel, Palestine, Saudi Arabia, Iraq, and Iran are obviously very much in play in addition to the ever-present concern about the stability of the region that contains most of the world's oil reserves.

Although we should not underestimate the progress made toward the fuller pricing of energy (or energy converters) during the past two generations, we must realize that most of the challenge is still ahead and that there are no easy ways to proceed. Communist central planning had, I hope, utterly discredited the route of comprehensive government intervention in energy, or any other, pricing. Many former Communist countries combined some of the world's most stringent emission standards and public health norms with enormous energy conversion inefficiencies and horrendous environmental destruction. Performance criteria imposed after an agreement between legislatures and industries, akin to the U.S. CAFE limits, have been useful but that practice is open to exceptions. The most egregious exemption is classifying all SUVs, although they are clearly used as passenger cars, as light trucks and thus making them exempt from the 27.5-mpg requirement that has to be met by cars. Equally indefensible is the now more than a decade-long pause in requiring higher fleet car efficiencies: CAFE has not been strengthened since 1987!

And even when all interested parties agree on the need for a much more inclusive pricing there is only a rare possibility of readily acceptable quantifications. Moreover, while internalized costs could promote better market transactions leading to more rational uses of energies and thus help to avoid environmental degradation, there are only a very few markets for environmental goods and services, and this reality raises the chance of further governmental interventions in introducing more inclusive pricing. An obvious counterargument is that states are already intervening in a large number of indirect, and less effective, ways, and that they get invariably involved in managing the consequences of environmental abuse—and hence it would be more rational if they would try to minimize both the frequency and the extent of such degradation.

Perfectly inclusive pricing is an elusive goal as there are always considerable uncertainties in the ranges of plausible cost estimates (Viscusi et al. 1994). And it will be particularly difficult to internalize precisely those hidden burdens of energy use that appear to have the highest external costs: rapid global warming comes naturally first to mind. The intertwined problems of scale (what to include) and complexity (how to account for it) will continue to be the most intractable obstacles in the quest for more inclusive energy pricing, but incremental improvements could bring us closer to rational internalizations. Because of these systemic uncertainties concerning the real cost of various energy conversions I am disinclined to put much trust into often resolutely phrased arguments about relative costs of individual processes or about comparative advantages of new techniques. There are, of course, many instances where standard accounting helpfully reveals indisputable advantages or drawbacks of a particular choice but I believe that the best possible evaluations of energy conversions must go beyond the narrow, and obviously imperfect, bounds of usual assessment of profitability.

Energy and the Quality of Life

All energy conversions undertaken by humans share the same raison d'être: they are just means toward a multitude of ends. All of the commonly used measures of energy use—be it conversion efficiencies, energy costs, per capita utilization levels, growth rates, consumption elasticities, or output ratios—are just helpful indicators of the performance and the dynamics of processes whose aim should not be merely to secure basic existential needs or to satisfy assorted consumerist urges but also to enrich intellectual lives and to make us more successful as a social and caring species. And, given the fundamental necessity to preserve the integrity of the one and only biosphere we inhabit, all of those aims should be accomplished in ways that are least disruptive to the maintenance of irreplaceable environmental services. High quality of life, physical and mental, is the goal; rational energy use is the means of its achievement.

Assessment of average national quality of human life cannot rely on a single surrogate variable. Quality of life is obviously a multidimensional concept that embraces attributes of narrow physical well-being (these, in turn, reflect wider environmental and social settings) as well as the entire spectrum of human mental development and aspirations. Foremost in the first category is the access to adequate nutrition and to health care, as well as the capacity to address effectively a large array of natural and man-made risks (ranging from air pollution to violent crime): only a successful record in these matters can assure a full span of active life. The second key component of human well-being starts with the universal delivery of good-quality basic education and should obviously include the exercise of personal freedoms.

I will chose one or two critical measures in each of these categories and correlate them with average per capita energy consumption in the world's 57 most populous nations. These countries, each having more than 15 million inhabitants, account for nearly 90% of the world's population. As with the links between the already explored relationship between energy use and the economic performance there is no doubt about some significant correlations between rising levels of per capita energy utilization and the higher physical quality of life characterized by such variables—but a closer look brings some surprising conclusions and it also reveals clear saturation levels.

Infant mortality and life expectancy are perhaps the two best indicators of the physical quality of life. The first variable is an excellent surrogate measure as it is a highly sensitive reflection of complex effects of nutrition, health care, and environmental exposures on the most vulnerable group in any human population. The second measure encompasses long-term effects of these critical variables. During the late 1990s the lowest infant mortalities were in the most affluent parts of the modern world, in Japan (a mere 4 deaths/1,000 live births), Western Europe, North America, and Oceania (5–7), and the highest rates (in excess of 100, and even of 150) were in nearly 20 African countries (mostly in the sub-Saharan part of the continent) as well as in Afghanistan and Cambodia (UNDP 2001).

Leaving the anomalously low Sri Lankan rate aside, acceptable infant mortalities (below 30/1,000 live births) corresponded to annual per capita energy use of at least 30–40 GJ. But fairly low infant mortalities (less than 20/1,000 live births) prevailed only in countries consuming at least 60 GJ a year per capita, and the lowest rates (below 10) were not found in any country using less than about 110 GJ (fig. 2.10). However, increased energy use beyond this point is not associated with any further declines of infant mortality, and the correlation for the entire data set of 57 countries is −0.67, explaining 45% of the variance.

In every society female life expectancies at birth are, on the average, 3–5 years longer than the rates for males. During the late 1990s global minima of average female life expectancy were below 45 years in Africa's poorest countries and maxima over 80 years in Japan, Canada, and in a dozen European nations. As in the case of infant mortality the correlation with average per capita energy use is less than 0.7 (the sign being positive in this case), which means that the relationship explains less than half of the variance. And, once again leaving the Sri Lankan anomaly aside, female life expectancies above 70 years are seen in countries consuming no more than 45–50 GJ of energy per capita, the seventy-five-year threshold is surpassed at about 60 GJ, but the averages above 80 years are not found in any country consuming less than about 110 GJ/capita (fig. 2.11).

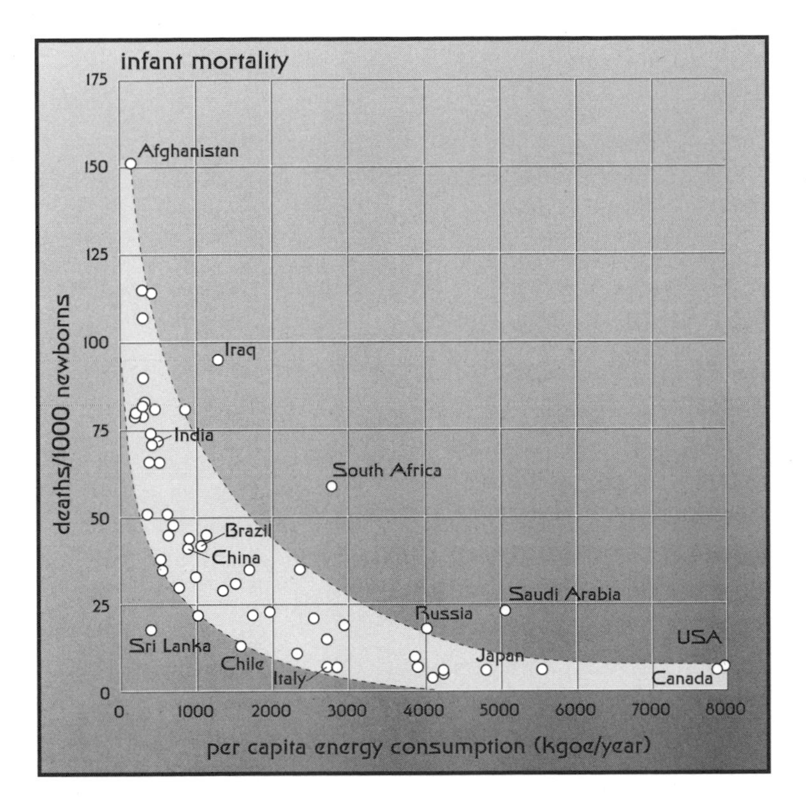

Figure 2.10
Comparison of infant mortality with average annual per capita use of commercial energy. Plotted from data in UNDP (2001).

Comparisons of average per capita energy consumption with average food availability are not particularly useful. Effective food rationing can provide adequate nutrition in a poor nation even as the variety of foodstuffs remains quite limited while high per capita food supplies in rich countries are clearly far beyond any conceivable nutritional needs and dietary surveys shows that as much 40% of all food available at retail level is wasted (Smil 2000c). This means that national means of food energy availability should not be used for "the higher the better" comparisons but rather as indicators of relative abundance and variety of food, two considerations associated with the notion of good life in virtually every culture. Minimum per capita availabilities satisfying the conditions of adequate supply (with comfortable reserves) and good variety are over 12 MJ/day, the rates corresponding, once again, to the average per capita consumption of between 40–50 GJ of primary energy per

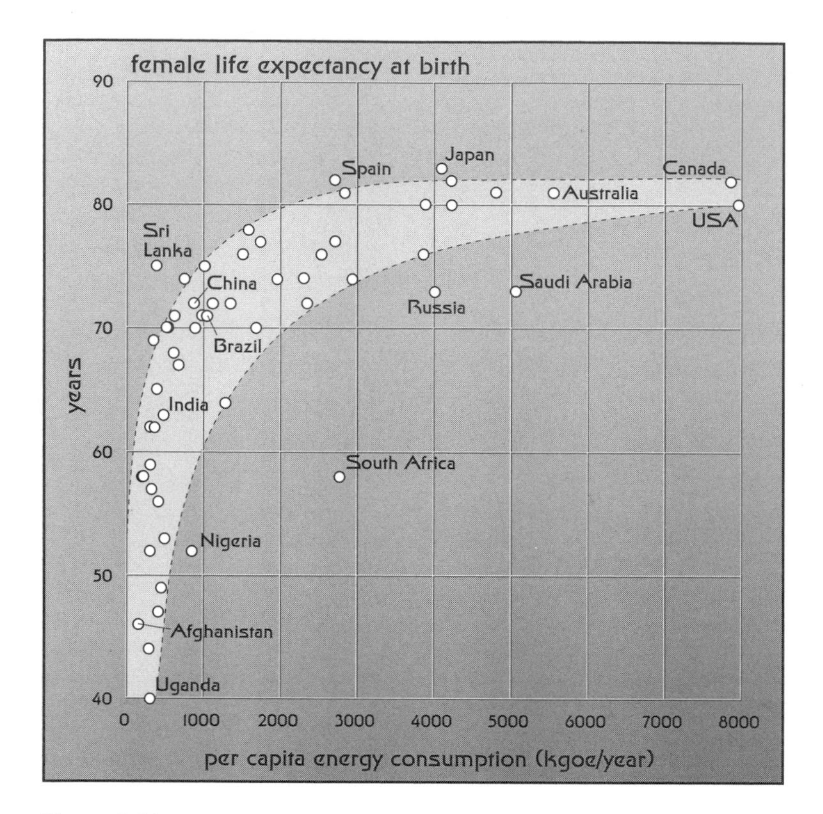

Figure 2.11
Comparison of average female life expectancy with average annual per capita use of commercial energy. Plotted from data in UNDP (2001).

year (fig. 2.12). Not only there is no benefit in raising food availabilities above 13 MJ/day but such surfeit of food also greatly increases the chances of overeating and widespread obesity.

National statistics on education and literacy levels may be readily available but their interpretation is not easy. Enrollment ratios at primary and secondary level tell us little about actual ability to read, comprehend, and calculate; they reflect mostly just the shares of pupils completing compulsory schooling but convey no qualitative distinctions between those who remain functionally illiterate and those who can continue postsecondary studies. And the rates of university attendance and graduation may reflect more the much-lowered admission and testing standards than any notable intellectual achievements. Many recent studies have shown dismal performance

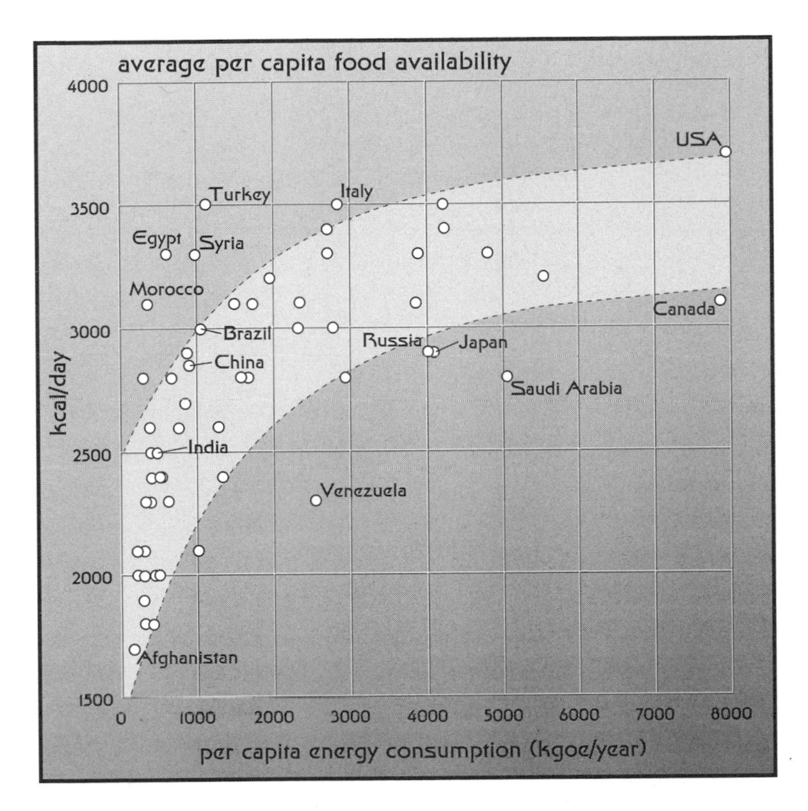

Figure 2.12
Comparison of average daily per capita food availability with average annual per capita use of commercial energy. Plotted from data in FAO (2001) and UNDP (2001).

of U.S. students; for example, during the late 1990s as many as 23% of high school seniors lacked even rudimentary reading skills (NCES 2001).

Both the literacy rates and combined primary, secondary, and tertiary enrollment ratios must then be seen as merely quantitative indicators of general availability of education and not as truly revealing qualitative measures. In any case, very high levels of the latter rate (at and above 80% of the population in the relevant age category) have been attained in some countries with primary energy per capita use as low as 40–50 GJ/year. As university-level education is obviously more expensive, its large-scale availability—judged, once again, just quantitatively by more than 20–25% of all young adults receiving postsecondary education—has been associated with per capita energy consumption of at least about 70 GJ.

UNDP uses four quality-of-life indicators (life expectancy at birth, adult literacy, combined educational enrollment, and per capita GDP) to construct the Human Development Index (HDI; UNDP 2001). The index differs little among the world's 20 best-off countries: in 2001 Norway topped the list with 0.939, followed by Australia, Canada, and Sweden (all with 0.936) but the value for Italy, in the twentieth place, was still 0.909. The lowest values, below 0.35 are shared by seven sub-Saharan countries, including the populous Ethiopia and Mozambique. At about 0.75 the correlation between the HDI and average per capita energy use is somewhat higher than for individual variables, but even so, 45% of the variance remains unexplained. Data plot shows, once again, a nonlinear trend with high level (above 0.8) reachable with as little as 65 GJ/capita and with minimal, or no gains, above 110 GJ (fig. 2.13).

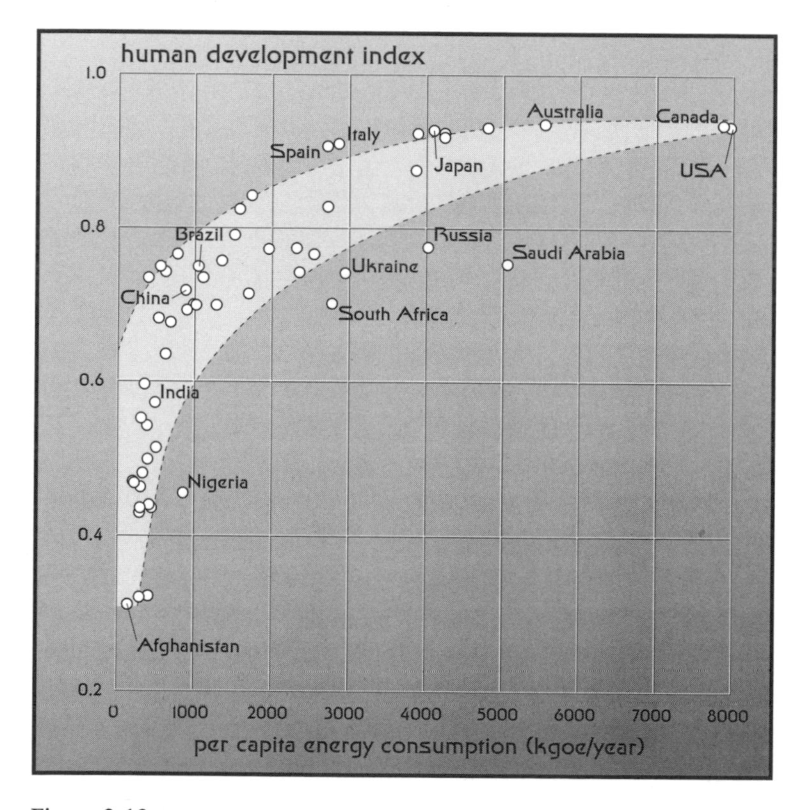

Figure 2.13
Comparison of the Human Development Index (HDI) with average annual per capita use of commercial energy. Plotted from data in UNDP (2001).

The weakest of all important links among the energy consumption and the quality of life concerns the political arrangements effectively guaranteeing personal freedoms. This is hardly surprising as soon as one recalls that all of the fundamental personal freedoms and institutions of participatory democracy were introduced and codified by our ancestors generations before the emergence of high-energy, fossil-fueled civilization when average per capita primary energy use was a mere fraction of the late twentieth-century levels. The only key exception is the suffrage for women. Although many U.S. states adopted that right during the nineteenth century the U.S. federal law was passed only in 1920, and an analogical British act only made it through the Parliament in 1928 (Hannam et al. 2000).

And the history of the twentieth century shows that suppression or cultivation of these freedoms was not dictated by energy use: they thrived in the energy-redolent United States as they did in energy-poor India, and they were repressed in the energy-rich Stalinist Soviet Union as they still are in energy-scarce North Korea. While it is indisputable that personal freedoms remain beyond the reach in most low-energy societies in Africa, Asia, and Latin America, a comprehensive worldwide survey of political rights and civil liberties indicate that such a link is not inevitable and that cultural and historical forces have been much more important in determining the outcome.

Consequently, correlating the annual comparative assessment of the state of political rights and civil liberties that is published annually by Freedom House (2001) with the average per capita energy use reveals only a weak linkage. The ranks of free countries (Freedom House ratings between 1–2.5) contain not only all high-energy Western democracies but also such mid- to low-level energy users as South Africa, Thailand, Philippines, and India. Conversely, countries with the lowest freedom rating (6.5–7) include not only energy-poor Afghanistan, Vietnam, and Sudan but also oil-rich Libya and Saudi Arabia. Correlation coefficient for 57 of the world's most populous nations is only −0.51, explaining just 27% of the variance. Personal freedoms are thus compatible with societies using as little as about 20 GJ of energy a year per capita (Ghana, India) and the populous nations with the highest freedom ranking (1–1.5) using the lowest amount of commercial fuels and electricity (less than 75 GJ) are Chile and Argentina (fig. 2.14).

These correlations allow some fascinating conclusions. A society concerned about equity and willing to channel its resources into the provision of adequate diets, availability of good health care, and accessibility to basic schooling could guarantee decent physical well-being, high life expectancy, varied nutrition, and fairly good

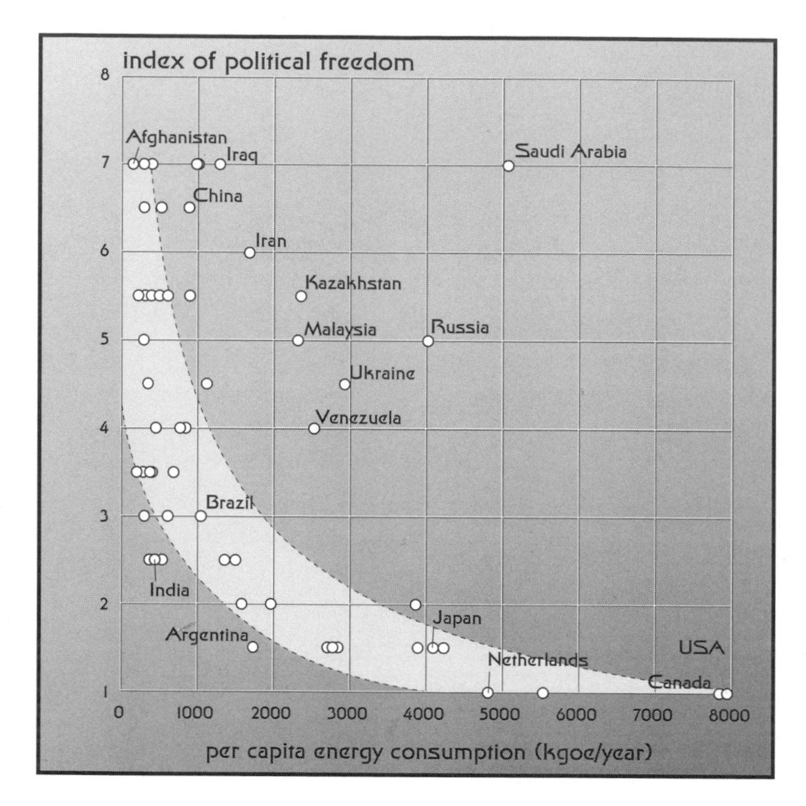

Figure 2.14
Comparison of the Freedom Index with the average annual per capita use of commercial energy shows that a high degree of personal freedom is not predicated on very high energy consumption. As the outliers right of the best-fit line show, just the reverse is true as far as a number of major oil producers is concerned. Plotted from data in UNDP (2001) and Freedom House (2001).

educational opportunities with an annual per capita use of as little as 40–50 GJ of primary energy. A better performance, pushing infant mortalities below 20, raising female life expectancies above 75 and elevating HDI above 0.8 appears to require at least 60–65 GJ of energy per capita, while currently the best global rates (infant mortalities below 10, female life expectancies above 80, HDI above 0.9) need no less than about 110 GJ/capita. All of the quality-of-life variables relate to average per capita energy use in a nonlinear manner, with clear inflections evident at between 40–70 GJ/capita, with diminishing returns afterwards and with basically no additional gains accompanying consumption above 110 GJ/capita. And prospects for a nation's political freedoms have little to do with any increases in energy use above existential minima.

Annual per capita energy consumption of between 50–70 GJ thus appears to be the minimum for any society where a general satisfaction of essential physical needs is combined with fairly widespread opportunities for intellectual advancement and with the respect for basic individual rights. I will explain some fascinating implications of this mean and some remarkable opportunities associated with its possible attainment in some detail in the book's last chapter. But before ending this brief exploration of linkages between energy the quality of life I must note the notable absence of correlation between the average economic well-being and energy use on one hand and the feelings of personal and economic security, optimism about the future, and general satisfaction with life on the other.

In 1999 annual average per capita energy use in Germany (175 GJ) was only half, and in Thailand (40 GJ) a mere one-eighth, of the U.S. rate (340 GJ), and the American PPP-adjusted GDP was 34% above the German mean and 5.2 times the Thai average (UNDP 2001). But the Gallup poll done in 1995 found that 74% of Germans and Thais were satisfied with their personal life—compared to 72% of Americans (Moore and Newport 1995). Such findings should not be surprising as a personal assessment of the quality of life involves strong individual emotions and perceptions that may be largely unrelated to objectively measured realities. In fact, many studies have shown little connection between subjective appraisals of quality of life and personal satisfaction on one hand and the objective socioeconomic indicators on the other (Nader and Beckerman 1978; Andrews 1986; Diener, Suh and Oishi 1997). The quest for ever-higher energy use thus has no justification either in objective evaluations reviewed in this section, or in subjective self-assessments.

Energy and the Environment

Environmental consequences of producing, moving, processing, and burning coals and hydrocarbons and generating nuclear electricity and hydroelectricity embrace an enormous range of undesirable changes (Smil 1994a; Holdren and Smith 2000). Spectacular failures of energy systems—such as the destruction of Chernobyl's unshielded reactor in 1986 (Hohenemser 1988), or a massive spill of crude oil from the *Exxon Valdez* in 1989 (Keeble 1999)—will periodically capture public attention with images of horrifying damage. But the effects of cumulative gradual changes—including those such as acidification or eutrophication of ecosystems whose causative factors are invisible inputs of reactive compounds—are far more worrisome as these processes are far more widespread and persistent. Also the need for long-term commitments, be they for a safe disposal of radioactive wastes or effective worldwide reductions of greenhouse gas emissions, is a far greater challenge than dealing with spectacular accidents.

Chemical energy of fossil fuels is released by their combustion and hence it is inevitable that the fossil-fueled civilization has had a particularly acute impact on the atmosphere. Combustion of fossil fuels has increased substantially the atmospheric burden of particulate matter (PM) and it has multiplied the global atmospheric flux of sulfur and nitrogen oxides (released mostly as SO_2 and NO and NO_2), hydrocarbons, and carbon monoxide (CO). These compounds were previously released only from biomass burning or as by-products of bacterial or plant metabolism.

Particulate matter includes all solid or liquid aerosols with diameter less than 500 μm. Large, visible particulates—fly ash, metallic particles, dust, and soot (carbon particles impregnated with tar)—that used to be released from uncontrolled coal and oil combustion in household stoves and in industrial and power plant boilers settle fairly rapidly close to their source of origin and are rarely inhaled. Very small particulates (diameters below 10 μm) can be easily inhaled, and aerosols with diameter 2.5 μm and smaller can reach alveoli, the lung's finest structures, and contribute to the extent and severity of chronic respiratory problems (NRC 1998). Small PM can also stay aloft for weeks and hence can be carried far downwind and even between the continents. For example, only 7–10 days after Iraqi troops set fire to Kuwaiti oil wells in late February 1991, soot particles from these sources were identified in Hawai'i, and in subsequent months solar radiation received at the ground was reduced over an area extending from Libya to Pakistan, and from Yemen to Kazakhstan (Hobbs and Radke 1992; Sadiq and McCain 1993).

SO_2 is a colorless gas that cannot be smelled at low concentrations while at higher levels it has an unmistakably pungent and irritating odour. Oxidation of sulfur present in fossil fuels (typically 1–2% by mass in coals and in crude oils) is its main source (smelting of color metals, petroleum refining, and chemical syntheses are the other major emitters). U.S. emissions of the gas peaked in the early 1970s at nearly 30 Mt/year, and have been reduced to less than 20 Mt by the mid-1990s, with the 1980–1999 decline amounting to about 28% (Cavender, Kircher, and Hoffman 1973; EPA 2000; fig. 2.15). Global emissions of SO_2 rose from about 20 Mt at the beginning of the twentieth century to more than 100 Mt by the late 1970s; subsequent controls in Western Europe and North America, and collapse of the Commu-

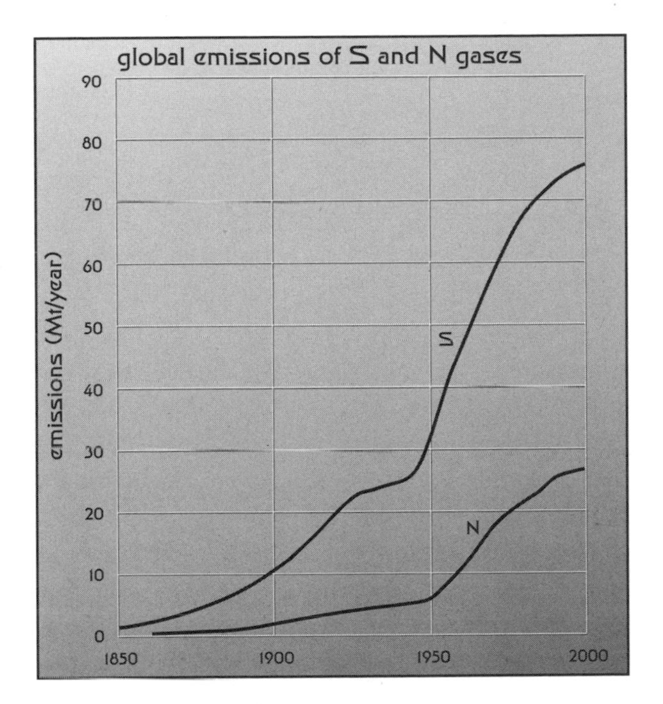

Figure 2.15
Global emissions of sulfur and nitrogen oxides during the twentieth century. The rates are expressed in Mt N and Mt S per year. Based on figures in Smil (2002), EPA (2000), and Cavender, Kircher, and Hoffman (1973).

nist economies cut the global flux by nearly one-third—but Asian emissions have continued to rise (McDonald 1999).

Nitrogen oxides are released during any high-temperature combustion that breaks the strongly bonded atmospheric N_2 and combines atomic N with oxygen. Electricity-generating plants are their largest stationary sources, vehicles and airplanes the most ubiquitous mobile emitters. Anthropogenic hydrocarbon emissions result from incomplete combustion of fuels, as well as from evaporation of fuels and solvents, incineration of wastes, and wear on car tires. Processing, distribution, marketing, and combustion of petroleum products are by far the largest source of volatile hydrocarbons in all densely populated regions. In spite of aggressive control efforts total U.S. emissions of NO_x had actually increased by about 1% between 1980 and 1999, remaining at just above 20 Mt/year, but during the same time average concentrations in populated areas declined by about 25% (EPA 2000).

Colorless and odorless CO is the product of incomplete combustion of carbon fuels: cars, other small mobile or stationary internal combustion engines (installed in boats, snowmobiles, lawn mowers, chain saws), and open fires (burning of garbage and of crop residues after harvest) are its leading sources. Foundries, refineries, pulp mills, and smouldering fires in exposed coal seams are other major contributors. Emission controls (using catalytic converters) that began on all U.S. vehicles in 1970 have been able to negate the effects of a rapid expansion of car ownership and of higher average use of vehicles: EPA estimates that by the late 1990s U.S. CO emissions fell by about 25% compared to their peak reached in 1970 (Cavender, Kircher, and Hoffman 1973; EPA 2000).

Effects of these emissions differ greatly both in terms of their spatial impact and duration as well as in their impact on human health, ecosystems, and materials. Air pollution from combustion of coals and hydrocarbons was the twentieth century's most widespread energy-related environmental degradation. Combination of uncontrolled emissions of PM and SO_2 created the classical (London-type) smog, which was common in Europe and in North America until the 1960s (Stern 1976–1986). Its marks are greatly reduced visibility, higher frequency of respiratory ailments and, during the most severe episodes occurring during the periods of limited atmospheric mixing (as in London in 1952 or in New York in 1966), increased mortality of infants and elderly with chronic lung and cardiovascular diseases.

Laws to limit ambient air pollution (British Clean Air Act of 1956, the U.S. Clean Air Act of 1963 and Air Quality Act of 1967), gradual replacement of coal by hydrocarbons, and widespread uses of electrostatic precipitators (they can remove in excess of 99% of all PM) that began in the 1950s combined to virtually eliminate visible black smoke from Western cities and industrial regions. For example, in 1940 the U.S. combustion and industrial processes released almost 15 Mt of particulates smaller than 10 μm but the total was only about 3 Mt during the late 1990s (Cavender, Kircher, and Hoffman 1973; EPA 2000).

Epidemiological evidence assembled since the late 1980s indicates that increases in human mortality and morbidity have been associated with particulate levels significantly below those previously considered harmful to human health (Dockery and Pope 1994). This effect has been attributed to particles smaller than 2.5 μm that are released from mainly motor vehicles, industrial processes, and wood stoves. For this reason in 1997 the EPA introduced new regulations to reduce concentrations of such particles (EPA 2000). Once implemented, this new rule might prevent as many as

20,000 premature deaths a year and reduce asthma cases by 250,000—but these claims have been a highly controversial, and appropriate control measures are to be phased in gradually.

Anthropogenic SO_x and NO_x are eventually (in a matter of minutes to days) oxidized and the resulting generation of sulfate and nitrate anions and hydrogen cations produces precipitation whose acidity is far below the normal pH of rain (about 5.6) acidified only by carbonic acid derived from the trace amount of CO_2 (about 370 ppm) constantly present in the atmosphere (fig. 2.16). Beginning in the late 1960s the phenomenon was noted to cover large areas up to about 1,000 km downwind from large stationary sources in Western and Central Europe as well as in Eastern North America (Irving 1991; see also fig. 2.9). After 1980 an increase in acid deposition became evident over large areas of South China (Street et al. 1999).

Its effects include loss of biodiversity in acidified lakes and streams (including disappearance of the most sensitive fish and amphibian species); changes in soil

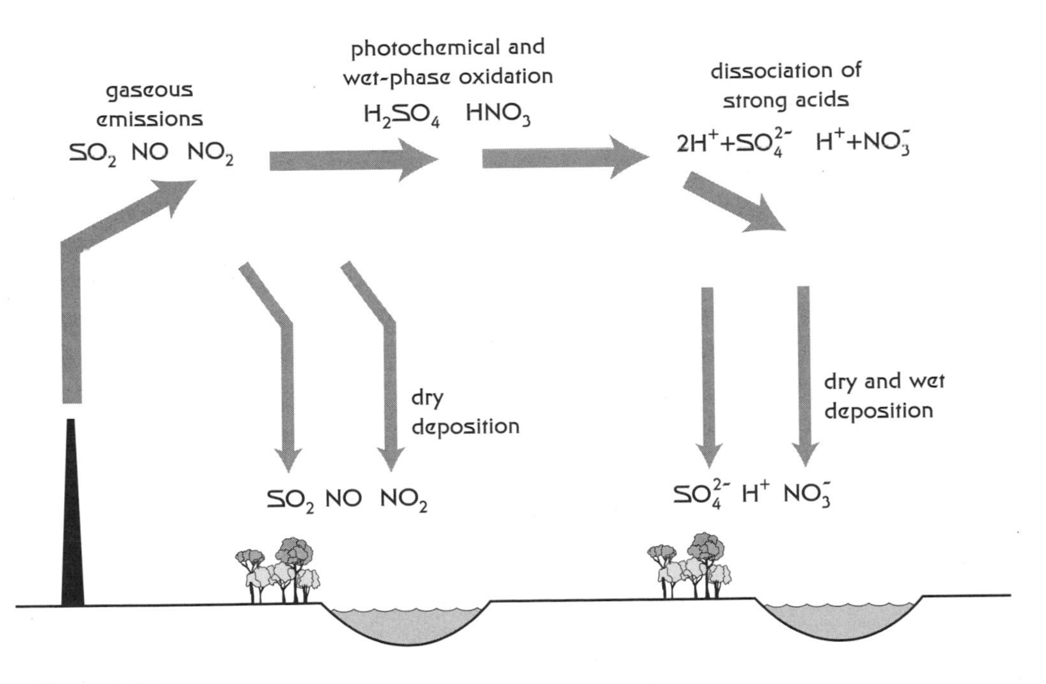

Figure 2.16
Dry and wet acidifying deposition originates from emissions of sulfur and nitrogen oxides generated by combustion of fossil fuels.

chemistry (above all the leaching of alkaline elements and mobilization of aluminum and heavy metals); and acute and chronic effects on the growth of forests, particularly conifers (Irving 1991; Godbold and Hütterman 1994). Acid precipitation also increases the rates of metal corrosion, destroys paints and plastics, and wears away stone surfaces. Some ecosystemic consequences of acid deposition are short-lived and relatively easily reversible, others may persevere for decades even after such emissions were sharply reduced or nearly eliminated.

As already noted, commercial FGD of large stationary sources, introduced for the first time during the early 1970s, has dramatically lowered SO_2 emissions in the United States, Japan, and parts of Europe (NAPAP 1991; Stoddard et al. 1999; EPA 2000). Together with switching to less sulfurous fuels these improvements stopped, or even reversed, decades-long processes of environmental acidification whose impact has been particularly severe on sensitive aquatic ecosystems (Stoddard et al. 1999; Sullivan 2000). But in China, now the world's largest emitter of SO_2, the area of acid precipitation is still expanding in the country's southern part (Street et al. 1999).

Progress has been also achieved in controlling the precursors of photochemical smog, a mixture of air pollutants generated in the presence of sunlight by complex atmospheric reactions of NO_x, CO, and volatile organic compounds. These reactions produce ozone, an aggressive oxidant causing higher incidence of respiratory diseases in both people and animals, reduced crop yields and damaged forests and other materials (Colbeck 1994). This form of air pollution, observed for the first time in Los Angeles during the mid-1940s, is now a semipermanent, or a seasonal, presence in large cities on every continent. Expansion of megacities and of intercity land and air traffic is now creating regional, rather than just urban, photochemical smog problems (Chameides et al. 1994).

Stationary sources of NO_x have been harder to control (for more see chapter 4) but since the early 1970s automotive emissions of the three smog precursors have been greatly reduced by a combination of redesigned internal combustion engines and by mandatory installation of three-way catalytic converters removing very large shares of CO, NO_x, and hydrocarbons (Society of Automotive Engineers 1992). As a result, by the year 2000 average U.S. car emissions (measured as g/km) were cut by 97% for hydrocarbons, 96% for CO, and 90% for NO_x when compared to the precontrol levels of the late 1960s (Ward's Communications 2000). Few other pollution control efforts can show such impressive gains, and new low-emission vehicles will reduce the remaining hydrocarbon emission by an additional 75% and will halve the existing NO_x rate.

Only as we were achieving these successes in reducing the levels of major outdoor pollutants did we come to appreciate that indoor air pollution poses often higher health risks than does the ambient air (Gammage and Berven 1996). High levels of fine PM and carcinogens are especially common in rural areas of poor countries where inefficient combustion of biomass in poorly ventilated rooms leads commonly to chronic respiratory diseases (World Health Organization 1992). For example, in China the rural mortality due to chronic obstructive pulmonary diseases is almost twice as high in rural areas than in cities: rural ambient air is cleaner but villagers using improperly vented stoves are exposed to much higher levels of indoor air pollution (Smith 1993; Smil 1996). The effect on children younger than five years is particularly severe: in poor countries 2–4 million of them die every year of acute respiratory infections that are greatly aggravated by indoor pollutants.

Two kinds of global concerns caused by fossil fuel combustion emerged during the last two decades of the twentieth century: global climate change and interference in the nitrogen cycle. During the 1980s, after a century of on-and-off studies of anthropogenic global warming, it became evident that the emissions of greenhouse gases, at that time dominated by CO_2, have become a factor in global climate change. Basic consequences of the process were outlined rather well by Arrhenius (1896) just before the end of the nineteenth century: a geometric increase of CO_2 producing a nearly arithmetic rise in surface temperatures; minimum warming near the equator, maxima in polar regions; and a lesser warming in the Southern hemisphere. The current period of concern was relaunched in 1957 when Roger Revelle and Hans Suess concluded that

> human beings are now carrying out a large scale geophysical experiment of a kind that could not have happened in the past nor be reproduced in the future. Within a few centuries we are returning to the atmosphere and oceans the concentrated organic carbon stored in sedimentary rocks over hundreds of millions of years (Revelle and Suess 1957, p. 18).

When the first systematic measurements of rising background CO_2 concentrations began in 1958 at two American observatories, Mauna Loa in Hawai'i and at the South Pole, CO_2 concentrations were about 320 ppm but the average for the year 2000 at Mauna Loa was just a fraction below 370 ppm (Keeling 1998; CDIAC 2001; fig. 2.17). Expanding computer capabilities made it possible to construct the first three-dimensional models of global climatic circulation during the late 1960s, and their improving versions have been used to forecast changes arising from different future CO_2 levels (Manabe 1997). During the 1980s global climate change studies became increasingly interdisciplinary, and their extensions included a delayed

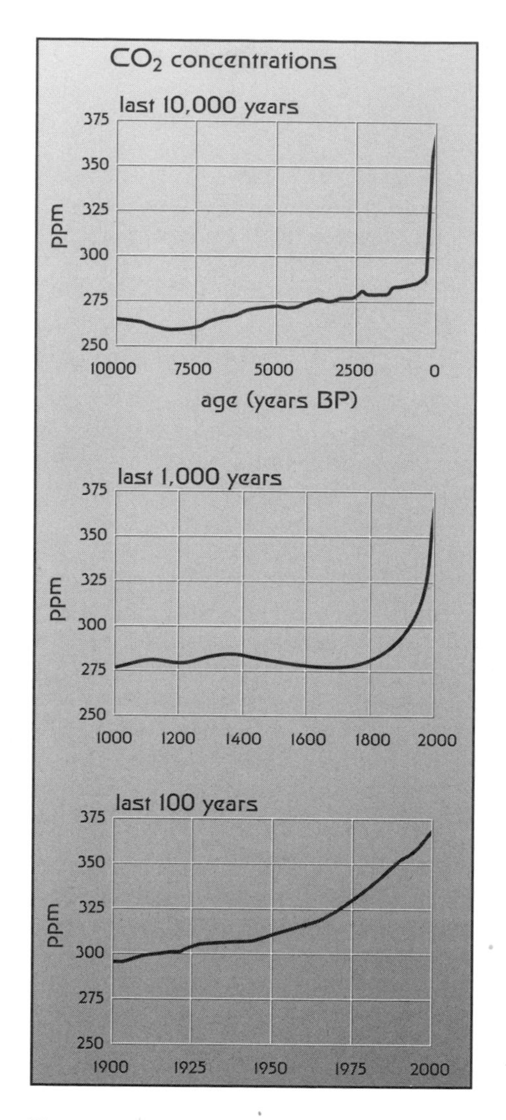

Figure 2.17
Atmospheric CO_2 concentrations are shown on three time scales: during the past 10,000 years, during the last millennium, and during the last century. Based on figures in Houghton et al. (2001).

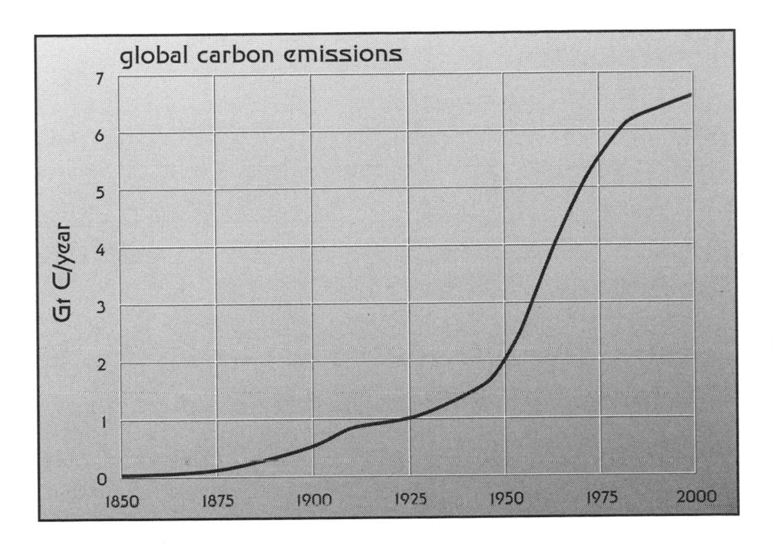

Figure 2.18
Global anthropogenic carbon emissions from fossil fuel combustion, cement production, and the flaring of natural gas, 1850–2000. Plotted from data in Marland et al. (2000).

recognition of the importance of other greenhouse gases (CH_4, N_2O, CFCs), and attempts to quantify both hydrospheric and biospheric fluxes and stores of carbon (Smil 1997).

Annual CO_2 emissions from the combustion of fossil fuels have been surpassing 6 Gt C since 1989, with about 6.7 Gt C emitted in the year 2000 (Marland et al. 2000; fig. 2.18). This is a small fraction of the atmosphere-biosphere exchange as about 100 Gt C are withdrawn from the atmosphere every year by terrestrial and marine photosynthesis and plant and heterotrophic respiration promptly returns nearly all of this carbon to the atmosphere, leaving behind only 1–2 Gt C/year in the slowly expanding biomass stores in forests (Wigley and Schimel 2000). But almost half of all carbon released every year from fossil fuels remains in the atmosphere and as the main absorption band of CO_2 coincides with the Earth's peak thermal emission the more than 30% increase of the gas in 150 years has already increased the energy reradiated by the atmosphere by 1.5 W/m^2.

Moreover, the warming effect of other greenhouse gases now roughly equals that of CO_2, bringing the total anthropogenic forcing to about 2.8 W/m^2 by the late 1990s (Hansen et al. 2000; IPCC 2001; fig. 2.19). This is equivalent to little more than 1% of solar radiation reaching the Earth's surface, and a gradual increase of

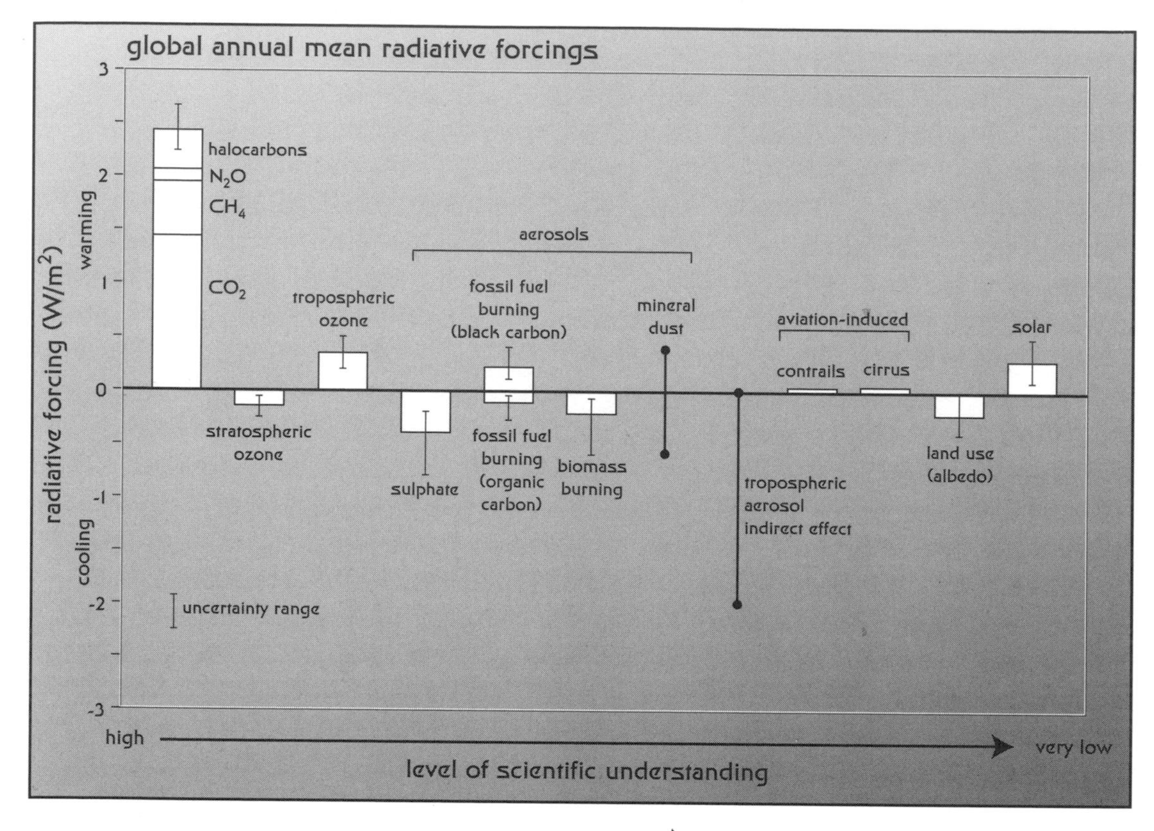

Figure 2.19
Global mean of radiative forcing of the Earth's atmosphere for the year 2000 relative to the year 1750. The overall effect has been nearly 2.5 W/m^2, with CO_2 accounting for about 60% of the total. Based on a figure in IPCC (2001).

this forcing would eventually double preindustrial greenhouse gas levels (about 280 ppm in 1850) and, according to the latest consensus report (Houghton et al. 2001), raise the average tropospheric temperatures by anywhere between 1.4–5.8 °C above today's mean. Key points of broad scientific consensus on all of these matters are best summarized in a series of reports by the Intergovernmental Panel on Climatic Change (Houghton et al. 1990; Houghton et al. 1996; Houghton et al. 2001).

Although higher atmospheric CO_2 levels would have some beneficial consequences for the biosphere (Smil 1997; Walker and Steffen 1998), both the public and the research attention has been concentrated on possible, and often indefensibly exaggerated, negative impacts. These would arise above all from the accelerated global water cycle and from three kinds of increases during winters in high Northern latitudes: higher surface warming, greater surface warming of the land than of the sea, and increased precipitation and soil moisture. More disturbingly, practical results in actively moderating greenhouse gas emissions have been meager. While it is true that the world's primary energy supply at the end of the twentieth century contained about 25% less carbon than at its beginning—declining from about 24 tC/TJ in 1900 to about 18 tC/TJ in 2000—this decarbonization has been a consequence of gradual long-term substitutions of coal by hydrocarbons and primary electricity rather than of any concerted action driven by concerns about the global climate change (Marland et al. 2000).

Concerns about rapid global warming have spawned endless rounds of ever more expensive scientific studies, as well as arcane negotiations aiming at eventual global agreements on moderating the emissions of greenhouse gases—but no resolute actions. Even the rather feeble reductions of greenhouse gases mandated by the 1997 Kyoto agreement proved impossible to realize. This is indefensible, as there is little doubt that major reductions of greenhouse gas emissions can be achieved with minimal socioeconomic costs in affluent countries. I will return to this critical concern in the last chapter.

During the latter half of the twentieth century the high-energy civilization also began interfering to an unprecedented degree in the global nitrogen cycle (Galloway et al. 1995; Smil 1997). Applications of synthetic nitrogen fertilizers and NO_x emissions from fossil-fuel combustion are the two largest sources of the anthropogenic enrichment of the biosphere with what is normally a key growth-limiting macronutrient. Anthropogenic nitrogen flux now rivals the natural terrestrial fixation of the nutrient. Nitrogen from combustion is the smallest of the three principal fluxes: about 85 Mt N come from nitrogen fertilizer, between 30–40 Mt N a year originate in the planting of leguminous species symbiotic with N-fixing bacteria, and about 25 Mt N come from fossil-fuel combustion. But reactive nitrogen compounds produced by atmospheric reactions can be transported over considerable distance, contributing either to acid deposition or excessive nitrogen enrichment of sensitive ecosystems. Obviously, this problem will also require a much more effective management.

A long array of other environmental impacts arising from extraction, transportation, and conversion of fossil fuels and from the generation of primary electricity has many local and regional consequences. Among the most objectionable changes are the visual assaults caused by large-scale destruction of plant cover and reshaping of landscapes that accompanies surface coal extraction, particularly in mountainous or hilly areas, such as the Appalachia (NRC 1981) and the accidental oil spills from crude oil tankers. These spills are particularly objectionable and damaging to wildlife and recreational potential when they take place in coastal waters and affect large areas of sandy or rocky beaches (Burger 1997).

Invisible but no less worrisome alterations range from excessive silting of reservoirs that impound water for electricity generation to accidental releases of excessive levels of radioactivity from nuclear power plants. Reservoirs are also often surprisingly large sources of greenhouse gases, and concerns about radioactive contamination will not end with the decommissioning of existing power plants as some of the accumulated wastes will have to be monitored for 10^2–10^3 years. I will address these, and many other environmental concerns regarding renewable energies and nuclear generation while assessing the future of these conversion techniques in chapter 5.

Energy and War

The first draft of this chapter, including the remarks about the impossibility of any clear-cut attribution of military costs to the real price of oil, was finished just after the attacks of September 11, 2001 (fig. 2.20). A few weeks later, after rereading the chapter, I decided that at least a brief survey of links between modern wars and energy would be appropriate: after all, wars were among the key events shaping the twentieth century and the use of Boeing 767s as projectiles on September 11, 2001 demonstrated that it will not be any naïve "end of history" (Fukuyama 1991) but rather new violent conflicts that will shape the twenty-first century.

New weapons produced and energized by abundant and cheap fossil fuels and electricity transformed the twentieth-century war making an extraordinarily rapid and frightening fashion, with horrifying consequence for military and civilian casualties. As with so many other twentieth-century developments foundations of this new destructive power were put in place by the discoveries of a new class of chemicals prepared by nitration of such organic compounds as cellulose, glycerine, phenol, and toluene during the latter half of the nineteenth century (Urbanski 1967). Much like the venerable gunpowder these compounds, including Alfred Nobel's dynamite and ballistite, and Hand Henning's cyclonite, the most powerful prenuclear explosive, were self-oxidizing, but they produced a far more powerful blast able to create shock waves.

More destructive delivery of these high-energy explosives has been made possible by new machines whose design and production has invariably demanded much more

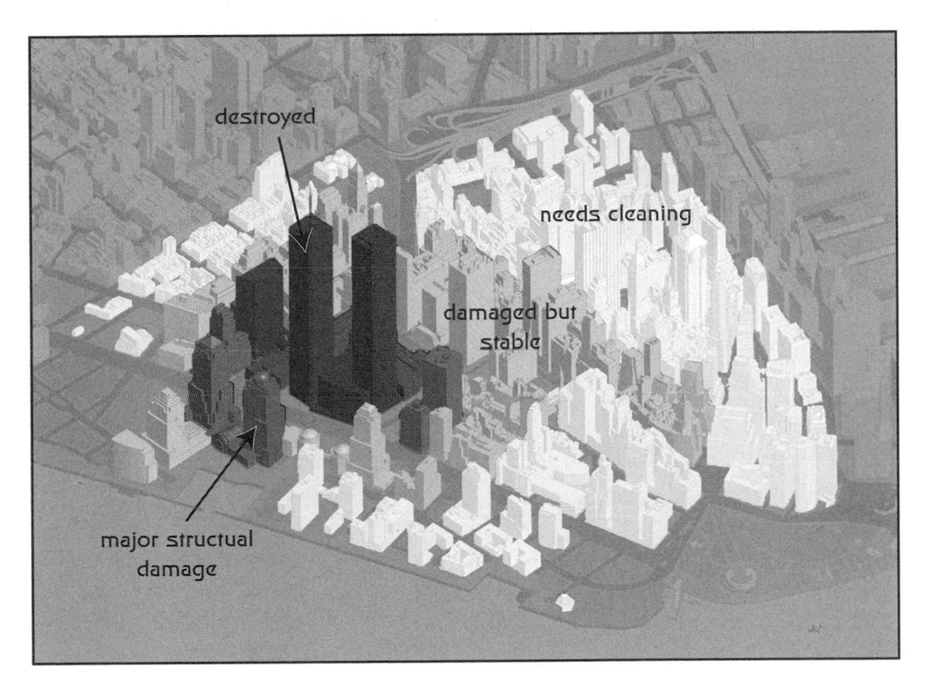

Figure 2.20
Structures destroyed and damaged by the terrorist attack on the World Trade Center on September 11, 2001. Based on an image prepared by the City of New York and Urban Data Solutions and posted at <http://www/cnn.com/SPECIALS/trade.center/damage.map.html>.

embedded energy. These deadly advances can be seen in every class of weapons on land and sea and in the air, by comparing rifles with machine guns, horse-drawn field guns with heavily armored tanks, battery-powered submarine with nuclear vessels, cruisers with aircraft carriers and first wood-canvas-wire planes of World War I with the titanium-clad stealth machines of the 1990s. The last example is a perfect illustration of the escalating energy cost: materials needed to make the pioneering airplanes (wood, cloth, iron, and steel) required no more than 5–25 MJ/kg to produce, those going into the latest stealth machines need more than 200 MJ/kg for composite materials and special alloys.

High-energy explosives delivered by more effective means raised the battle toll by an order of magnitude. Combatant death toll during the two major nineteenth-century conflicts that involved major military powers of the day powers, the Crimean War of 1853–1856 and the Franco-Prussian War of 1870–1871, was less than 200 fatalities per 1,000 men of armed forces fielded at the beginning of a conflict. This death ratio surpassed 1,500 during World War I and 2,000 during World War II, with the staggering total of 4,000 for Russia (Singer and Small 1972). Germany lost about 27,000 combatants per million people during World War I, but more than 44,000 during World War II. And civilian casualties grew even faster, reaching more than 70% of the total death total during World War II.

Conventional bombing of large cities produced repeatedly huge noncombatant losses within days or mere hours (Kloss 1963). German fatalities totalled nearly 600,000, and about 100,000 people died during nighttime raids by B-29 bombers on Japan's four principal cities between March 10–20, 1945. These effects resulted from a massive release of incendiary bombs rather than from dropping any extraordinarily powerful explosives. In contrast, the two nuclear bombs that killed at least 100,000 people in August 1945, released energy of, respectively, 52.5 and 92.4 TJ (Committee for the Compilation . . . 1981). In retrospect, these were toy weapons compared to ten independently targeted warheads placed on a single submarine-launched intercontinental ballistic missile whose explosion would release up to 800 times more energy than did the Hiroshima bomb. And, to cite just one more of those unimaginably huge totals, by the Cold War's end the total explosive power of U.S. and Soviet nuclear warheads was equivalent to more than 800,000 Hiroshima bombs.

The other unmistakable feature of modern wars that is closely linked with high energy consumption is the degree of economic mobilization required for conducting major armed conflicts. The trend started during World War I and it is readily illustrated with the growth of aircraft manufacturing. In August 1914 Britain had just 154 airplanes but four years later the country's aviation industry was employing 350,000 people and turning out 30,000 planes a year (Taylor 1989). Similarly, U.S. manufacturers delivered just 514 aircraft to the country's air force during the last quarter of 1940 but before the war was over they produced more than 250,000 planes (Holley 1964).

At the same time, the link between energy use and success in war is—very much like the previously discussed links between energy consumption and the quality of life—far from being a simple matter of strong positive correlations. There is, of course, no doubt that the excessive stockpile of nuclear weapons, whose energy cost amounted to appreciable shares of superpower energy consumption (my conservative estimate is that it took about 5% of all U.S. and Soviet energy use between 1950 and 1990), has been the main reason why they did not fight a war. But while a rapid and enormous mobilization of American energy was clearly instrumental in overcoming Japan and Germany—the expenditure of more explosives and the deployment of more sophisticated weapons in Vietnam did not lead to the U.S. victory. And the attacks of September 11, 2001 illustrate the perils and the penalties of the aptly named asymmetrical threats: a coordinated operation launched with a minimal energy cost produced global repercussions necessitating the deployment of considerable military power and leaving behind economic damage of at least $2 trillion.

Finally, a few paragraphs on energy resources as the cause of war. Historians of the twentieth century like to point out to the Japanese quest for oil supply as an explicitly stated reason for the country's surprise attack on Pearl Harbor in December 1941 (Sagan 1988). Indeed, during the summer and fall of 1941 there were repeated warnings by top Japanese military leaders that the navy was running out of oil and that they wanted a quick decision to act. But a more accurate reading of the events is not to deny a proximate role that Japan's declining oil supplies (following the U.S. export embargo) played in launching the attack on the Pearl Harbor but also to acknowledge a long history of expansive Japanese militarism (so clearly demonstrated with the 1933 conquest of Manchuria and the 1937 attack on China) as well as a peculiarly self-inflicted nature of the entire confrontation with the United States (Jansen 2000). And there is certainly no disagreement in concluding that neither Hitler's aggression against Poland nor the decisions of major European powers to enjoin World War I could be ascribed to any quest for energy resources.

Post–World War II military interventions, directly or by a proxy, and sales of arms designed to bolster friendly regimes include a number of involvements where a nation's oil endowment had clearly played a leading, if not the dominant, role. These actions started with the Soviet attempt to take over northern Iran in 1945 and 1946 and continued with decades of Soviet arms sales to Egypt, Syria, and Iraq and concurrent American arm shipments to Iran, Saudi Arabia, and the Gulf states. During the 1980s they included strong—and as it turned out very soon an utterly misplaced—Western support of Iraq during its long war with Iran (1980–1988), and they culminated in the Desert Shield/Storm operation of 1990–1991, whose monetary costs were noted earlier in this chapter.

By invading Kuwait, Iraq not only doubled crude oil reserves under its control, raising them to about 20% of the global total, but it also directly threatened the nearby Saudi oilfields, including al-Ghawar, the world's largest reservoir of liquid fuel, and indeed the very survival of the monarchy that alone possesses one-fourth of the world's oil reserves. Yet even in this seemingly clear-cut case there were, as previously described, other compelling reasons to check the Iraqi expansion. Critical assessment of modern conflicts lends a clear support to Lesser's (1991) conclusions that resource-related objectives have been generally determined by broader strategic aims, and not vice versa.

At the time of this writing we can only guess how the linkage between energy and war could be affected by the combination of declining availability of the cheapest conventional oil in countries outside of the Middle East (these prospects are assessed in some detail in chapters 3 and 4) and further advances of fundamentalist and aggressively anti-Western Islam. Plausible scenarios span an uncomfortably wide range of possibilities, from a manageable transition toward more open and at least quasi-democratic regimes throughout the region thus far dominated by dynastic (Saudi Arabia), clerical (Iran), or military (Iraq) autocracies—all the way to visions of a global Armageddon unleashed by the suicidal armies of the faithful against the infidels.

3

Against Forecasting

Why do we not forecast? And how should we look ahead without forecasting? I will answer the first question by looking back and demonstrating that for more than 100 years long-term forecasts of energy affairs—no matter if they were concerned with specific inventions and subsequent commercial diffusion of new conversion techniques or if they tried to chart broad sectoral, national, or global consumption trends—have, save for a few proverbial exceptions confirming the rule, a manifest record of failure. I see no reason to perpetuate such repeatedly failing, unproductive and, as they may engender false feelings of insight, actually counterproductive endeavors. Repeated forecasting failures call for a radical departure: dismal record of long-range forecasts of energy affairs demonstrates convincingly that we should abandon all detailed quantitative point forecasts. But what is a rational substitute? Replacing detailed quantitative predictions by exploratory forecasts clearly outlining a range of plausible alternative options appears to be a much more sensible approach—but such exercises have often presented fans of plausible outcomes that are obviously too broad to serve as useful guides of effective action.

A great deal of evidence shows that we should not expect any fundamental improvements of this unhelpful record by substituting simpler forecasts of any kind by increasingly complex computerized models. Consequently, I will argue that only two kinds of looking ahead are worthwhile, indeed essential. The first kind consists of contingency scenarios preparing us for foreseeable outcomes that may deviate substantially, even catastrophically, from standard trend expectations or from consensus visions. Unusually deep and prolonged global economic depression that would set back average living standards by a generation or two is one such possibility, as is a Middle Eastern conflagration that would result in a loss of at least one-third of the world's crude oil supply and all of the attendant repercussions for the world's economy. The terrorist attack on the United States on September 11, 2001 had widened tragically the scope of such plausible catastrophic events. Now we have to contemplate also the use of chemical, bacterial, or viral weapons against civilian populations in large cities or the use of jetliners as missiles to attack nuclear power plants, all undertaken by terrorist groups whose eradication presents enormous challenges for any modern open society.

The second kind of forecasts encompasses no-regret normative scenarios that should be prepared to guide our long-term paths toward the reconciliation of human aspirations with the biospheric imperatives. While it may not be easy to reach consensus on every major issue that should form a part of such normative outlines it is quite realistic to expect broad agreement on many key desiderata. After all, does anybody favor a rapid global warming with unpredictable consequences for the biosphere, or an even greater inequality of access to basic energy services than is experienced in today's poor, populous nations? Naturally, a careful examination of various limits that circumscribe our options and of numerous uncertainties that may derail our plans must precede the formulation of any broad or specific goals.

But before making the arguments in favor of the normative forecasting and before detailing the limits and uncertainties whose understanding must inform their formulation I will review first the two opposite tendencies commonly encountered among forecasts of technical advances. On one hand there are many fascinating (and depressing) failures of imagination regarding the invention and subsequent commercialization of important energy conversion techniques. On the other hand there is excessive confidence in the potential of particular technical fixes that are seen to hold (often near-magical) solutions to our problems and whose early commercialization is forecast to bring prosperous futures. I will follow these reviews by similarly disheartening looks at failed forecasts of energy requirements, prices, intensities, and resource substitutions.

Failing Endeavors

Widespread forecasting of particular technical, social, or economic developments is, as with so many other developments, a product of industrial and postindustrial eras with their rapidly changing techniques and mores. Explicit forecasts began appearing more frequently during the closing decades of the nineteenth century. They became ubiquitous only since the late 1960s as the computerization of business decision making, technical design, and academic research has made it easy to deploy many quantitative forecasting methods. More importantly, computers have made it possible to construct increasingly more complex, and hence seemingly more realistic, models of the real world. An expanding array of forecasting tools—ranging from simple algebraic formulas to intricate computer simulations, and from expert consensus to elaborate scenario building—has been used not only for technical and economic prognoses but also for forecasts of national political trajectories or trends of the global environmental change.

Most of the formal, institutional forecasts, particularly those concerning economic matters, are exercises in very short-term vision. The long-standing obsession of modern economists with high rates of economic growth and the recent infatuation of affluent populations with high-flying (and nose-diving) stock markets has generated an unprecedented demand for economic forecasting but most of these forays look only months, or a year or two, ahead and a decades-long horizon might as well be in eternity. The same is true about countless supply-and-demand forecasts that are being prepared constantly by all kinds of businesses. In contrast, advances in transportation (ranging from new kinds of private cars to space travel), medical breakthroughs (elimination of specific diseases, timing of particular cures), and diffusion of new consumer items (when will everybody have a PC?) have been among the most common subjects of longer-range forecasts extending up to a decade, or even two, ahead.

As expected in the world dominated by short-term concerns, most of the forecasts of overall primary energy and electricity requirements and sectoral energy needs span only months to a few years. Still, there have been many forecasts extending for more than a decade, and up to half a century, and recent concerns about global warming have brought energy consumption forecasts spanning an entire century. Long-range forecasts of energy matters became more common only during the 1960s, and they turned into a small growth industry a generation ago, after OPEC quintupled the price of crude oil by the beginning of 1974. They now cover an enormous spectrum ranging from fairly narrow exercises focusing on capacities and performances of individual exploration, production, and conversion techniques to ambitious, and highly disaggregated, demand and price models of national, regional, and global fuel and electricity futures. Some of these models are available at no cost from their authors; others, such as DRI/McGraw-Hill World Energy Projections, touted for their unique modelling and forecasting capabilities, have required their subscribers to pay every year the sum several times as large as the average global per capita income.

One commonality is shared by virtually all of these forecasts: their retrospective examinations show a remarkable extent of individual and collective failure to predict both the broad trends and specific developments. Lack of imagination has been a common failure when appraising technical breakthroughs, but excessive faith in new techniques, particularly by people involved in their development and promotion, has been no less common when assessing future commercial diffusion of these innovations. With rare exceptions, medium- and long-range forecasts become largely worthless in a matter of years, often just a few months after their publication. Moreover, these failures appear to be unrelated to the subject of a particular forecast or to specific techniques used. Routinely performed long-range forecasts of national energy demand tied to GDP growth rates and consumption elasticities have not fared better than the appraisals of unprecedented technical innovations by the world's leading experts, and intricate econometric exercises have not been more successful than simple demand models.

Conversion Techniques

Predictions of technical developments became quite common during the second half of the nineteenth century as new inventions began changing the world that in terms of leading energy sources and principal conversion techniques had been generally stagnant for millennia. The best way to review this record of forecasting failures is to let the original, irresistibly poor, judgments speak for themselves. I will thus quote several notable examples concerning new energy conversion techniques introduced since the 1880s, beginning with the epochal invention of commercial electricity generation and distribution.

In 1879, just three years before T. A. Edison began selling electricity for lighting in both London and New York (fig. 3.1), the Select Committee on Lighting by Electricity of the British House of Commons (1879) heard an expert testimony that there is not "the slightest chance" that electricity could be "competing, in a general way, with gas." Exactly a decade later Edison (1889, p. 630) himself was making a big blunder:

> My personal desire would be to prohibit entirely the use of alternating currents . . . I can therefore see no justification for the introduction of a system which has no element of permanency and every element of danger to life and property . . . I have always consistently opposed high-tension and alternating systems . . . because of their general unreliability and unsuitability for any general system of distribution.

Burying and insulating AC wires, Edison argued, "will result only in the transfer of deaths to man-holes, houses, stores and offices." But Edison's extraordinary feel for innovation did not desert him for long and this emotional "battle of the systems"

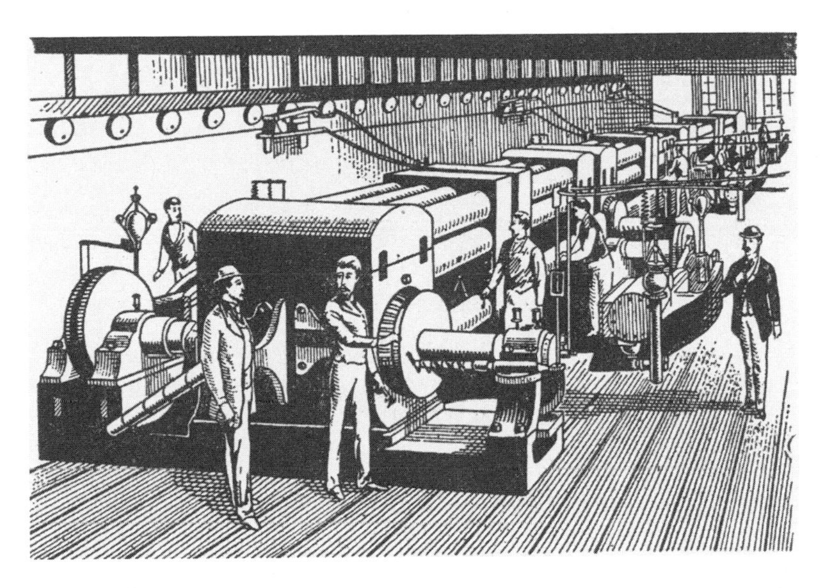

Figure 3.1
Six jumbo dynamos of Thomas Edison's first American electricity-generating station located at 255–257 Pearl Street in New York and commissioned on September 4, 1882. The station's direct current was initially supplied to only 85 customers and it energized just 400 light bulbs. Image reproduced from *Scientific American* (August 26, 1882).

was short-lived. Alternating current transmission and distribution, favored by the other three great pioneers of the electric era—Nikola Tesla, George Westinghouse, and Sebastian Ferranti—prevailed during the 1890s. Edison, realizing his error, quickly abandoned his advocacy of direct current and steered his company into the AC world (David 1991).

Edison's second miscalculation concerning energy conversion came a decade later. Before the end of the nineteenth century many people saw the internal combustion engine as only a short-lived choice for a transportation prime mover. In his biography Henry Ford reminisced that his employer objected to his experiments with the gas engine, believing that electric car was the coming thing. And the Edison Company offered him the general superintendency but "only on the condition that I would give up my gas engine and devote myself to something really useful" (Ford 1929, pp. 34–35). Although they cost more than the internal combustion cars, by 1900 electric cars dominated the emerging U.S. automotive market (fig. 3.2). In 1901 six recharging stations in New Jersey made it possible to drive electric cars from New York to Philadelphia, and in 1903 Boston had 36 recharging sites (McShane 1997).

Figure 3.2
Early electric cars (Immisch & Co. model built in 1888 for the Sultan of Turkey is shown here) looked very much like the contemporary gasoline-fueled vehicles. There were 24 small batteries in the pictured car, enough to power it at 16 km/hour for 5 hours. Reproduced from Tunzelmann (1901).

But these were ephemeral gains as cars powered by internal combustion engines were steadily gaining commercial ground. In spite of this clear trend Edison spent nearly the entire first decade of the twentieth century trying to develop a high-density battery that could compete with gasoline (Josephson 1959). This was a costly, and ultimately futile, quest: Edison's solution, a nickel–iron–alkaline battery introduced in 1909, proved to be a dependable standby source of electricity for uses ranging from miners' lights to ship-board propulsion, but not a competitive prime mover for automobiles (fig. 3.3). Yet a century later the lure of the electric car is still with us. New forecasts keep promising that these vehicles will capture a significant share of the automotive market within a decade—but once that decade is over we find that, mirage-like, the promised share is still another decade ahead.

Recent legislative forcing of electric cars in California has not helped either. In 1995 the California Energy Commission (CEC) decided that by 1998, 2% of all new vehicles (or about 22,000 cars) sold in the state will have to be electric and that the share of zero-emission vehicles (ZEVs) will rise to 10% of the state's car sales (or close to 150,000 units) by the year 2003 (Imbrecht 1995). But no commercial electric cars went on sale during the late 1990s and in January 2001 California's Air Re-

Figure 3.3
Thomas Edison believed that electric cars, such as the one that he is seen here inspecting in 1913, will prevail over the unclean gasoline-powered cars. From the Smithsonian Institution's photo set *Edison After 40* at <http://americanhistory.si.edu/edison/ed_d22.htm>.

sources Board redefined the goal for the year 2003: at least 10% of newly sold vehicles must have low emissions but only 2% must be ZEVs, i.e., electric cars (Lazaroff 2001). Subsequently, this requirement was put on hold at least until 2005.

The very idea of an airborne internal combustion engine was even harder to accept. Three years before the Wright brothers accomplished the first flight with a heavier-than-air machine above the dunes at Kitty Hawk in North Carolina on December 17, 1903 (fig. 3.4), Rear Admiral George W. Melville (1901, p. 825) concluded that

> Outside of the proven impossible, there probably could be found no better example of the speculative tendency carrying man to the verge of chimerical than in his attempts to imitate the birds . . . Should man succeed in building a machine small enough to fly and large enough to carry himself, then in attempting to build a still larger machine he will find himself limited by the strength of his materials in the same manner and for the same reasons that nature has.

And in the year when the brothers actually took off, Octave Chanute (1904, p. 393), their enthusiastic supporter and himself a designer of excellent gliders, called on first principles when he argued that airplanes

Figure 3.4
This was not supposed to happen: one of the twentieth century's most memorable photographs shows the Wrights' machine lifting off above the sands of the Kitty Hawk, North Carolina at 10:35 AM on December 17, 1903. Orville is piloting and Wilbur, who ran alongside the plane on its slow takeoff, watches as the machine rises. Image available at the Smithsonian National Air and Space Museum site at <http://www.nasm.edu/galleries/gal100/wright_flight.gif>.

will eventually be fast, they will be used in sport, but they are not to be thought of as commercial carriers . . . the sizes must remain small and the passengers few, because the weight will, for the same design, increase as the cube of the dimensions, while the supporting surface will only increase as the square . . . The power required will always be great, say something like one horse power to every hundred pounds of weight, and hence fuel can not be carried for long single journeys.

Chanute's assumption works out to about 17 W/kg of the total airplane mass but the rate for the Boeing 747, whose maximum takeoff weight is close to 400 t and whose four turbofan engines develop about 80 MW during takeoff, is over 200 W/kg, an order of magnitude higher. This is, of course, primarily the consequence of lighter yet more powerful prime movers. The Wrights' heavy homemade four-stroke engine (with aluminum body and a steel crankshaft) weighed 8 g/W, while two generations later the best piston engines of World War II needed just 0.8 g/W. Today's large gas turbines, high-bypass turbofans installed on wide-bodied jets, weigh less than 0.1 g/W (Smil 1994a).

Technological miscasting of energy matters has continued to thrive during the second half of the twentieth century but in contrast to the pre-WWII period the principal failure has not been due to timid imagination but rather to excessive faith in the practical potential of new techniques. Certainly the most flagrant example of this overblown confidence was the uncritical faith in nuclear fission-powered electricity generation. This blindness was not common only during the late 1940s and the early 1950s when excuses could be made because of the pioneering stage of the fission's development, and when nuclear energy was expected to open a future of abundance where "consumption, not production, will be a problem" (Crany et al. 1948, p. 46). A generation later Spinrad (1971) predicted that by the late 1990s almost 90% of new electrical generating capacity everywhere except in Africa will be nuclear, and that fission will supply over 60% of the world's electricity generation.

Also in 1971, Glenn Seaborg, a Nobelian (in 1951 for his research on chemistry of transuranium elements) and at that time the Chairman of the U.S. Atomic Energy Commission, predicted that by the year 2000 nuclear energy will bring "unimagined benefits" that will directly improve the quality of life for most of the world's population (Seaborg 1971, p. 5). Fission reactors were to generate not only nearly all electricity for households and industrial uses but also to transform the world's agriculture by energizing food production complexes producing fertilizers and desalting sea water. The concept of large 'nuplexes' was first presented by Richard L. Meier in 1956 and it was later elaborated by the Oak Ridge National Laboratory (Meier 1956; ORNL 1968; Seaborg 1968). These complexes—centered on large nuclear plants (eventually huge breeder reactors) and located in coastal desert areas and providing energy for desalinization of sea water, production of fertilizers, industrial plants, and intensive crop cultivation (fig. 3.5)—were to make many areas of the world's deserts habitable and productive.

And there were no doubts about nuclear energy's indispensability: Seaborg and Corliss (1971) thought that without it civilization would slowly grind to a halt. But it would be unfair to single out just one illustrious forecaster whose visions failed as so many scientists and engineers, including a number of other Nobel Prize winners, foresaw a world shaped by ubiquitous and inexpensive nuclear energy. Indeed, another Nobelian, Hans Bethe (1967 prize in physics, for his work on nuclear energy production in stars), concluded that "the vigorous development of nuclear power is not a matter of choice, but of necessity" (Bethe 1977, p. 59).

Bold forecasters also foresaw that the increasingly nuclear world, deriving its electricity from numerous large stationary reactors, will have nuclear-powered cargo ships and passenger planes, that it will use directed nuclear explosions to uncover mineral riches, reverse river flows, open new canals, and dig new harbors in Alaska

Figure 3.5
An agro-industrial complex centered around a large nuclear power plant sited on a desert coast. During the late 1960s nuclear enthusiasts envisaged that before the century's end these nuplexes—combining electricity generation, desalination, chemical syntheses, industrial production, and intensive cropping—will feed humanity and convert most of the world's deserts into habitable zones. Their actual number as of the year 2000: 0. Reproduced from ORNL (1968).

and Siberia, and that it will install propulsion reactors in rockets that were to ferry men to Mars. Nuclear fission was to produce energy for thermochemical splitting of water producing abundant hydrogen for urban and intercity transportation. In other visions of the nuclear future, people were to move into underground cities leaving the surface to return to wilderness: all it would take to reconnect with nature was to take an elevator. In such a world classical fission reactors were to be just the magic for the beginners, merely a temporary fix before being supplanted by fast breeders.

In 1970 the Nixon administration scheduled the completion of the U.S. prototype Liquid Metal Fast Breeder Reactor (LMFBR) for 1980. When outlining long-range possibilities of the global energy supply Weinberg (1973, p. 18) captured the hopes many invested into the technique by concluding that there is not "much doubt that a nuclear breeder will be successful" and that it is "rather likely that breeders will be man's ultimate energy source." Not surprisingly, Westinghouse Electric, the company contracted to build the U.S. breeder, was confident that "the world can reap tremendous benefits in terms of greatly increased energy resources" (Creagan 1973, p. 16). General Electric expected that commercial fast breeders would be introduced by 1982, and that they would claim half of all new large thermal generation markets in the United States by the year 2000 (Murphy 1974; fig. 3.6). And in 1977 a consortium of European utilities decided to build Superphénix, a full-scale breeder reactor at Creys-Malville as the prototype of the continent's future nuclear plants (Vendryes 1977). In reality, this inevitable, ultimate technique was abandoned long before the century's end.

The U.S. breeder program amounted to a string of failed promises. In 1967 the first demonstration reactor was proposed for 1975 completion at a cost of $100 million; by 1972 the completion date advanced to 1982, and cost estimates reached $675 million (Olds 1972). The entire project was abandoned in 1983, and the country's only small experimental breeder reactor (EBR-II, operated at Argonne National Laboratory since 1964) was shut down in 1994. As Superphénix was nearing its completion Vendryes (1984, p. 279) thought that the age of LMFBR "is now at hand, with all necessary safety guarantees," but it was precisely because of safety concerns that the reactor was shut down in 1990.

And breeders have not been actually the only ultimate energy promise: ever since the early1950s there has been an enormous interest and commensurate investment in that truly endless source of clean energy, nuclear fusion (Fowler 1997). Until the late 1960s, with no fundamental breakthroughs after 15 years of intensive research, these hopes were vague and distant but the first experimental successes with the toroidal magnetic-confinement device of the late 1960s (Artsimovich 1972) brought a great deal of optimism. Seaborg (1971) was confident that fusion experiments would show positive energy production before 1980. By 1972 it was predicted that electricity from commercial fusion will be generated by the year 2000, and during congressional hearings a witness pushed for a rapid boost in research funds in order to make commercial fusion available in 20 rather than in 30 years (Creutz 1972).

Group consensus of energy experts gathered for a long-range forecasting Delphi study I conducted in the early 1970s was that not just the median probability, but

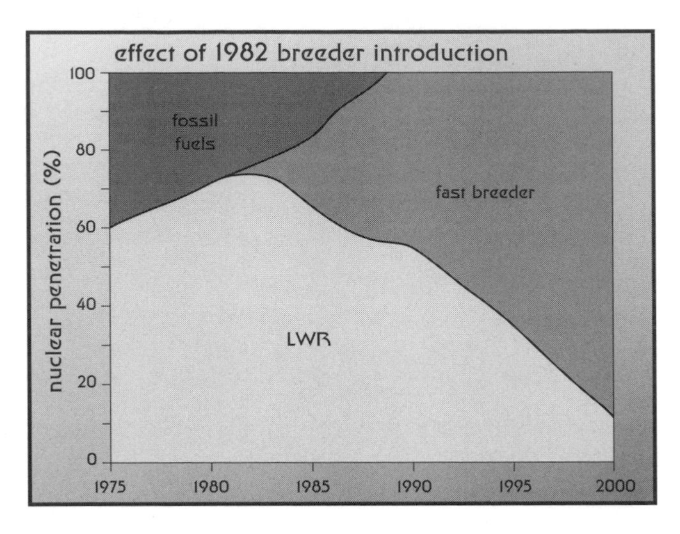

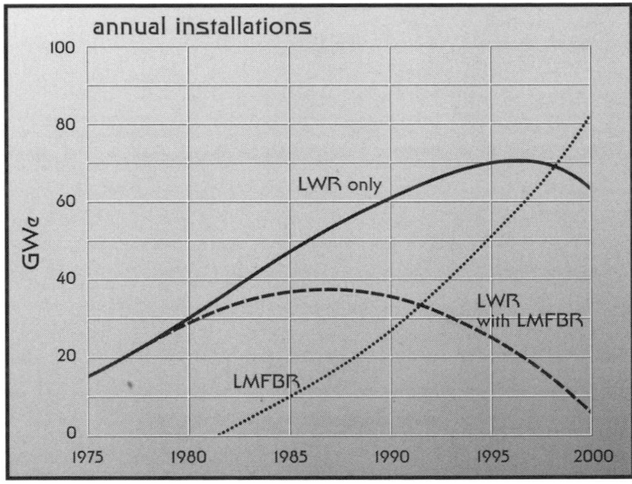

Figure 3.6
In 1974 General Electric expected no new fossil-fueled electricity generation in the United States after 1990 and LMFBRs to dominate new installations before the year 2000. Based on Murphy (1974).

also the seventy-fifth-quartile estimate for commercial introduction of thermonuclear generation was in the year 2000 (Smil 1974). Less than a decade after I completed my study the International Fusion Research Council (1979) concluded that "the net generation of fusion energy within fifteen years and the practical demonstration of an electricity-producing plant ten years later are reasonable goals for research and development in fusion." But after looking back at 40 years and $20 billion of fusion research an OTA (1987) panel concluded that the elusive goal of commercial fusion power was at least 50 years away. Five years later Colombo and Farinelli (1992) concurred that the goal of commercial energy from fusion appears to be 40–50 years away. That appears to be an eternally receding goal. In a November 1999 meeting, scientists involved in the International Thermonuclear Experiment Reactor project insisted that "electricity from fusion could be real in 50 years" (Ariza 2000, p. 19).

But fusion timetables were always purely speculative because, so far, no experimental reactor has even achieved break-even conditions. What is then actually even more surprising is that as the twenty-first century began there were no fission-powered food complexes, no nuclear islands containing plants with capacities of 10 GW, and no LMFBRs. To suggest now—as former President Eisenhower (urged by Lewis Strauss, former Chairman of the U.S. Atomic Commission) did in 1967, and as Weinberg still tried to do fifteen years later (Weinberg 1982)—that the best way to defuse the Arab–Israeli conflict is to build huge nuclear-power desalinization plants in the Sinai and the Gaza strip is utterly surrealistic. There is no shortage of true believers among the members of the once so large nuclear community who are waiting for the resurrection of fission and for an eventual breakthrough for fusion. I will return to these dreams in chapter 5. Meanwhile the reality is that although nuclear fission has become a relatively important contributor to the world's energy supply—providing about 7% of the world's primary demand and 17% of all electricity—it is withering away in all affluent countries where 85% of its capacity is located (most of the rest is in the states of the former Soviet Union).

Except for France and Japan, no new nuclear plant has been ordered in any affluent country since 1980, and neither any European country nor the United States or Canada have any plans to replace their existing ageing nuclear reactors (fig. 3.7). Instead, some countries are examining ways to shorten their reliance on nuclear generation, all are troubled by the prospect of decommissioning scores of reactors, and no nation has been able to solve the problem of long-term disposal of radioactive waste. Most of the world's new generating capacity installed during the last quarter of the twentieth century has not been filled by multigigawatt nuclear stations sited

Figure 3.7
Aerial view of Saint-Laurent-des-Eaux nuclear power plant on the Loire between Blois and Orléans. Two small graphite reactors, built in 1969 and 1971, were shut down in 1990 and 1992, two 900-MW PWRs were completed in 1981. Photo courtesy of Electricité de France.

offshore on artificial energy islands but by fossil fuel-fired units of less than 300 MW, and by even smaller (less than 100 MW) gas turbines (Williams and Larson 1988). As far as I know there were even no hints of such a trend in the technical literature of the 1960s and the early 1970s when expectations were for the dominance of large thermal units with installed capacities in excess of 1 GW (Smil 1974).

Before leaving nuclear matters I cannot resist a few brief citations from an editorial in *The Economist* whose bold and assured arguments appeared in print on March 29, 1986. Its opening sentence claimed that "only by investing heavily in nuclear power today can the world be sure of avoiding high-cost energy in the 1990s and beyond," and one of the key arguments for doing this was that "while several hundred people a year are accidentally killed digging coal from the earth, the nuclear-power industry remains as safe as a chocolate factory" (*The Economist* 1986, p. 11). The Chernobyl accident took place exactly four weeks later. Repeated, and very correct, explanations given by Western engineers, namely that the accident was not an inevitable outcome of generating electricity by fission but that it happened because of a poor Soviet design and incompetent operation procedures, did nothing to assuage the already deep public distrust of nuclear power. Less explicable is the fact that the weekly that relishes exposing rigid thinking editorialized in such a dogmatic way, affirming that only nuclear power can provide the abundance of clean, cheap energy the world will need in the twenty-first century.

As the nuclear mania began fading during the late 1970s, the emerging void of technical salvation filled rapidly with rising interest in hydrocarbons derived from nontraditional sources and in renewable energies. The first delusion was based on the undeniably enormous amount of such fuels in the Earth's crust. Even conservative appraisals show the total energy content of solid fuels being a multiple of liquid and gaseous fuels. Coal from huge open-cast mines could be liquefied and gasified, and liquid fuels could be extracted from oil shales and tar sands. IIASA's high scenario (Häfele et al. 1981) had an equivalent of 5 Gtoe (sic!) coming from these sources by the year 2030. Even its low version anticipated about 2.5 Gtoe from liquefied coal, shales, and tars, or an equivalent of about 70% of the annual crude oil output in the year 2000.

Even before it got spooked by the Iranian *coup d'etat* and humiliated by 14 months of the Tehran hostage affair, the Carter administration was a particularly eager promoter of coal liquefaction and gasification (Hammond and Baron 1976; Krebs-Leidecker 1977). By the century's end the United States was to be producing synthetic fuels equivalent to more than 21 EJ or about one-fifth of actual primary energy use in the year 2000. This huge total was to be dominated by coal-derived, high-energy synthetic gas and shale-derived synthetic crude. In 1980 the Energy Security Act established the Synthetic Fuels Corporation (SFC), allotted a stupendous $88 billion to the program and stipulated that "in no event" can it be terminated prior to September 1992.

The program's delusionary nature becomes clear when one recalls that no large-scale coal gasification or liquefaction facilities existed at that time—and yet by 1992 these untested conversions were to be in place at 40 plants each producing an equivalent of 2.5 Mt of oil a year and costing $3.5–4.5 billion (Landsberg 1986). As not a single one of these monsters (each was to occupy about 2.5 km^2, consume 7 Mt of coal/year, and have enormous environmental impacts) was ever built we will never know how much these hasty back-of-the-envelope cost estimates were off. President Reagan signed a bill disbanding the SFC as of mid-April 1986. The only completed gasification plant, the Great Plains Project in North Dakota built with $1.5 billion of federally guaranteed loans, began converting lignite into gas in 1984, defaulted on its payments in 1985, and it was sold by the government in 1988 for $85 million (Thomas 2001). *Sic transit gloria mundi.*

At the beginning of the twenty-first century these conversions are either nonexistent (massive coal gasification or liquefaction, tar oil recovery) or they make a marginal contribution to the world oil supply (extraction from oil sands). In the year 2000 oil extracted from Athabasca oil sands in north-central Alberta amounted to less than 5% of Canadian and less than 0.2% of global crude oil output (Suncor Energy 2001; BP 2001). Conversions of renewable energy flows have fared only marginally better. As breeders went out of fashion and biogas digesters became new low-tech favorites during the early 1980s the new infatuation brought a new wave of uncritical expectations and a new crop of forecasts extolling the near-miraculous potential of renewable energy conversions. Too many people—admirers of Schumacherian smallness, environmental and anticorporate activists, nostalgic ruralists, and do-it-yourself backyard fixers—put their faith in salvation through simple (always only a few moving parts), rugged, decentralized, user-friendly gadgets whose cost will only keep rapidly falling with time so they will soon dominate the Earth.

These nirvana techniques were to operate on land as well as in the ocean. Cultivation of common reeds (*Phragmites*) were to energize Sweden (Björk and Granéli 1978). Giant kelp (*Macrocystis*) plantations in Pacific waters off California's coast (people spoke about 'arable surface water') were to be harvested by a special fleet of ships, and kelp fronds were to be chopped or ground to make a slurry that was to be anaerobically fermented to supply all of America's gaseous fuels (Show et al. 1979). Thousands of wind turbines were to dot windy sites of America's Great Plains and gigantic arrays of wind towers were to rise above coastal waters while wave-energy machines, incessantly devouring the North Sea's swell, were to generate a surfeit of electricity for the United Kingdom (Salter 1974; Voss 1979). Anaerobic fermentation of organic wastes, the idea of economies running on excrement, had a particular appeal to deep conservationists, and biogas was to be the fuel on which the world's most populous countries in Asia were to base their modernization.

In the early 1980s I felt strongly that these unrealistic claims called for a clear response and I devoted an entire book to detailing the weaknesses of such feebly thought-out propositions for biomass energy futures (Smil 1983). Just imagine cultivating, harvesting, and gasifying billions of tonnes of kelp in offshore waters to produce all of America's gas! Or running America's inefficient cars on corn-derived ethanol, most likely at a net energy loss! But I should not have bothered. As conceived and retailed during the 1970s nearly all of these "soft energy" schemes were so poorly thought out, relied on such inefficient techniques, and were so impractical that they were bound to meet the same fate as their extreme "hard energy" (that is mostly nuclear) counterparts. As I will explain later in this chapter, perhaps the most famous forecast of market penetration by "soft energy" in the United States (Lovins 1976) missed its target for the year 2000 by no less than 90%. I will address realistic prospects for biomass energy conversions, and for other kinds of renewable energy sources, in chapter 5.

In our progress-obsessed society, expectations of new experimental and technical breakthroughs never end: as old candidates lose their luster new possibilities emerge. Which ones will be soon forgotten and which one, or two, may change the world generations from now? Here are just three recently announced advances. Self-organized discotic solar cells, simple constructs made from a crystalline dye and a liquid crystal, were found to convert radiation near 490 nm with efficiency of more than 34% (Schmidt-Mende et al. 2001). Will we see the world full of efficient and durable organic solar cells? Sound waves in thermoacoustic engines and refrigerators can replace pistons and cranks now built into this machinery (Garrett and Backhaus 2000). Will we see many new machines without moving parts? New solid-state thermal diodes operating at 200–450 °C can produce electricity from a variety of energy sources without a turbine or a similar generator (MIT 2001). Will this latest advance in thermionics, whose concept is more than a century-old, lead to ubiquitous nonpolluting generation of electricity?

Primary Energy Requirements

Forecasting total primary energy consumption would seem to be a much easier task than glimpsing the fortunes of particular energy conversion techniques. After all, future demand for commercial energy is clearly tied to population growth and to the overall economic performance. Near-term population forecasts (up to 10 years) are fairly accurate, and GDPs, particularly those of mature economies, are advancing in a relatively orderly fashion. Moreover, only a few energy-related subjects have received as much research attention during the second half of the twentieth century as the elasticities of fuel and electricity consumption and, more recently, changing energy intensities of national economies. Yet even when they appear to be right on, the exercises tying energy needs to population and economic growth do not really succeed much better than the purely technical predictions: two of my own attempts illustrate perfectly this counterintuitive outcome.

In 1983 predictions of global primary commercial energy consumption in the year 2000 submitted by the participants in the International Energy Workshop ranged from the low of 5.33 Gtoe by Amory Lovins to the high of 15.24 Gtoe by the International Atomic Energy Agency; my forecast was 9.5 Gtoe (fig. 3.8; Manne and Schratenholzer 1983). Actual global primary energy consumption in the year 2000 amounted (with both hydroelectricity and nuclear electricity converted by using the average efficiency of fossil-fueled generation, the way I did it in 1983) to about 9.2 Gtoe (BP 2001), or 73% above Lovins's 1983 projection and 40% below the IAEA forecast.

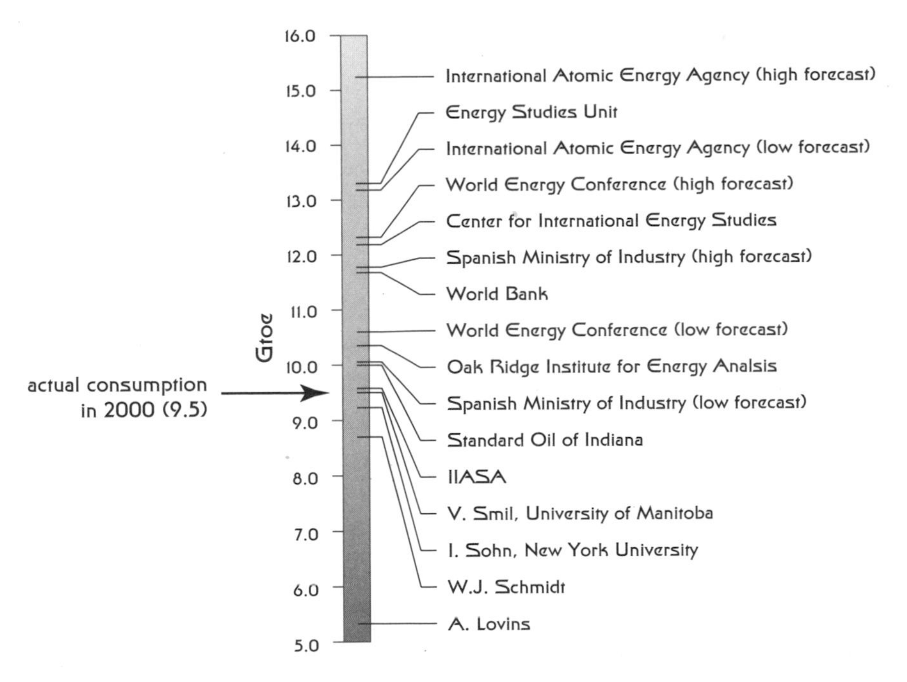

Figure 3.8
Forecasts of the global TPES in the year 2000 made by the participants in the 1983 International Energy Workshop. The predicted extremes were about 45% lower and 60% higher than the actual rate. Plotted from a table in Manne and Schrattenholzer (1983).

My forecast is thus off by a mere 3%—but I do not use this near-perfect hit in order to congratulate myself for having done better than both the "hard" and the "soft" energy protagonists, but rather in order to demonstrate how wrong I was. While I had virtually nailed the overall demand, I was much less successful in forecasting the makeup of the world's primary energy consumption. I underestimated both the use of natural gas and crude oil (by, respectively, 17% and 12%), and I overestimated the contributions of coal (by 14%) and renewable energies. If my breakdown would have been used for calculating the capital requirements for natural gas or coal industries, the future emissions of CO_2, the principal greenhouse gas, or releases of SO_2 and NO_x, the two gases responsible for nearly all anthropogenic acid deposition, the errors would have been considerable.

But I have an even better example of an aggregate demand forecast whose numbers turned out very well—but where the overall setting and countless details have changed beyond anybody's expectations. Median forecasts of China's primary commercial energy consumption for the years 1985 and 1990 that I made in my first book on the country's energy written in 1975 turned out to have errors of, respectively, a mere 2% and 10% (Smil 1976; Fridley 2001). I was certain that major changes would follow Mao's death (he died in September 1976), and a long-range forecasting Delphi study I conducted at that time anticipated many of these changes (Smil 1977). But I could not have predicted either the speed or the extent of China's post-1979 modernization with all of its complex implications for energy demand, economic expansion and environmental degradation which I traced ten and twenty years later (Smil 1988, 1998a). Between 1980 and 2000 the GNP of China's rapidly modernizing economy increased roughly sixfold, or almost twice as fast as I anticipated in the mid-1970s—but the country has been also dramatically reducing its relative need for energy.

During the last six years of Mao's rule, the energy intensity of China's economy actually rose by 34%. But then Deng Xiaoping's modernization resulted in closing of the most inefficient enterprises, large-scale modernization of energy-intensive processes, and gradual restructuring of industrial output in favor of higher-value added manufactures. Between 1980 and 1990 the average energy intensity of China's economic product fell by about 40% (Fridley 2001). Consequently, my excellent forecast of China's total energy needs in 1985 and 1990 was the result of being doubly wrong as the real economic growth was nearly twice as high as I anticipated but its energy intensity was almost halved. While I failed to anticipate its extent, the United

Nations forecasters failed to notice the very existence of this declining intensity even in 1990, after a decade of major efficiency improvements.

In their 1990 report they expected China's energy demand to increase at the average GDP growth rate of about 5% (UNO 1990)—but instead of being around 1.0 the actual elasticity was close to 0.5 (Fridley 2001). And the continuing decline of the average energy intensity of China's economy means that even forecasts less than a decade old are already substantially off. Six foreign and domestic forecasts issued between 1994 and 1999 put the country's total primary energy consumption in the year 2000 at 1.491–1.562 Gtce (mean of 1.53 Mtce) while the actual use was 1.312 Gtce, making those forecasts 12–19% too high in a matter of just a few years (China E-News 2001).

To see numerous examples of long-range projections of national or global energy demand that have badly missed their targets requires nothing more than consulting just about every major institutional forecast done since the 1960s. Figure 3.9 plots the totals of some notable forecasts of the U.S. primary energy consumption in the year 2000 that were released between the years 1960 and 1979 (Battelle Memorial Institute 1969; Perry 1982; Smil 2000d). Most of them ended up at least 40–50% above the actual demand of 2.38 Gtoe. Perhaps the most precious example of failed forecasting of long-range national energy demand is the goal of U.S. energy independence charted right after OPEC's first round of crude oil price increases by the Nixon administration for the 1980s (Federal Energy Administration 1974). Its excessive ambition made the document a paragon of wishful thinking. Alvin Weinberg, who at that time headed the Office of Energy Research and Development in the White House, recalled how this expensive ($20 million) blueprint was flawed because its predictions of future oil demand and supply depended largely on a single number, on the poorly known price elasticity of demand for liquid fuel (Weinberg 1994).

Yet many experts believed that this goal was not only achievable but also compatible with further huge increases of total primary energy consumption. For example, Felix (1974), in a classical example of blind exponential forecasting, put the U.S. energy demand in 1985 at 3.1 Gtoe and the year 2000 consumption at 5.1 Gtoe. And in spite of this huge increase he still thought that the self-sufficiency could be realized by the year 1985 and maintained this position afterwards, thanks largely to the combination of growing coal extraction and dramatically expanded generation of nuclear electricity (fig. 3.10). A reality check: in spite of the fact that in the year 2000 the U.S. primary energy use was, at 2.38 Gtoe, about 55% lower than Felix's forecast, the country imported almost 30% of its primary energy demand in

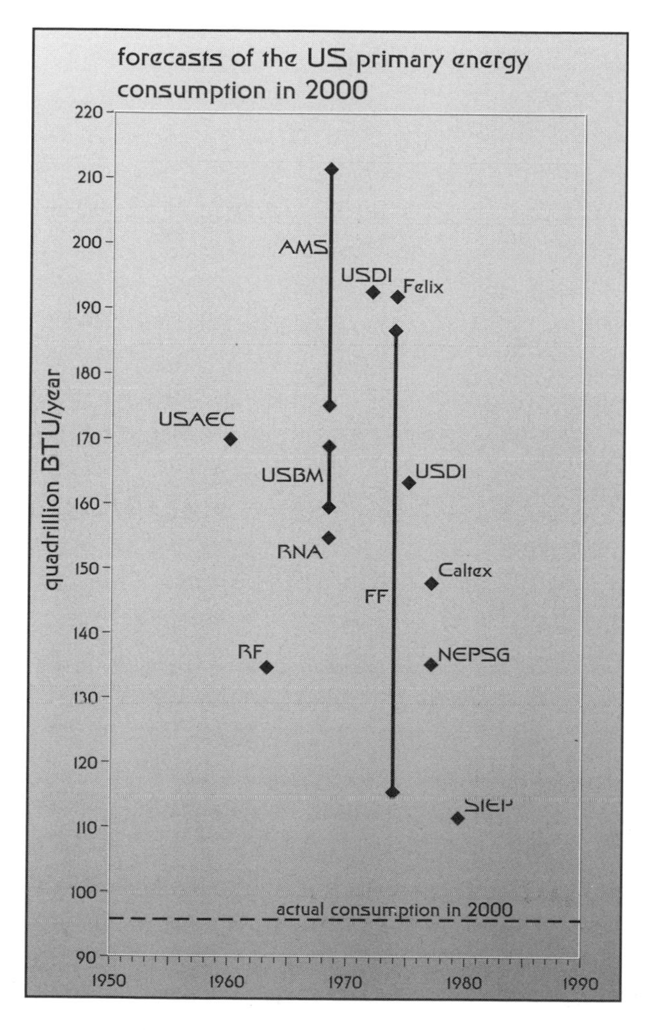

Figure 3.9
Long-range forecasts of the U.S. primary energy consumption in the year 2000 published between 1960 and 1980. Their average value was nearly 75% above the actual demand. Plotted from data in studies listed in the graph.

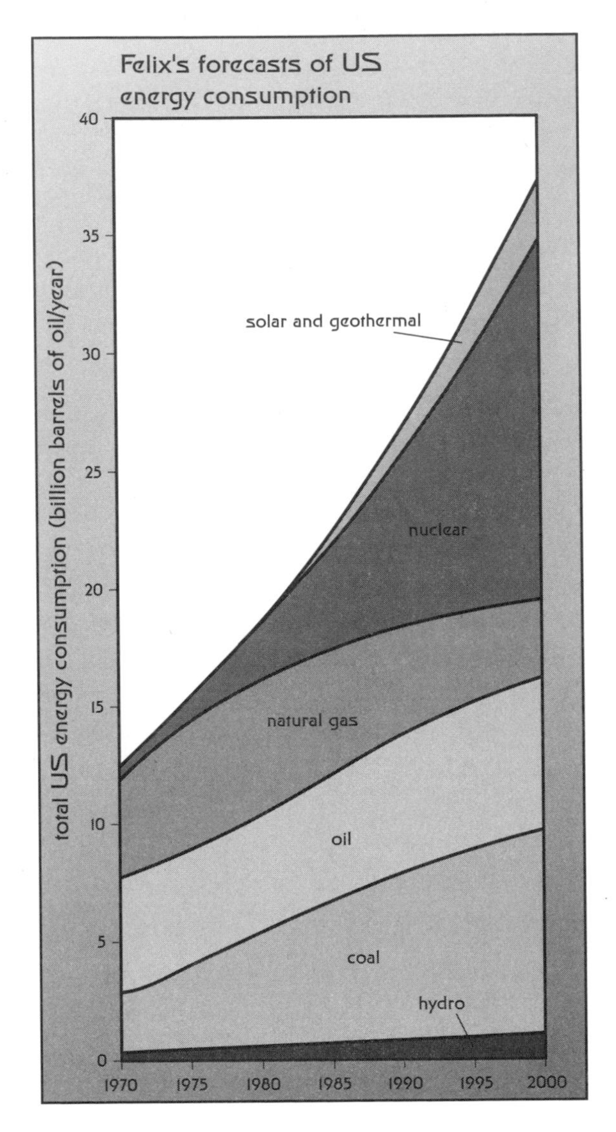

Figure 3.10
Felix's forecasts of the U.S. energy consumption turned out to be spectacularly wrong not only because of its prediction of self-sufficiency by the year 1985 but, as shown in this graph, also because of its expectations of continued exponential growth of demand and its 40% share of nuclear electricity by the year 2000. Simplified from Felix (1974).

that year, and about 55% of its consumption of liquid fuels (EIA 2001e)! Craig, Gadgil, and Koomey (2002) offer another retrospective examination of failed long-term energy forecasts for the United States. Substantial exaggerations have also characterized long-term forecasts of global energy requirements.

A generation ago the two most widely cited forecasts were those of the World Energy Conference (1978) and of the International Institute for Applied Systems Analysis (Häfele et al. 1981). The first forecast was produced by one of the world's oldest international bodies (established 1924) devoted to the study of energy systems and its low and high totals of global primary energy needed in the year 2000 were 12.3 and 13.9 Gtoe. The latter forecast involved contributions of more than 140 scientists from 20 countries; it subdivided the world into seven regions and used alternative scenarios of GDP growth and primary energy substitution to arrive at low and high annual consumption estimates of, respectively, about 10.2 and 12.7 Gtoe for the year 2000. Except for an unrealistically low total of 5.5 Gtoe by Lovins (1980), forecasts published a generation ago were too high (fig. 3.11). They were commonly 30–40%, or even more than 50% above the actual global primary energy consumption of 8.75 Gtoe (when converting hydroelectricity using its energy content and nuclear electricity using the equivalent of thermal generation) or 9.2 Gtoe (when using the equivalent of thermal generation for both).

Because of rapid price and demand changes even some short-term forecasts, and particularly those for individual fuels, can become almost instantly obsolete. OECD's forecast prepared just a year before the OPEC's first price hike put the world's primary energy consumption at 8.48 Gtoe in 1980, with crude oil supplying 4.05 Gtoe (OECD Oil Committee 1973). The actual 1980 global consumption of primary commercial energy was 6.01 Gtoe, with oil supplying 3.02 Gtoe; OECD's forecasting errors thus amounted, respectively, to 30% and 25% in just eight years. And the assumptions of world oil prices used by the Workshop on Alternative Energy Strategies (WAES)—an elaborate international, intergovernmental effort conducted between 1974 and 1977—to construct its long-range scenarios of energy demand in affluent countries were made obsolete by the second round of OPEC's price hikes just two years after this widely publicized study came out (Basile 1976; WAES 1977).

And just to dispel any belief that the only recent rapid shifts in energy demand were those following OPEC's unpredictable actions, I will note a recent unpredicted drastic reduction of China's coal extraction. This shift accounts for most of the already noted decline of average energy intensity of China's economy. Several factors explain this change: a weaker market for coal caused by slower economic growth

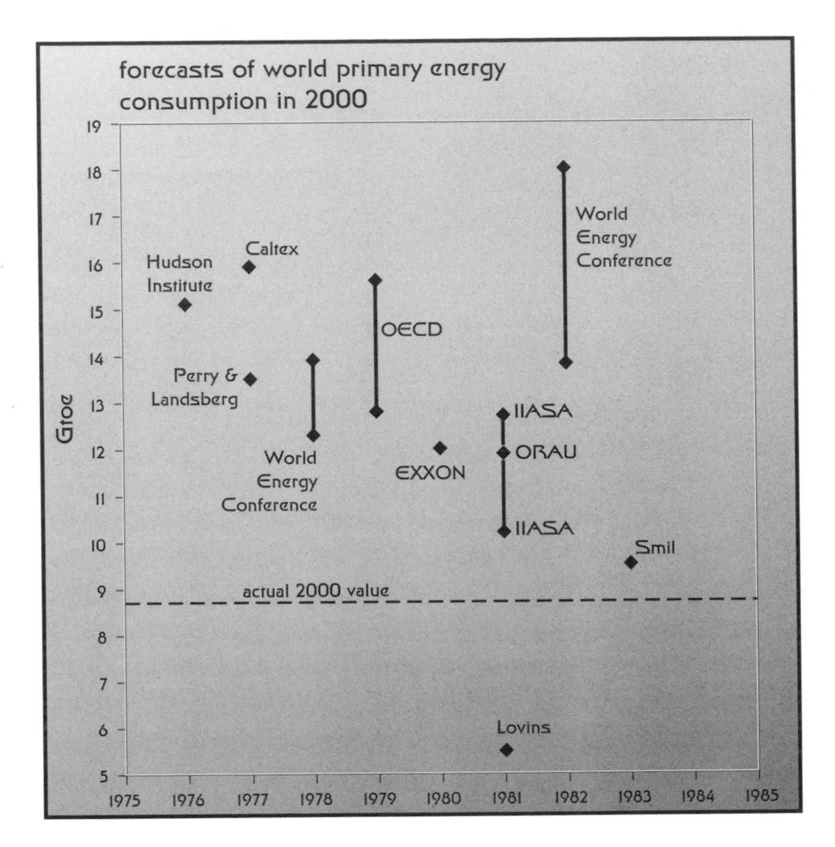

Figure 3.11
Long-range forecasts of the world's primary energy consumption in the year 2000 published between 1960 and 1980. Plotted from data in studies listed in the graph.

as well as by efforts to reduce high levels of urban air pollution originating in coal combustion; rising availability of higher-quality liquid and gaseous fuels (China's oil imports doubled in 2000); poor quality of coal extracted from China's primitive small coal mines (sold as run-of-the-mine fuel, without any cleaning and sorting); and concerns about the high rate of fatalities and injuries in these almost entirely unregulated enterprises.

After a total 26% increase in just five years, China's raw coal output reached 1.374 Gt in 1996 (or almost 1.04 Gtce) and it was expected to rise at a similar rate for the remainder of the 1990s. But as China's energy demand fell and the energy intensity of its economy declined by one-third during the late 1990s, the authorities

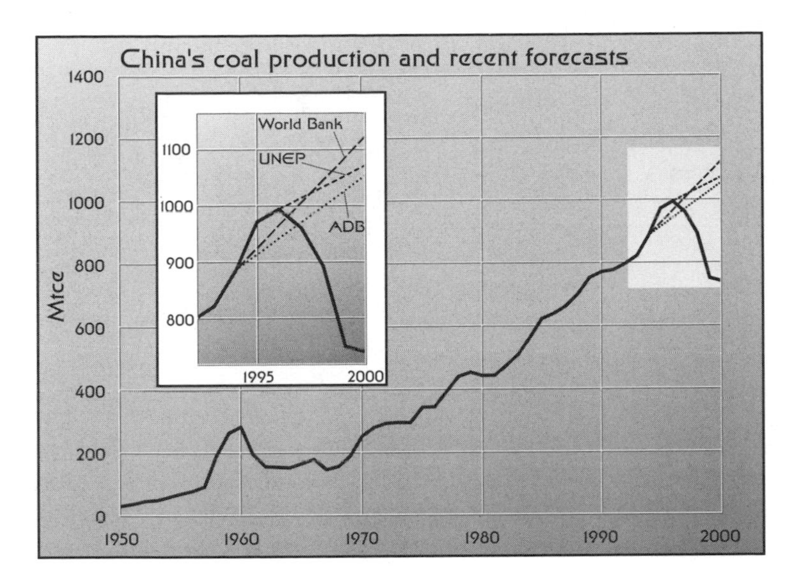

Figure 3.12
China's coal production 1950–2000, and some recent forecasts for the year 2000. Plotted from data in Smil (1976), Fridley (2001), and *China E-News* (2001).

closed tens of thousands of inefficient and dangerous small rural coal mines (both privately and collectively owned) and the actual extraction in the year 2000 was just 740 Mtce compared to forecasts of 1–1.1 Gtce published as recently as the mid-1990s (fig. 3.12; Fridley 2001; NBS 2001).

Demand for Electricity

Long lead times needed to put online new generating capacity make it imperative that the utilities engage in repeated long-range forecasting of electricity demand. Moreover, given electricity's irreplaceable role in modern societies (millions of electric motors, lights, and computers cannot use any substitute source of energy) and continuously expanding variety of its final uses, it could be expected that even the slightest signals of possible demand shifts will be looked at both promptly and closely. Yet the long-term forecasting success in this critical sector has not been any better than the predictions of total primary energy use. This poor record can be best illustrated by looking at the experience of the world's largest electricity producer. U.S. utilities, singly and collectively, did not anticipate a major downturn in demand during the 1970s and a generation later, they missed, almost inexplicably, the need for installing new generating capacity.

After two decades (1950–1970) of 7–10% annual growth (that is doubling every 10 to 7 years), virtually every forecaster in the United States (as well as in Canada) expected the identical, or even a bit faster, growth in decades ahead. Such expectations yielded incredibly large aggregate requirements. In 1970, in his opening speech at a meeting on environmental aspects of nuclear generation, Glenn Seaborg, at that time the Chairman of the U.S. Atomic Energy Commission, made the following forecast of the U.S. electricity needs (Seaborg 1971, p. 5):

> The projected growth of electrical generating capacity in the United States encompasses a range of estimates the median of which is 1600 million kilowatts by the year 2000. The upper limit of this range is currently about 2,100 million kilowatts, which I am inclined to believe may be more realistic. In this event, it would be necessary to build in each of the next three decades the equivalent of about 300, then 600, and then 900 power plants of 1,000 megawatt capacity each.

In reality, the U.S. generating capacity in the year 2000 was about 780 GW, less than 40% of Seaborg's "more realistic" estimate. Such exaggerated forecasts were not a U.S. oddity. In 1974 the United Kingdom Atomic Energy Authority believed that the country would need no less than 200 GW of generating capacity in the year 2000 and that some 120 GW of it will be nuclear (Bainbridge 1974). In reality the total British capacity at the century's end was only about 70 GW (just 35% of the forecast total) with less than 13 GW in nuclear stations, or a miss of nearly an order of magnitude.

Falling demand invalidated these forecasts almost as soon as they were made. Decadal increments of U.S. electricity sales fell from 202% during the 1960s to 50% during the 1970s, 30% during the 1980s, and 23% during the 1990s (EIA 2001e). This unexpected trend was inevitably reflected in gradually slumping forecasts of annual growth rates of electricity generation as U.S. producers had consistently exaggerated short- to medium-term growth of electricity demand: their repeatedly adjusted forecasts still kept ahead of real growth rates. In 1974 the first ten-year forecast of peak electric generating capacity in the United States prepared by the North American Electric Reliability Council envisaged the annual average growth of 7.6%; by 1979 the estimate was down to 4.7%, and by 1984 it was just 2.5% (fig. 3.13). Only at that point did the forecast begin, at least temporarily, matching the reality. Actual average growth rates of the U.S. electricity consumption were 2.5% during the 1980s and just 2% during the 1990s (EIA 2001e).

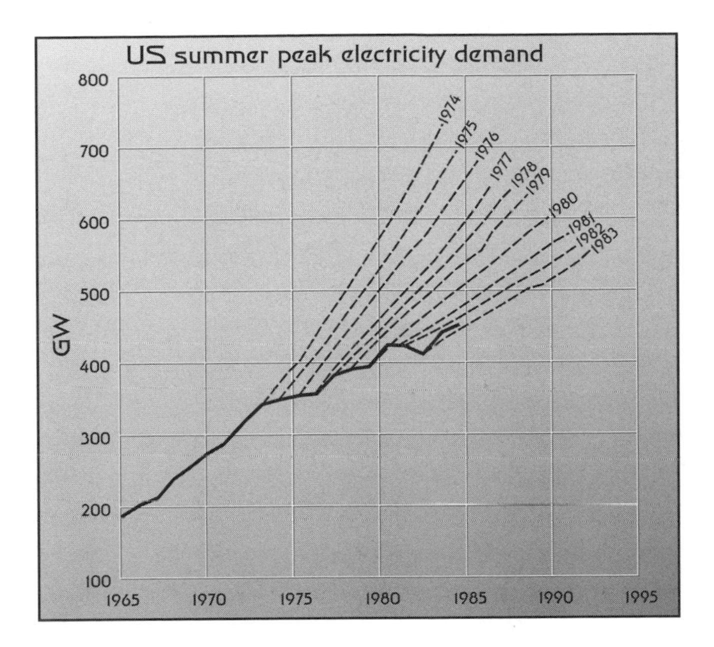

Figure 3.13
Successive ten-year forecasts of average annual growth rates of the U.S. electricity generation issued by the North American Electricity Reliability Council between 1974 and 1983 showed repeatedly excessive expectations of future demand.

California was no exception to the falling growth forecasts. Ten-year forecasts of load growth of one of its largest utilities, Southern California Edison Company, declined from 9% in 1965 to 8% in 1970, 5% in 1975, 3% in 1980, and to just 2% by 1985 (Southern California Edison Company 1988). Slow growth continued during the early 1990s when the state's electricity demand rose by less than 5%. But then the combination of rapid population and economic growth of the late 1990s (2.5 million people added between 1995 and 2000 as the bubble of the "new" electricity-intensive economy kept expanding) lifted the demand by about 20%. At the same time, the net addition of new generating capacity, in what is perhaps the world's most NIMBY-conscious jurisdiction averse to building any new generating plants, amounted to a mere 1.4% of the 1995 total (fig. 3.14; CEC 2001a; Pacific Gas & Electric Company 2001). Add the ineptly conceived deregulation of the state's generating industry—and the result has been an unprecedented spell of rotating blackouts, crisis management, and fears about the security of future supply.

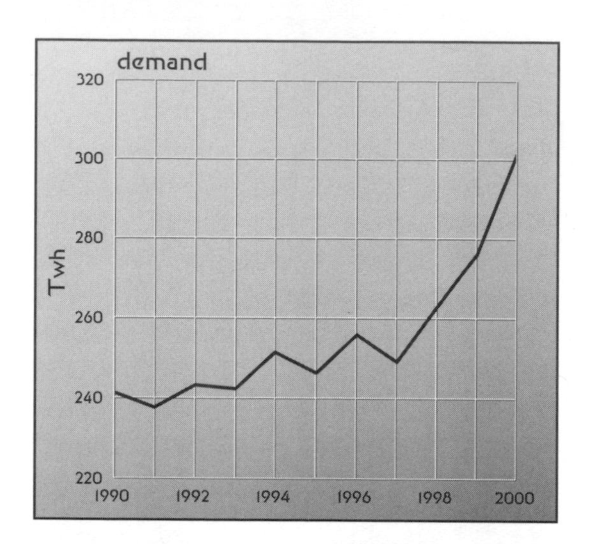
demand
Twh
320
300
280
260
240
220
1990
1992
1994
1996
1998
2000

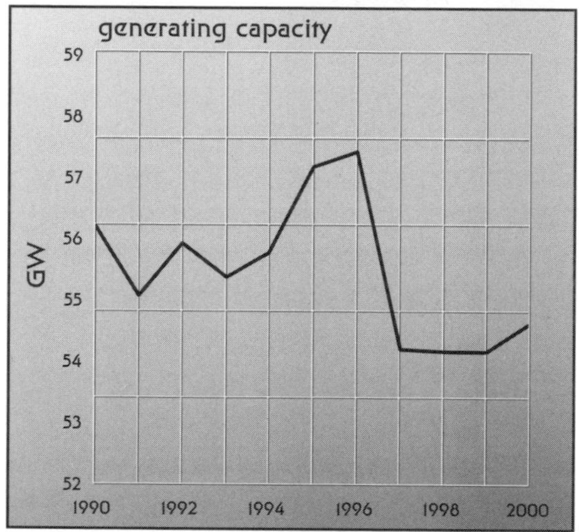
generating capacity
GW
59
58
57
56
55
54
53
52
1990
1992
1994
1996
1998
2000

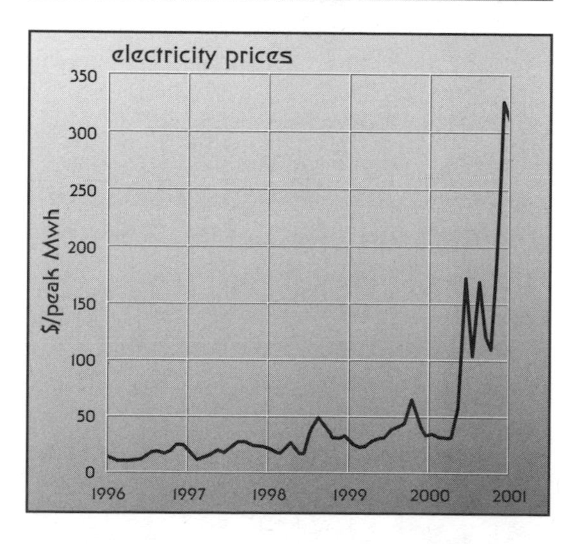
electricity prices
$/peak Mwh
350
300
250
200
150
100
50
0
1996
1997
1998
1999
2000
2001

Interestingly, while many forecasters were surprised by the growth of the 1990s, the California Energy Commission was actually fairly successful in anticipating the resurgent demand. In its May 1988 report it put the state's peak demand in the year 2000 at 56.673 GW. Adjusted peak demand for the year 2000, taking into account statewide voluntary curtailments in the use of electricity, was 53.257 GW, which means that the Commission's total was about 6.5% higher (CEC 2001a). This is an excellent result for a twelve-year-old forecast and an uncommon example of getting it right, at least over a relatively short period of time.

An opposite, albeit temporary, shift was under way in China from the late 1990s. For decades the country labored under chronic shortages of reliable electricity supply (Smil 1976, 1988). Massive post-1990 construction of thermal and hydroelectricity plants combined with more efficient industrial production not only reversed the situation, it has actually resulted in surplus generating capacity. As a result, some of China's largest new hydro stations, including Ertan (3.3 GW, completed in 1998) in Sichuan (which operated at 40% of its full capacity in 2000) and Xiaolangdi (1.8 GW, completed in 2001) in Henan, have not been able to conclude long-term contracts for sales of their relatively expensive (compared to small and rapidly built thermal stations) electricity (Becker 2000).

Energy Prices and Intensities

No energy commodity attracted more attention and generated more price forecasts during the last three decades of the twentieth century than did crude oil, the world's most important fossil fuel. History of the nearly 140 years of recorded crude oil prices (beginning with the Pennsylvania oil boom of the early 1860s) shows more than a century of general decline. When expressed in constant monies, the average price of crude oil in 1970 was less than one-quarter what consumers had to pay in 1870. This is why the OPEC-driven price spike between the years 1973–1982 came as such a shock, even when bearing in mind that the rapid inflation rate of the 1970s and early 1980s made the increase much more impressive than a comparison using constant prices.

◀ **Figure 3.14**
Three graphs illustrate California's electricity problem during the year 2000: lack of new generating capacity and rising demand by the rapidly expanding economy of the 1990s combined with poorly executed deregulation processes to produce a sharp spike in prices. Based on data in CEC (2001a).

As already noted in chapter 2, the short-lived rise was followed by return to levels not as low as those prevailing between the early 1920s and the late 1960s but still lower than they were a century ago (BP 2001). During the four decades preceding the spike, between 1930 and 1970, oil prices were remarkably constant because a very effective cartel in the fuel's largest producer, the Texas Railroad Commission (TRC), controlled the U.S. output, and with it the world price, through prorated production quotas (Adelman 1997). And when the Middle Eastern output began to rise after World War II producing companies precluded any fluctuations by using the system of posted crude oil prices. OPEC was set up in 1960 by Iran, Iraq, Kuwait, Saudi Arabia, and Venezuela (eight more nations joined later) basically in order to prevent any further reductions of posted prices and hence to protect the revenue of its members.

In 1971 came the first nationalizations of foreign oil companies in Algeria and Libya. As the nationalizations continued throughout the Middle East, OPEC's posted price (for the Saudi light crude) rose from $2.59 per barrel at the beginning of 1973 to $3.01 on October 1. Five days later, on Yom Kippur, an Egyptian army crossed the Suez Canal and broke through the Bar Lev line set up after the 1967 Israeli victory in the Six-Day War. Israel eventually regained the initiative but the Arab-dominated OPEC saw the conflict as a great opportunity to press its own advantage. Prices went to $5.12 per barrel on October 16, and a day later the organization's Arab members decided to use "oil weapon" and embargo oil exports to the United States, and a few days later also to the Netherlands, and to begin gradual reductions of oil production until Israel's withdrawal from occupied Arab lands (Smil 1987; EIA 2001f).

Predictably, the embargo was ineffective: it ended on March 18, 1974, after multinational oil companies simply redirected their tankers, but the oil price rose to $11.65 per barrel on January 1, 1974, quadrupling within just six months (fig. 3.15). Although there was never any threat of massive physical oil shortages, production cuts by OPEC's Arab members (ironically, in view of later events, only Iraq did not join this action) caused a panicky reaction in the West. Queues in front of gasoline pumps were a comparatively minor, although occasionally a violent, irritant. Most importantly, continued high oil prices were widely seen to usher in a new era of low economic growth, widening trade deficits for affluent importers, and the end of growth for modernizing low-income economies.

The second round of OPEC's price hike, a consequence of the fall of the Iranian monarchy and the beginning of the Iraq–Iran war, came just five years after the first

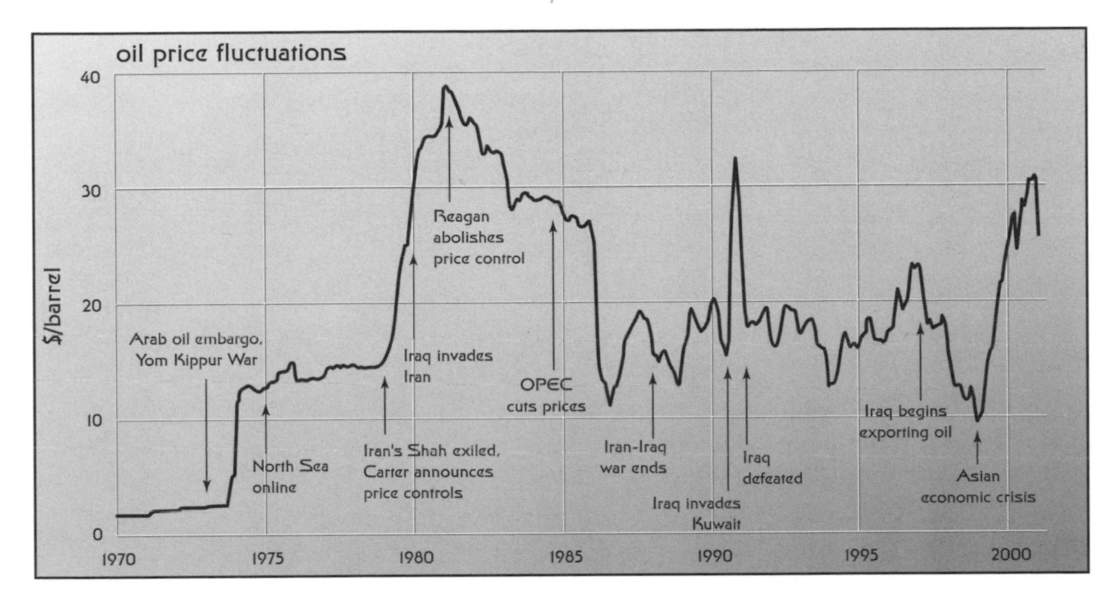

Figure 3.15
Oil price fluctuations, 1970–2000. Based on a graph in EIA (2001f).

one. As Shah Mohammed Reza Pahlavi was leaving Tehran for exile on January 16, 1979, the oil price stood at $13.62 per barrel; in November, as Khomeini's young disciples took over the U.S. embassy, the price of the Saudi light crude was $24.20. On September 23, 1980, Iraq invaded Iran and the oil price ended the year 1980 at $32.95 per barrel, finally peaking in March 1981 at $34.89, with spot market prices near $50 in Rotterdam (fig. 3.15; EIA 2001f).

Not surprisingly, enormous resources were expended, and recurrent anxieties were endured at that time in order to fathom the future of the world oil price. Expectations of future large oil price increases were nearly universal. This, in turn, lead to widespread fears that all of the fungible wealth of the Western world would soon end up in the treasuries of OPEC nations and that the fabulously rich Arabs would also gradually buy up any worthwhile European and North American real estate. Numbers for the 1970s were stunning: in 1970 OPEC countries got $8 billion for their oil; by 1979 their oil revenues topped $300 billion (Smil 1987). But this trend did not extend beyond the 1970s and it became clear very soon that these worries were vastly exaggerated. Marginal downward adjustments of posted crude oil prices took place in 1981 and 1982 and as falling consumption resulted in obvious oil glut the OPEC cut its overall output and a new official marker price of $28.74 was set up in March 1983. This price held, more or less, until August 1985 when the Saudi's, with revenues cut by declining demand, decided to regain their lost market share by doubling their crude oil output during the next five months.

The price held only for another three months, then it dropped below $20 per barrel in January 1986 and in early April it sank briefly below $10 per barrel before settling into a fluctuating pattern between $15–20. Energy analysts who just a few years ago feared that there is a 30%, even 50% chance of a massive shock to the Western world's economies from losing 10 Mbd of the Middle Eastern production for one year (Lieber 1983) were now worried that "prices could fall so much more that it's scary" (Greenwald 1986, p. 53). This great oil price countershock of 1986 meant that in inflation-adjusted monies crude oil prices returned almost to where they were before the end of 1973. But as the price chart shows (fig. 3.15), this, and every subsequent price dip, was followed by a relatively large, but short-lived rebound. The highest upswing, to nearly $33 per barrel in October 1990, came after the Iraqi invasion of Kuwait in August 1990—but on the day the United States began its air attacks on Iraq (January 16, 1991) the price fell by $9–10 per barrel and it did not spike even after the retreating Iraqis ignited Kuwaiti oilfields creating what amounted to perhaps the nearest image of inferno on the Earth (Sadiq and McCain 1993).

The next deep trough came in February 1994 (below $13 per barrel), and the highest point of the mid-1990s, reached in October 1996 (over $23 per barrel). This was followed by a steady slide as renewed Iraqi sales coincided with Asia's economic downturn. By the end of 1998 the price slid below $10 per barrel but almost immediately a cold northern winter, stronger demand, and low oil stocks began pushing prices sharply higher, and they tripled between January 1999 and September 2000. The month's mean cost to refiners was $30.53, and on September 20, NYMEX trades were as high as $37.80 per barrel (EIA 2001f). Not surprisingly, during the early months of 2001 predictions were for skyrocketing prices later in the year, but by July, in a sudden reversal of price trend OPEC ministers were considering yet another production cut as oil price settled down yet again well below $20 per barrel (fig. 3.15).

Rather than citing numerous examples of failed long-range oil price predictions I will offer, once again, the International Energy Workshop forecasts of the early 1980s as a perfect illustration of two common forecasting errors: the spell cast by recent events, and the herd instinct of forecasters. Unlike the case of total primary energy forecasts for the year 2000 (where I came within 3% of the actual

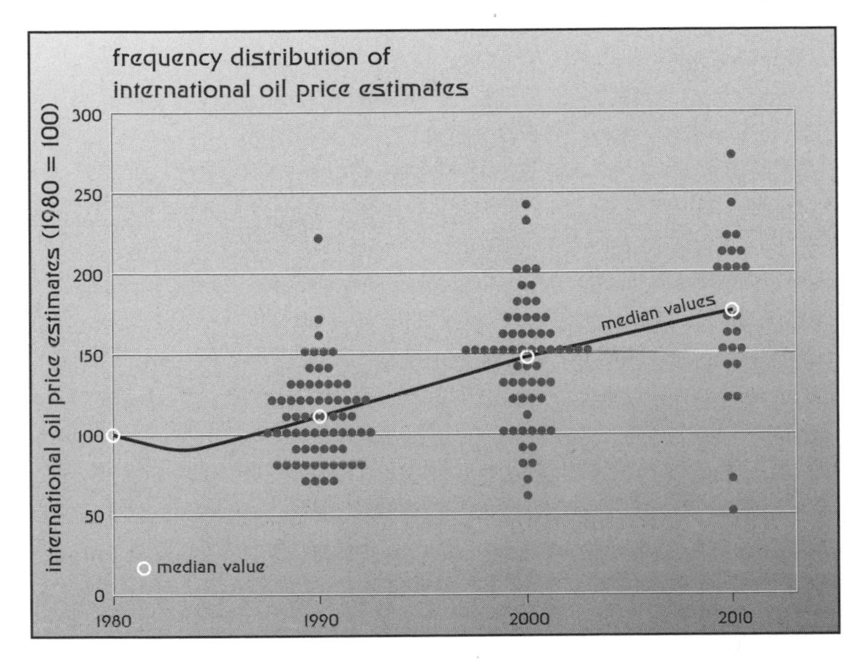

Figure 3.16
Frequency distribution and median values of international oil price forecasts by the participants in the 1983 International Energy Workshop (Manne and Schrattenholzer 1983). Even the most conservative predictions turned out to be excessive.

aggregate) my 1983 crude oil price forecast now looks quite ridiculous. I put the price of crude oil in the year 2000 at 30% above the 1980 level (Manne and Schrattenholzer 1983; fig. 3.16). Tenor of the time, the common bane of long-range forecasters, exercised its usual influence: in 1983 oil prices were just slightly off their historic peak they reached in 1980–81, and anticipation of further oil price increases was the norm, with most forecast tightly bunched. My forecast means that, adjusted for inflation, a barrel of crude oil should have averaged about U.S.$75 per barrel in the year 2000. But, as already explained, less than two years after that forecast was made OPEC lost the control of the world oil market and the world entered a new period of widely fluctuating prices whose mean remained below $20 per barrel during the 1990s.

By the mid-1990s the crude oil price was, in inflation-adjusted terms, about as low as in 1920 and although it rose in 2000 above $20, and shortly even above $30 per barrel, it still remained far below my 1983 forecast (fig. 3.15). The world where crude oil would wholesale for U.S.$75 (2000) per barrel in 2000 would have been a very different place from the one where it was selling, for most of the year, for about one-third of that price! The only small consolation I can draw from this failure is that my 1983 oil price forecast was less ridiculous than that of the World Bank's chief economist (he had it 54% above the 1980 level). I should also note that those institutional projections of the 1980s that erred on the high side of future energy prices also erred on the high side of energy demand, and hence their demand forecasting error was even higher than is suggested by comparing them with actual energy requirements.

If longer-term forecasts have been notable failures, the near-term ones have not been any more perspicacious. For example, none of the nine world oil price projections for the year 2000, published during 1995 and 1996 in the United States and United Kingdom (EIA 1996) and averaging $18.42 (1994) per barrel, turned out to be correct. That year's average price was $27.84 (2000) which translates into inflation-adjusted $25.77 (1994), 40% higher than the group's mean! And as the rapid succession of post-1985 troughs and peaks on the world oil price chart testifies, there tend to be relatively large shifts once the turning points are passed and forecasts can be spectacularly wrong (compare figs. 3.15 and 3.16). But temptation to forecast is simply too great even for people who should know better. In February 1999 the oil price was barely above $10 per barrel and *The Economist* forecast a month later that the price would soon slide to $5—but at the year's end it was five times as high and the weekly's lead article had to admit "We woz wrong!"

A perceptive reader (Mellow 2000, p. 6) had an excellent appraisal and a good advice applicable not just to the journal's editors:

> A greater element of "shamanism" has crept into the forecasting business as modelling has increased in esotericism. The decision maker anoints his shaman for life, for better or worse, or until a better one comes along. *The Economist* should avoid the shaman's role.

But shamanism is here to stay. Many new oil price forecasts are being issued constantly, some of them looking as far ahead as the year 2020. A group of them, published between 1997 and 2001, shows that most forecasts, undoubtedly under the influence of generally fairly stable and low-to-moderate oil prices prevailing during the 1990s, predicted a rather rapid decline from the highs of the year (that, indeed, took place, and even faster than they predicted) to around $20 per barrel followed by 10–15 years of fluctuations within a narrow band (EIA 2001e; fig. 3.17).

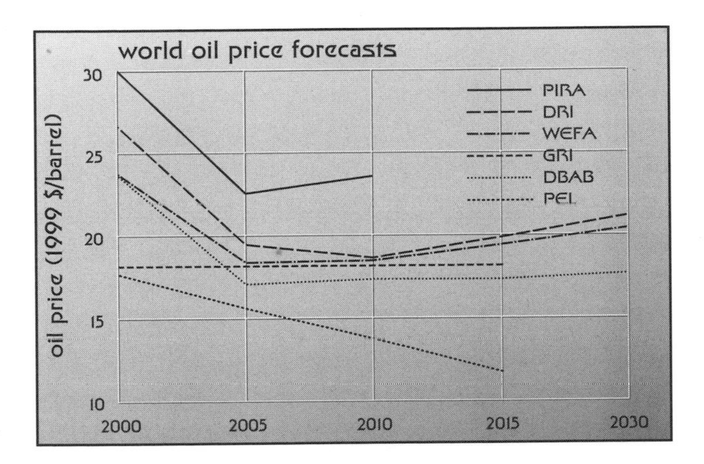

Figure 3.17
World oil price forecasts published between 1997–2001 stay overwhelmingly within a narrow band ($15–22) until the year 2015. Plotted from data in EIA (2001e).

This would be, I suppose, OPEC's dream scenario of moderately high and stable prices, with only one forecast of the price rising to $30 per barrel by 2015, and another one falling to close to $10 per barrel. The best advice I can give to these incorrigible shamans is to look carefully at fig. 3.18 that charts relative year-to-year swings of the world oil price since 1974: these shifts have been obviously quite random and hence nothing appears to be more incorrect where the future of oil prices is concerned than charting largely horizontal lines.

And OPEC's draining of the Western riches did not come to pass. After peaking in 1980 the cartel's revenues fell during the next two years while total expenditures kept rising and by 1982 OPEC's aggregate current account balance turned negative (just over $10 billion) and by 1985 expenditures were ahead of revenues by about $50 billion (Smil 1987). Between 1973 and 1983 the world's seven largest oil importing nations boosted their exports to OPEC countries faster than their purchases of crude oil and their deficit with the cartel was actually smaller than it was before 1973! And although the volume of traded crude oil rose by 30% during the 1990s, the decade's rapid increase in foreign trade (from $3.4 to $5.6 trillion, or 1.64 times, between 1990 and 1999) reduced the global oil exports to only about 4.5% of the total value (compared to about 8% in 1990). By 1999 crude oil sales were worth only about half as much as all food exports and equalled less than 6% of the burgeoning manufacturing trade (WTO 2001).

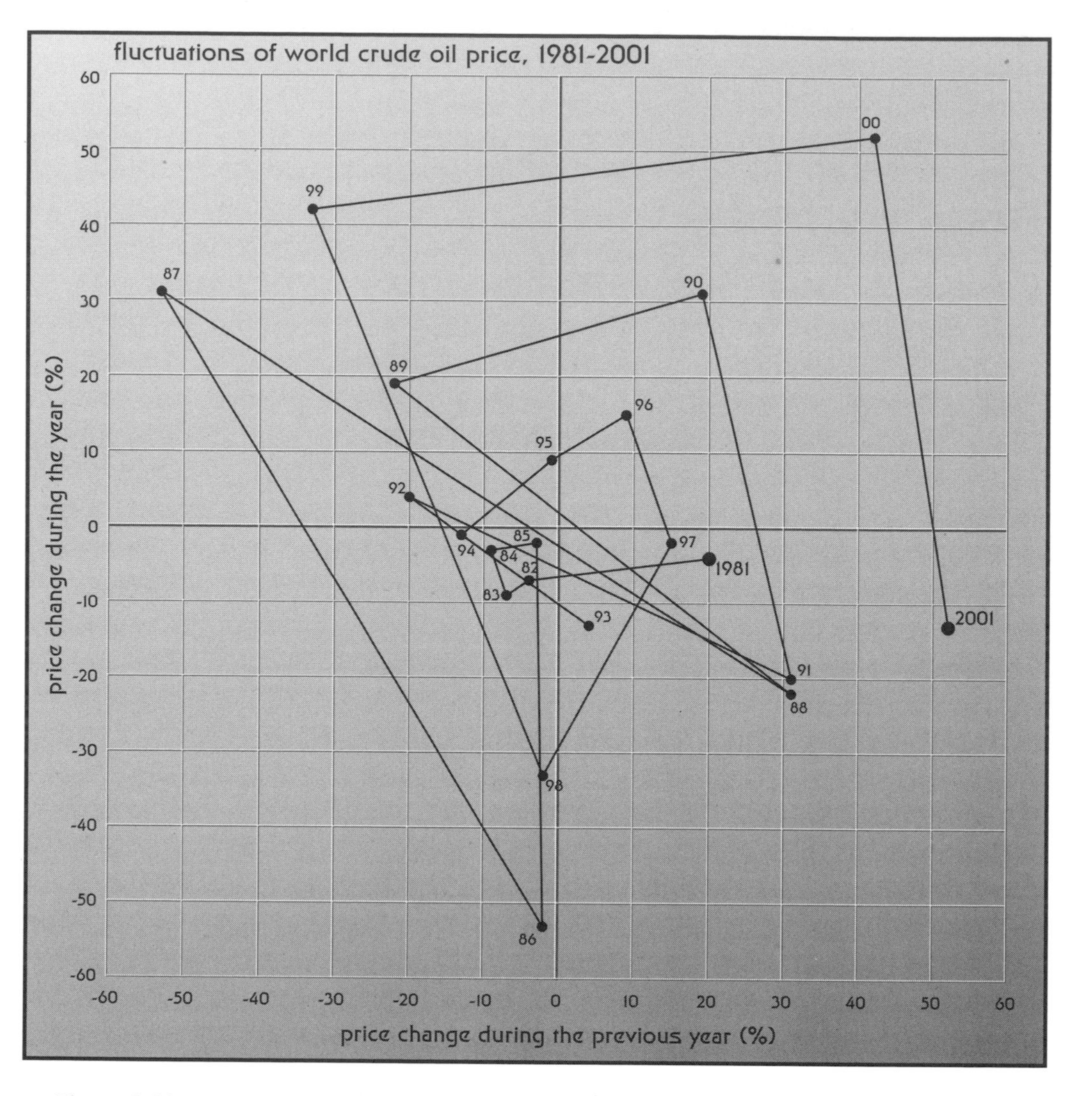

Figure 3.18
Graph of year-to-year changes in world oil prices since 1981 is an impressive way to show that random swings make it impossible to forecast what comes next. Plotted from average annual price data listed in BP (2001).

Changing energy intensities, be they for sectoral production or for entire national economies, were usually ignored by pre-1975, long-term energy forecasts. Given the U.S. dominance in energy matters there is a very plausible explanation for this crucial omission. Energy intensity of the U.S. economy peaked rather sharply around 1920, and by 2000 it was down by nearly 60%; but between 1952–1974 it remained remarkably stable, fluctuating by less than 5% around the period's mean and ending up only about 7% lower (fig. 3.19; EIA 2001e). This constancy suggested, wrongly, an immutable long-term link between energy consumption and wealth creation and it accounts for many exaggerated forecasts of the U.S. energy demand prepared during the 1970s.

OPEC's 1973–74 sharp crude oil price hike jolted the system back with astonishing, and at that time entirely unpredicted, consequences: between 1974 and 2000 the U.S. energy intensity declined by about 40% (fig. 3.19). But high energy prices that prevailed between 1973 and 1985 made surprisingly little difference

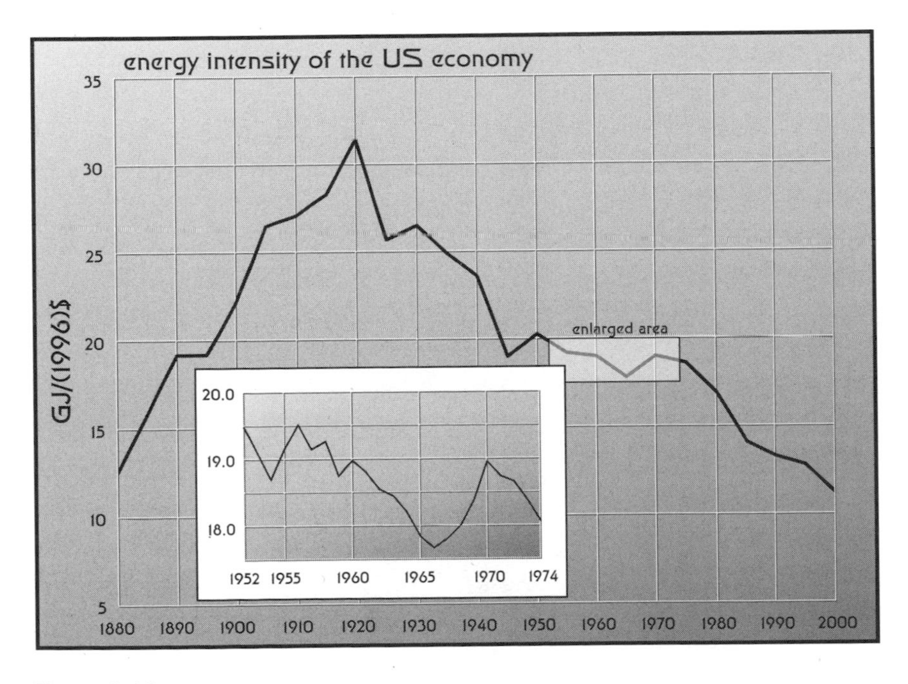

Figure 3.19
Energy intensity of the U.S. economy, 1880–2000. Post-1920 decline was interrupted by two decades of relative stability and it resumed after 1974. Plotted from data in Schurr and Netschert (1960), the U.S. Bureau of the Census (1975), and EIA (2001a).

for energy savings in manufacturing. Meyers and Schipper (1992) found that, when holding sectoral structure fixed, the aggregate energy intensity of U.S. manufacturing declined at nearly the same rate as it did between 1960 and 1973 when energy prices were essentially stable (fig. 3.20). These findings indicate that autonomous technical advances, rather than prices, are the key reason for higher efficiency in manufacturing. A very similar trend was found also in Japan—but the Japanese history of the overall energy intensity of its national economy is very different from the U.S. record.

Between 1885 (the first year for which relevant statistics are available) and 1940, energy intensity of the rapidly modernizing Japanese economy rose about 2.7-fold; by 1960 it fell by about 30% but by 1975 it was back to the early WWII level (fig. 3.21; IEE 2000). Another impressive period of efficiency gains reduced it by nearly one-third during the next 15 years but a curious, and again entirely unpredicted, shift took place during the 1990s. Japan, the paragon of high post-1974

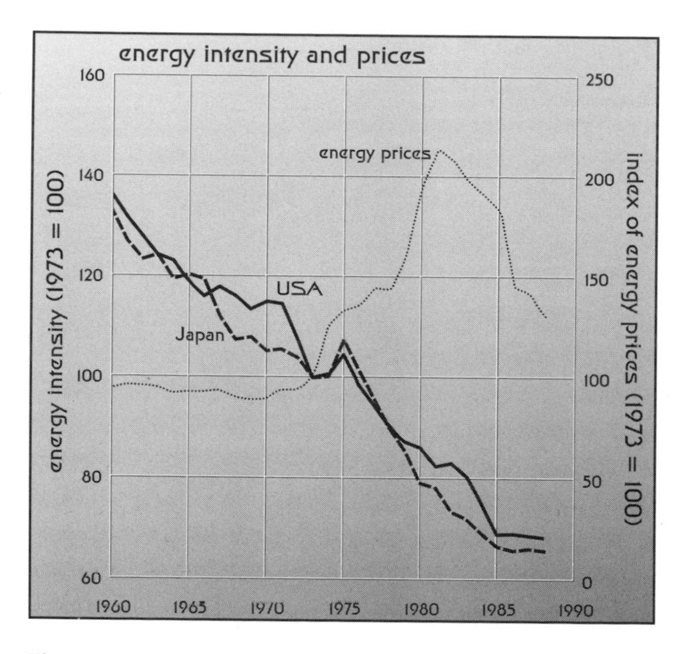

Figure 3.20
Energy intensity of manufacturing (structure-adjusted) in the United States and Japan contrasted with the average real cost of energy. Remarkably, improvements in energy use show very similar rates of decline regardless of the price trend. Based on Meyers and Schipper (1992).

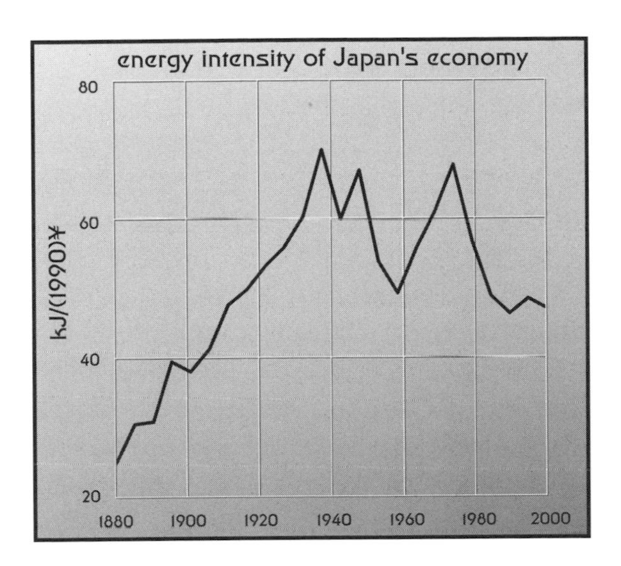

Figure 3.21
Energy intensity of Japan's economy, 1880–2000. Plotted from data in IEE (2000).

efficiency gains, consumed about 22% more energy in 2000 than it did in 1990 while its slow-growing, or stagnating, economy generated GDP only about 13% higher. Consequently, between 1990 and 1995 the country's average energy intensity actually increased by about 8% (fig. 3.21).

The energy intensity of China's grossly inefficient Maoist economy rose about fourfold between 1950 and 1977. Not surprisingly, IIASA's long-term forecast (Häfele et al. 1981) assumed, as I did at that time (Smil 1976), that China's energy consumption would continue growing faster than its GNP for several decades to come (i.e., energy elasticity slightly above 1.0). But three years after Mao's death Deng Xiaoping changed the country's course and, in yet another example of an entirely unpredicted shift, a combination of structural changes, technical innovation, and conservation cut the average energy intensity of China's economy by almost 60% between 1980 and 1995 (Fridley 2001; fig. 3.22). And the remainder of the 1990s was even more surprising. In spite of large efficiency gains achieved during the previous 15 years, average energy intensity of China's economy fell by another 35%, and even the aggregate primary energy consumption fell by nearly 5% while the country's inflation-adjusted GDP rose by 45% (NBS 2001)!

And so by looking just at the United States, Japan, and China we have excellent examples of tight coupling of economic growth and energy use (United States 1952–

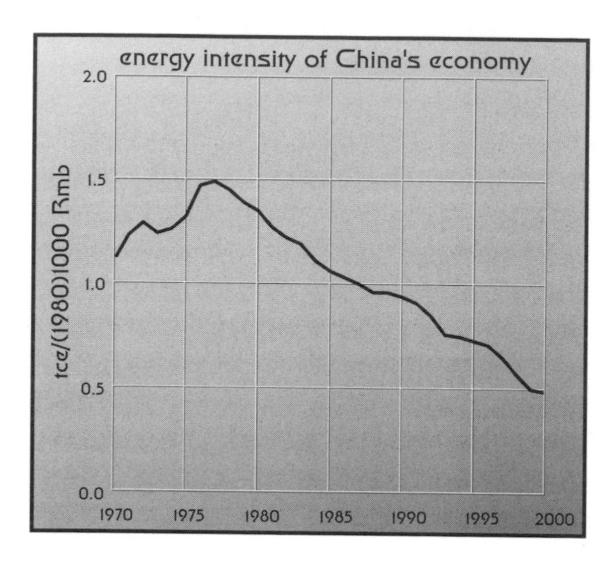

Figure 3.22
Energy intensity of China's economy, 1970–2000. Plotted from official GDP and TPES data listed in various issues of *China Statistical Yearbook* and also available in Fridley (2001). Using adjusted (i.e., lower) GDP values would result in a less dramatic decline in energy intensity.

1974, Japan until 1940 and again between 1960 and 1975, China 1949–1978), renewed decoupling (post-1974 United States, Japan between 1975–1995), dramatic, unprecedented, and still ongoing decoupling (post-1978 China) and surprising trend reversals (Japan between 1960 and 1975, and again since 1995). These examples demonstrate that long-term forecasts of national energy intensities remain inherently very difficult as they are products of complex combinations of both concordant and countervailing national trends that follow a broadly universal pattern of transitions but do so at country-specific times and rates.

While it is generally correct to conclude that energy intensities rise during the early stages of industrialization, peak, and then decline as economies use energy more efficiently it is also obvious that national peculiarities abound. Ascents and declines will have very different slopes, peaks can be both sharp or they can appear as extended plateaux, reversals are not impossible, and even impressive gains in industrial and agricultural energy efficiency may be substantially counteracted, or even swamped, by increased energy consumption by affluent households. The only solid conclusion is that energy intensities are not preordained and that they can be reduced both by gradual technical change and by determined conservation policies.

Substitutions of Energy Resources

There is one kind of energy forecasting that was supposed to be virtually error-free. Cesare Marchetti, IIASA's resident long-range energy forecaster, had studied histories of energy substitutions and found that these transitions are remarkably orderly as full cycles of shares of the TPES delivered by individual energy sources resemble fairly closely normal (Gauss–Laplace) curves (Marchetti 1977; fig. 3.23). The pattern's inexorability looks particularly impressive when the market shares (f) of primary commercial energy consumption are plotted as logarithms of [f/(1-f)]. Their ascents and descents then appear as straight lines and all that is needed to forecast the future is to look at the revealing chart of trends that prevailed between the mid-nineteenth century and 1975 (fig. 3.23). The process is very slow, with every new source taking about a century to penetrate half of the market share, and it is surprisingly regular.

In spite of many possible perturbations, including wars and periods of economic stagnation and rapid growth, penetration rates remained constant during the first

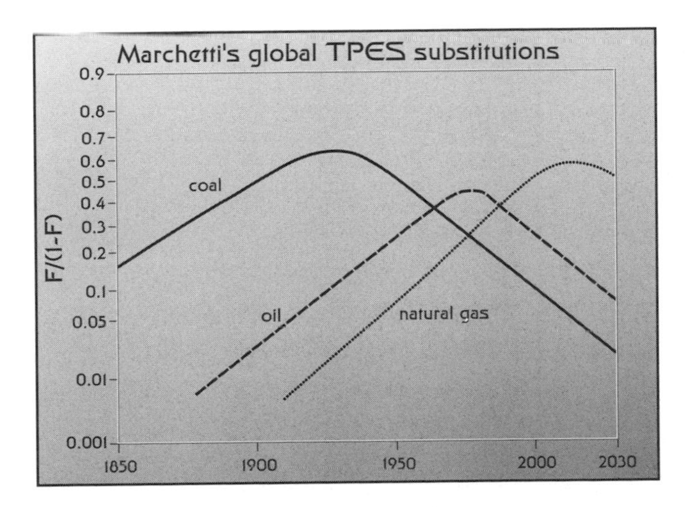

Figure 3.23
Marchetti's model of global TPES substitutions working as if the system "had a schedule, a will, and a clock." Based on Marchetti (1977).

three-quarters of the twentieth century. That is why Marchetti and Nakićenović (1979, p. 15) concluded that "it is as though *the system had a schedule, a will, and a clock,*" and that "all perturbations are reabsorbed elastically without influencing the trend." This regularity would offer an outstanding forecasting tool: as coal displaced wood and oil displaced coal, so will economic and technical imperatives ensure that natural gas and nuclear and solar energies will displace oil on a highly predictable global schedule. According to Marchetti, trying to change the course of these developments is futile: we are not decision makers, at best we are optimizers, and it is the system that is making the decisions. But Marchetti was wrong in concluding that the system's dynamics cannot be influenced. After the oil price hikes of 1973–74 many forces began reshaping the global energy system on a massive scale and the result has been a shift from a regime of seemingly preordained energy substitutions to one of surprisingly stable energy shares with relatively little structural change.

Only a decade after Marchetti made his predictions the actual share of oil in global energy consumption was well ahead of the forecast value. And Marchetti's model and reality were far apart by the year 2000. In that year crude oil supplied 40% of the world's primary commercial energy needs, nearly 60% above Marchetti's prediction of 25%; and coal and natural gas delivered, respectively, 23% and 25% of all primary energy. This was a pattern very much unlike Marchetti's prediction of, respectively, a mere 10% and about 50% (fig. 3.24). Oil, in spite of numerous predictions of imminent production decline (see chapter 4 for details) is not retreating rapidly, natural gas is advancing much more slowly than predicted, and coal is still claiming a share more than twice as large as was irrevocably preordained by Marchetti's clockwork mechanism. Only the share of nuclear electricity in the world's total primary energy consumption (about 7.5%) was close to his forecast (6%).

I calculated all of the actual shares for the year 2000 solely on the basis of total commercial energy consumption as Marchetti's grand substitution scheme has wood falling to just a fraction of 1% after the mid-1990s. That is, obviously, another major error of the clockwork scheme. As I already noted in chapter 1, and will explain in more detail in chapter 5, biomass fuels (mostly wood, but also crop residues and dung) still provided about 10% of the world's primary energy in the year 2000, or more than did all electricity generated by nuclear fission. I hasten to add that this comparison of gross energy inputs, while not incorrect, is misleading: because of much lower conversion efficiencies 35–45 EJ of biomass fuels produced much less

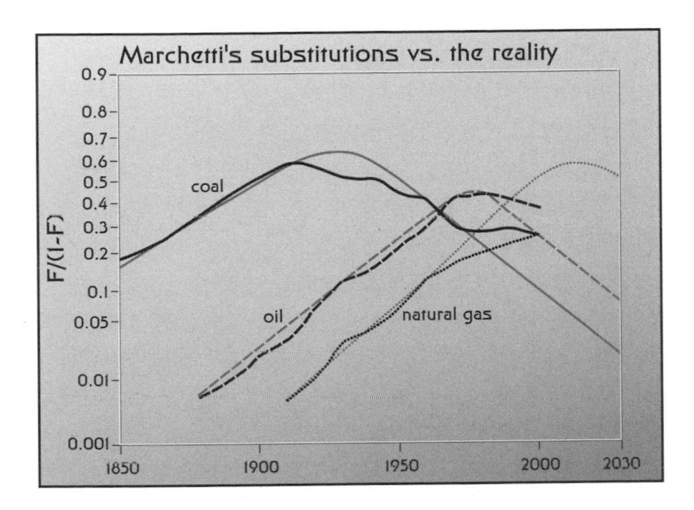

Figure 3.24
Real world versus a clock: after 1975 the actual shares of the three fossil fuels in the global TPES have shown major departures from expected trends, with both coal and crude oil consumption declining much more slowly, and natural gas use rising less rapidly than expected by Marchetti's substitution model. Plotted from global consumption data in UNO (1956; 1976; 1980–2001) and BP (2001).

useful energy than did 28 EJ of fission-generated electricity used by industries and households. However, this does not affect the matter at hand, namely the fact that the world still uses much more biomass energy than is dictated by Marchetti's clockwork.

Finally, Marchetti's inexorable mechanism requires a regular appearance of new sources of supply: without their sequential emergence the last entry in the substitution sequence, that is currently fission-generated electricity, would have to supply eventually all of the world's energy needs! But, as already noted, battered and contracting nuclear fission is not set to take over: it has been more than 20 years since any utility in the European Union or in North America ordered a nuclear reactor. Moreover, there are no plans to install more nuclear capacity in any affluent nation (for more on this see chapter 5). Decommissioning of ageing nuclear plants, rather than new construction, will be the industry's main concern in years to come (Farber and Weeks 2001). Marchetti's forecast of about 30% of the world's TPES coming from fission by the year 2020 is thus manifestly wrong.

And so is his forecast of the rising share of renewable conversions. At first look it appears that the share of 5% of all energy supplied by direct solar conversions by the year 2020 is a relatively modest requirement to fill—but only as long as one recalls that even with no increase in the world's energy use (surely a most unrealistic assumption) it would amount to an equivalent of almost 450 Mtoe, or more energy than was delivered in the year 2000 by hydroelectricity. Even when Marchetti's "solar" is interpreted broadly as all direct solar conversion and all wind- and wave-driven generation the task remains daunting. These three sources supply currently less than 0.01% of the world's primary commercial energy and so we would have to see their share of the global TPES growing by more than 30% a year for the next 20 years to reach Marchetti's goal for the year 2020. Unrealistic would be a charitable adjective to describe such a prospect.

Not surprisingly, latest IIASA global energy consumption forecasts do not refer to any inevitable clockwork substitutions and do not foresee any rapid end to the fossil fuel era. Instead, a study conducted jointly with the World Energy Council (WEC and IIASA 1995; Grübler et al. 1996) and forecasting global energy demand until the end of the twenty-first century presents three groups of scenarios. There are two cases of a high-growth future (one relying largely on both conventional and nonconventional hydrocarbons, the other one dominated by coal), a middle course, and two "ecologically driven" outcomes, one assuming a complete phase-out of nuclear energy, the other one an eventual introduction of inherently safe small reactors. But, most importantly, the scenarios make it clear that world of the late 1990s was perhaps only one-third of the way through the oil age and only one-fifth through the natural gas age (fig. 3.25). I will have more to say on scenario forecasts in the next section of this chapter.

Marchetti's forecasting failure has notable counterparts in many unfulfilled expectations on the extremes of the energy spectrum. While the real world has confounded his predictions that were pinned on a system with a supposedly immutable internal schedule, it also has failed to conform to unrealistic visions of excessive government-led tinkering promoted by advocates of both hard and soft energies. Massive worldwide resurgence of coal projected by the World Coal Study (nearly a tripling of its global output between 1977 and 2000) is an outstanding example in the first category (Wilson 1980). According to the study global coal extraction was to reach about 6.8 Gtce in the year 2000, an equivalent of 4.74 Gtoe, with the United States producing 25% of this huge total. Actual worldwide coal extraction in the year 2000 was less than half as large (about 2.2 Gtoe) and the U.S. output was less than one-third of the projected total (BP 2001). Global coal extraction thus has not conformed

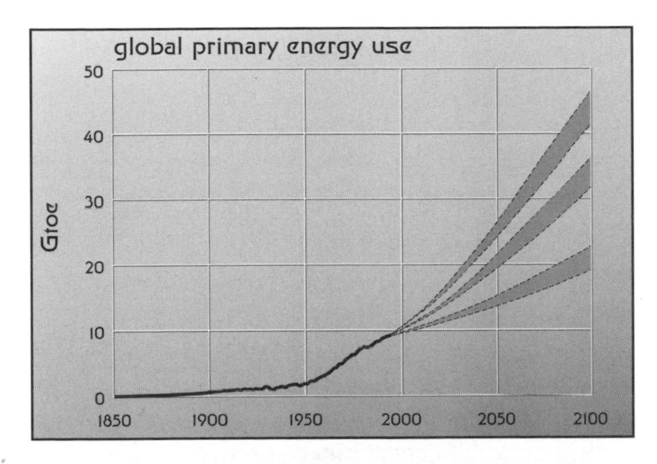

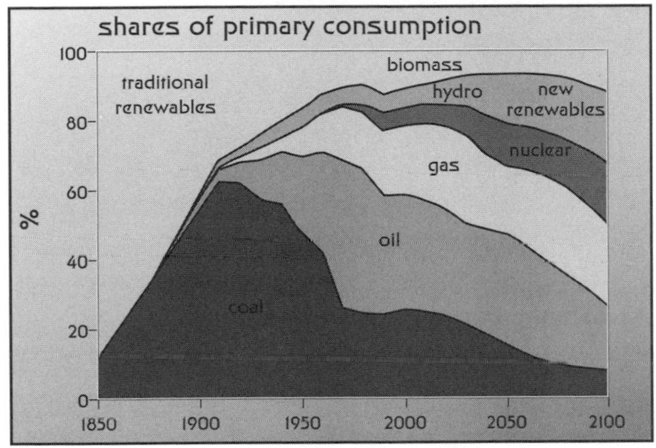

Figure 3.25
The range of forecasts of the global primary energy use by the WEC and IIASA and the shares of major sources in A1 scenario (high economic growth and a large future availability of hydrocarbons). Based on graphs in WEC and IIASA (1995).

either to Marchetti's predictions of rapid decline or to the World Coal Study's forecasts of a new prominence.

Renewable conversions have not fulfilled all those generation-old anticipations either. This point is best illustrated by comparing the actual performance of "soft energies" with their most touted forecast. In 1992 Amory Lovins looked back at his "soft path" prediction of aggregate energy consumption in the United States that was published in 1976 (Lovins 1976) and concluded that 15 years later his scenario stood the test of time far better than the conventional wisdom (Lovins 1992). True, his forecast is much closer to reality than all those simplistically exponential governmental predictions published during the 1970s. But it is a curious interpretation of reality when Lovins says that "the hard path hasn't happened and won't" (Lovins 1992, p. 9). We do not have giant nuclear islands—but neither do we have a new economy that is significantly dependent on renewable commercial energies, nor one that is poised to swing sharply in that direction.

In 1976 Lovins anticipated that the United States would derive about 750 Mtoe from soft techniques by the year 2000—but the actual total for renewables, including all hydro, biomass, and solar, in that year was about 175 Mtoe (Lovins 1976). After subtracting conventional large-scale hydro generation (clearly a kind of energy conversion that is neither small nor soft) renewables contributed just over 75 Mtoe, no more than about 10% of the share projected a generation ago by Lovins (fig. 3.26). Missing the target by 90% over a period of 24 years is hardly a noteworthy forecasting accomplishment. At least Lovins called for "just" around 30% of the U.S. total energy consumption to be delivered by renewables in the year 2000. In contrast,

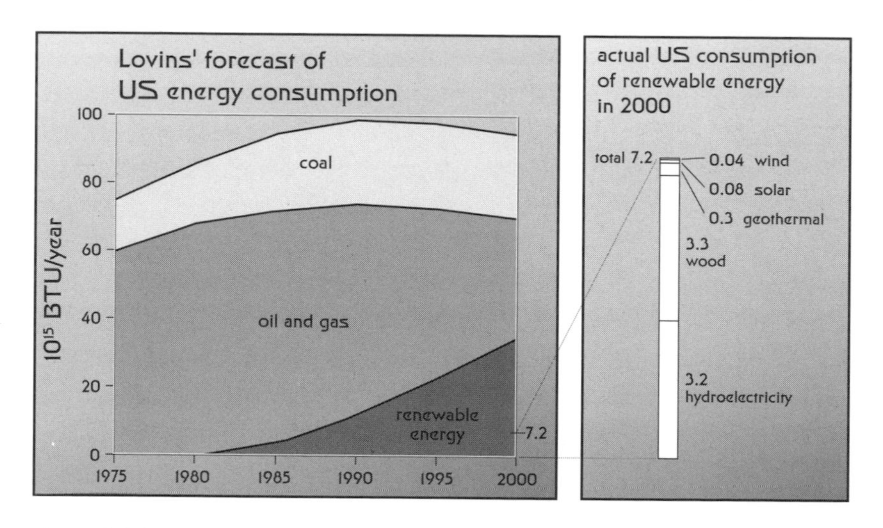

Figure 3.26
In 1976 Lovins forecast that "soft energies" will provide nearly one-third of the U.S. TPES by the year 2000. Actual contribution by all renewable sources was about 7%, and the share of small-scale, soft-energy techniques was no more than about 3%. Lovins thus missed his target by some 90%. Based on a graph in Lovins (1976) and consumption data reported by EIA (2001a).

Sørensen (1980) forecasted an American energy future where 49% of the country's energy use by the year 2005 were to originate from renewables, with biogas and wind supplying each 5%, and photovoltaics 11% of the total. As the actual U.S. consumption shares in the year 2000 were practically 0 for biogas, 0.04% for wind, and 0.08% for all forms of direct solar conversions (EIA 2001e), Sørensen's forecasts are off by anywhere between two orders of magnitude and infinity.

I cannot leave this section on resource substitution without citing opinions of the man who actually launched the modern obsession about running out of energy sources and about the consequences of such a development. In 1865 William Stanley Jevons (1835–1882), one of the leading economists of the Victorian era, published perhaps his most famous work, *The Coal Question,* in which he rightly connected the rise of British power with the widespread use of coal converted to the mechanical energy of steam. But Jevons also wrongly concluded that coal's exhaustion must spell an inevitable demise of national greatness. In forecasting coal demand he made the perennial error of vastly exaggerating future growth rates (see chapter 4). And after examining all supposed substitutes for coal (wind, water, tides, atmospheric electricity, peat, turf, and petroleum) he concluded that it is "of course . . . useless to think of substituting any other kind of fuel for coal" and that future advances in science "will tend to increase the supremacy of steam and coal" (Jevons 1865, p. 183, p. 188). Advance 135 years: as already noted (in chapter 1) in the year 2000 marginalized British coal industry extracted less than 20 Mt but the country was the world's ninth largest producer of crude oil and the fourth largest producer of natural gas—and a substantial exporter of both of these hydrocarbons!

Complex Models and Realities

Summarizing this dismal record of long-term energy forecasting is done easily with only three conjoined axioms. No truly long-range forecast can be correct in all of its key aspects. Most of these forecasts will be wrong in both quantitative and qualitative terms. Some forecasts may get a few quantities approximately right but they will miss the real qualities arising from subtly to profoundly altered wholes. The last of these three realities goes a long way toward explaining why it would be naïve to expect any radical improvements with the use of increasingly complex, and now invariably computerized, quantitative models.

Such a claim may seem unjustifiable as even some very complex and highly dynamic systems can be now simulated by increasingly sophisticated models whose calibrated runs offer reasonable replications of reality. Global climate modelling is a fine example of this slow, but indisputable, trend as it succeeded in replicating one of the most fundamental, and also the most intricate, planetary processes. The first three-dimensional models of global climatic circulation, prepared during the late 1960s, simulated well only the crudest features of planetary atmospheric dynamics. Although they are still too simple in order to accurately project regional and local conditions, today's best three-dimensional models do a much better job of replicating many complexities of global climate patterns and can take into account temperature forcing by natural and anthropogenic greenhouse gases and sulfates (GAO 1995a; Shukla 1998).

There is no shortage of other outstanding examples of complex realistic modeling ranging from photosynthetic productivity of crops to the flight aerodynamics of jumbo airplanes. All of these disparate modeling successes share a common denominator: they deal with systems that, although extraordinarily complex, lack complex interactions of social, economic, technical, and environmental factors that govern the course of energy production and use. Standard pre-1970 models of energy consumption not only failed to consider even basic interactions of these factors, they simply ignored most of these variables and used population and economic growth (often just predicted GDP or GNP rates with no allowance for changing efficiency of energy use) as the principal drivers of future energy consumption. Unfortunately, the first global model that attempted to incorporate all of the key variables and account for many of their complex interactions was a confusing mixture of laudable innovation and indefensible generalizations.

The Limits to Growth (Meadows et al. 1972), an outgrowth of Jay Forrester's work on dynamic systems (Forrester 1971), was easily the most widely publicized, and hence the most influential, forecast of the 1970s, if not the last one-third of the twentieth century. Energy consumption was, necessarily, an important part of this global exercise. But those of us who knew the DYNAMO language in which the simulation was written and those who took the model apart line-by-line quickly realized that we had to deal with an exercise in misinformation and obfuscation rather than with a model delivering valuable insights. I was particularly astonished by the variables labelled *Nonrenewable Resources* and *Pollution.* Lumping together (to cite just a few of scores of possible examples) highly substitutable but relatively limited resources of liquid crude oil with unsubstitutable but immense deposits of sedimentary phosphate rocks, or short-lived atmospheric gases with long-lived radioactive wastes, struck me as extraordinarily meaningless.

In spite of the fact that some publications identified major flaws of *The Limits to Growth* right after the book's publication (Nordhaus 1973; Starr and Rudman 1973), too many people took seriously this grotesque portrayal of the world. The report pretended to capture the intricate interactions of population, economy, natural resources, industrial production, and environmental pollution with less than 150 lines of simple equations using dubious assumptions to tie together sweeping categories of meaningless variables. But even if some very complex models were to succeed in conceptually bridging all of the relevant natural, technical, and socioeconomic realms they would not generally fare better than much simpler exercises. One of the two key reasons for this failure is that many critical variables cannot be either appropriately quantified or constrained by good probability estimates that are needed to generate restricted fans of possible outcomes to be used in confident decision making.

Naturally, this has not stopped many forecasters from supplying the needed quantities and I turn, once more, to examples provided by the IIASA experts who have been in the forefront of conjuring models of dubious complexity for more than two decades. I still remember my feelings of incredulity as I came across ludicrous manifestations of forecasting delusions in assorted acronymically named computer models such as MEDEE or MESSAGE (Sassin et al. 1983). Estimating, in the early 1980s, average GDP growth rates for a particular group of economies during the period 2015–2030 reveals one kind of misplaced confidence. Offering such numbers as the share of air-conditioned areas in service establishments, average bus ridership, or demolition rate of dwellings in a "region III," an entirely artificial unit made up of Western Europe, Japan, Australia, New Zealand, Israel, and South Africa (sic!) during the period 2025–2030 (as MEDEE-2 did) calls for another level of sibylline powers.

Chesshire and Surrey (1975, p. 60) identified the most obvious danger arising from this kind of modelling a long time ago:

> Because of the mathematical power of the computer, the predictions of computer models tend to become imbued with a spurious accuracy transcending the assumptions on which they are based. Even if the modeller himself is aware of the limitations of the model and does not have a messianic faith in its predictions, the layman and the policymaker are usually incapable of challenging the computer predictions . . . a dangerous situation may arise in which computation becomes a substitute for understanding . . .

Moreover, once the inherent uncertainties make the outcome fan too wide there is little point in building more complex models—and one might have obtained very similar results with a small electronic calculator and the proverbial back of an envelope. Recent energy models offer excellent examples of this forecasting genre. The already mentioned WEC and IIASA (1995) set of five scenarios offers global primary energy consumption forecasts ranging from 14–25 Gtoe in the year 2050, implying anywhere between 60–285% increase of the recent global energy use in two generations. Pretty much the same results can be obtained by simply assuming that the world's primary energy needs will continue growing at their recent (1980–2000) mean rate of 1.9% per year—or that a combination of continuing technical innovation, higher energy prices, and deliberate efforts to reduce energy use in order to minimize the impact of increasingly worrisome global warming will reduce that rate by at least one-third, to 1.25% per year. These simple, but clearly plausible assumptions, yield the range of about 16–23 Gtoe by the year 2050, the span very similar to 14–25 Gtoe obtained by elaborate WEC-IIASA scenarios.

The latest assessment of global warming uses a much wider fan of global energy consumption futures (Houghton et al. 2001). There are 40 different scenarios representing specific quantifications of one of the four principal story lines consider a huge range of population, economic, and technical factors that will drive future greenhouse gas and sulfur emissions (SRES 2001). These scenarios range from a world of convergent development with very rapid economic growth whose population peaks by 2050 and relies on new and more efficient conversion techniques to a world emphasizing local solutions whose continuously increasing population puts more stress of environmental protection and social equity. Extreme primary energy consumption values of these scenarios go from just 520 to 870 EJ in the year 2020 and from 510 to 2680 EJ in the year 2100 (fig. 3.27).

Again, it would have been so easy to get an almost identical fan of outcomes by using a few simple yet highly plausible assumptions explained in single-sentence scenarios. If seven billion people (the lowest total used by the IPCC scenarios for the year 2100) were to average 60 GJ of primary energy a year per capita (today's global mean) the total need in the year 2100 would be just 10 Gtoe. But if that energy would be used with average efficiency twice as high as it is today (a rather conservative assumption, we have done better during the past 100 years), then every person on this planet would have access to as much useful energy as does an average Italian today (who eats healthier and lives longer than just about anybody else). On the other hand, if today's global mean per capita energy consumption will rise as much as it did during the twentieth century (3.6 times) then the world of 10 billion

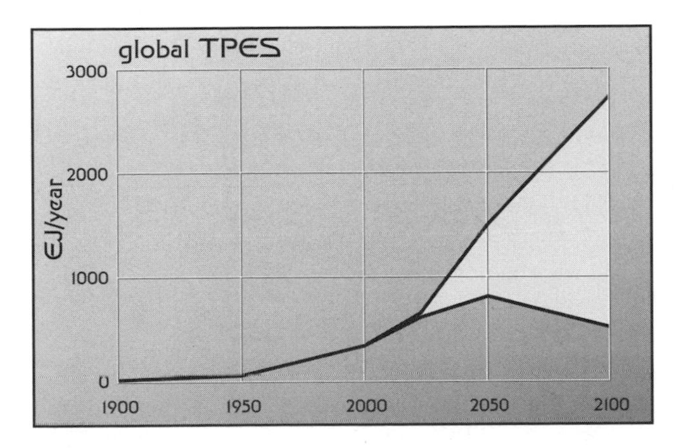

Figure 3.27
SRES forecasts demonstrate the questionable utility of century-long predictions: by the year 2100 the extreme global TPES values span more than a fivefold range. Plotted from data in SRES (2001).

people (the highest population total used by the IPCC) would need 57 Gtoe in the year 2100.

In a single short paragraph I outlined two highly plausible extreme scenarios bracketing the global energy use in the year 2100 (10–57 Gtoe) in an almost identical way (12–64 Gtoe) as did 40 elaborate IPCC scenarios stringing together multitudes of much more dubious assumptions. How can we even start guessing, as SRES (2001) did, what will be the per capita income or final energy intensities of economies in transition in the year 2100? What fundamentally new insights have been gained by those complicated exercises? As their authors note, their four featured marker scenarios were chosen from each of the scenario groups only because they best reflected the storyline and main attributes of specific models, not because they were seen more (or less) likely than any other scenario. For this reason Grübler and Nakićenović (2001) rejected Schneider's (2001) call for assigning subjective probabilities to their scenarios and continue to stress the fact that, in order to capture the continuing uncertainty about the future energy use, all illustrative scenarios should be seen as equally plausible.

This agnosticism may make logical sense (the level of future primary energy consumption and ensuing greenhouse gas emissions do remain highly uncertain)—but it is of no help in guiding sensible decisions: a fan of 10–60 Gtoe is simply too wide. And, to begin with, derivation of the plausible future range of global energy does not need any involved storylines, any strings of concatenated assumptions, or any computers. The exercise could be done, as just demonstrated, in a few minutes and as long as you know the formula for calculating exponential growth and remember that the current global energy consumption averages about 60 GJ/capita and that there are 42 GJ/t of crude oil all you need is a scientific calculator.

But there is nothing new about modellers' predilections for built-in complexity. They think they can do better by making their creations progressively more complex, by including more drivers and more feedbacks. They do not seem to realize that a greater complexity that is required to make such interactive models more realistic also necessitates the introduction of more questionable estimates (which can soon turn into pure guesses) and often also of longer chains of concatenated assumptions—and these necessities defeat the very quest for greater realism. As the variables become more numerous and time horizons more distant, quantifications becomes inevitably more arbitrary. And, even more importantly, no matter how complex a model might be and how lucky a forecaster can get in nailing a particular rate or share (or even a chain of assumptions) it is simply impossible to anticipate either the kind or the intensity of unpredictable events.

Unexpected Happenings

History is full of singularities, unpredictably rare events capable of changing destinies of nations, and now perhaps even of the entire humanity, for generations to come. No expert assemblies of the early-1970s would have identified the individuals whose unexpected decisions and actions changed the world's energy fortunes during the last quarter of the twentieth century and will continue shaping them in the near future. When I wrote the first version of this section during the summer of 2001, I believed that any list of such notable individuals should contain at least four men whose actions steered the world into utterly uncharted futures. The fifth name must be obviously added after September 11, 2001.

Decisions made by Shaikh Ahmad Zaki Yamani, oil minister of the Wahabite kingdom during the tumultuous period of sudden oil price rises, helped to create, and then to break, the era of OPEC's high oil prices as he changed the Saudi oil policy (Robinson 1988). In 1970 the very existence of Ayatollah Ruhollah Khomeini, an ascetic potato-eating Shi'ite mullah living in exile in Najaf in Iraq, was known only to a mere handful of Western experts but his creation of a powerful fundamentalist state in Iran after the overthrow of Shah Reza Pahlavi helped to nearly quadruple the world oil price less than a decade later and to change the geopolitics of the Middle East for generations to come (Zonis 1987). Deng Xiaoping, a veteran Chinese Communist Party leader, had barely survived Mao's Cultural Revolution, and in 1970 was living in internal exile in Sichuan. Eight years later his bold reforms began transforming China from an inefficient Stalinist economy with a subsistence standard of living to a rising world power (Ash and Kueh 1996). And in 1970 Mikhail Gorbachev—the man who 20 years later presided over an astonishingly swift and, even more astonishingly, peaceful dissolution of the Soviet Union—was still an obscure Communist Party secretary in Stavropol (Ruge 1992).

Who among the energy CEOs in 1970 actually worried about the ten-year-old OPEC, which was selling its oil at $1.85 per barrel but which, before the decade's end, was commanding up to 20 times as much? Who among energy economists would have guessed that this enormous price rise would be accompanied by increasing oil consumption? Whatever happened to demand-price elasticity? Who among the political analysts dissecting the 1973 Yom Kippur War, the proximate cause of the Arab oil embargo, would have predicted Sadat's and Begin's toasts in Jerusalem a mere four years later? Who among the East Coast liberals oozing disdain for the B-film credentials of Ronald Reagan would have believed that the President's 1987 challenge to Mikhail Gorbachev—"Tear down this wall!"—would become reality just two years later (Reagan 1987)? And who would have predicted that Reagan's policies would help to speed up that epochal opening? And who would have dared to predict that by the year 2001 the Evil Empire of 10,000 nuclear warheads, the perennial nightmare of the U.S. Cold War policy, would be gone, and that, in spite of its enormous natural riches its unraveling economy would make it poorer, in per capita terms, than Malaysia or Uruguay (UNDP 2001)?

Usama bin Ladin's eventual success—eight years after the failed, and curiously underestimated, 1993 attempt to bring New York's World Trade Center towers down by a truck bomb—in terrorizing the United States by a concerted crash of Boeing 767s into the iconic buildings of the world's most powerful economic and military power is not just another obvious addition to this list of men who have driven the world into uncharted futures (see also fig. 2.20). His murderous plan was clearly calculated to evoke enormous political, military, and economic dislocations and the costly short-term impact of economic damage and military and counterterrorism expenditures may be trivial compared to (perhaps imaginable but in their particulars utterly unknown) long-term consequences. Reduced global demand for crude oil is one of the certain short- to mid-term effects and renewed attempts to lessen the Western dependence on Middle Eastern energy are among the likely longer-term reactions.

And who among the U.S. commentators, whose articles and television appearances were gloomily predicting the end of the U.S. industrial dominance, did foresee the end of the Japanese economic miracle and a robust resurrection of U.S. productivity gains and economic growth during the 1990s when the country's inflation-adjusted GDP rose by 30%? Of course all of these unexpected events had enormous repercussions for the world's energy affairs. I will take a closer look at just two of these stunning changes, at Japan's economic reversal and at the generation of CO_2 from the combustion of fossil fuels as they offer very different, but equally excellent, examples of changes whose timing and intensity, and indeed the very occurrence, were not foreseen by any professional forecasters.

Japan's steady economic rise proceeded at a rapid rate from the end of WWII to 1989, and inflation-adjusted GNP growth averaged 10% during the 1960s, and more than 5% between 1970 and 1973 (Statistics Bureau 1970–2001; IEE 2000). The first oil shock caused the first postwar GNP decline (−0.7% in 1974), and it became easy to believe that a country importing 99.7% of its large, and rising, crude oil consumption would not fare well in the new world of high energy prices. But during the late 1970s the Japanese economy managed an average annual growth of more than 4% as the country's exporting manufacturers absorbed higher energy costs, improved the efficiency of their production, and increased their share of the global trade and their surplus with the United States (Statistics Bureau 1970–2001). Incredibly, five years after the first oil shock, and just as the second round of steep oil price rises was unfolding, Ezra Vogel (1979) envisaged Japan as the world's number one economy and the country's continuing solid performance engendered a period of intense U.S. worries and self-examination (Fallows 1989; Oppenheim 1991).

During the 1980s Japan's economy continued to grow at an average annual rate of 4%, and when the Plaza agreement of 1985 pushed the yen to the unprecedented height against the U.S.$, the Japanese investors began buying up America. The Nikkei index, below 7,000 in 1980, was closing on 40,000 a decade later (it stood at 38,916 on December 31, 1989) as the yen kept on soaring, from close to ¥250/U.S.$ in the early 1980s to ¥144/U.S.$ by the end of 1989 (fig. 3.28). If the expectations prevalent during the 1980s had turned out to be correct, North America and Europe would be by now destitute tributaries of the new Empire of the Rising Sun.

And then the bubble economy suddenly burst in the spring of 1990 and Japan's extraordinary boom turned into an unprecedented decline (Wood 1992). By the end

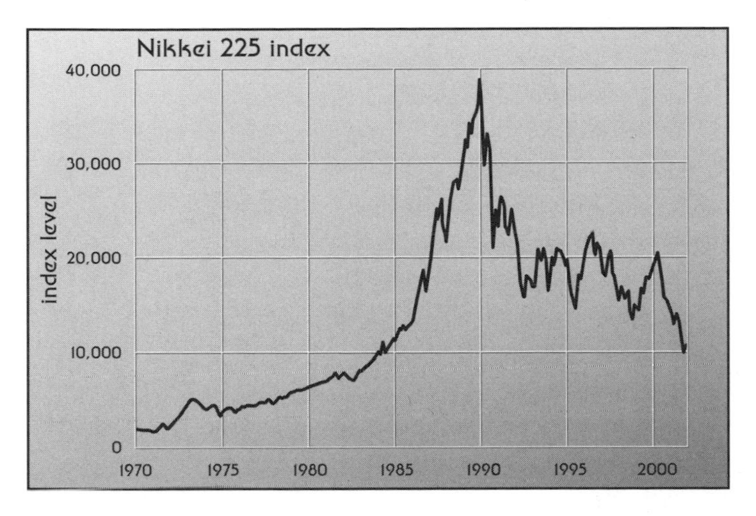

Figure 3.28
A quarter-century of the Nikkei Index (1976–2001) shows the gradual pre-1984 ascent, rapid rise during the late 1980s, steep post-1989 tumble, and recent stagnation. Plotted from data available at Nikkei Net Interactive, <http://www.nni.nikkei.co.jp>.

of 1990 the Nikkei was 40% below its peak and, scores of recurrent forecasts of an imminent recovery notwithstanding, the index has stayed down for more than a decade, spending most of the 1990s well below half of its peak value, and sinking to less than 10,000 in the year 2001 (Nikkei Net Interactive 2001; fig. 3.28). By the mid-1990s admirers of rapid Asian economic growth (IMF and the World Bank included) shifted their adulation to the continent's smaller tigers (Indonesia, Thailand, South Korea, and Taiwan) only to be stunned when those economies tumbled suddenly in 1997. Naturally, both the protracted post-1989 Japanese economic stagnation and the recent Asian economic downturn have had enormous immediate, and important long-term, effects on the global demand for traded fossil fuels, and hence on their world price: Asia's Pacific countries now buy more than one-third of the world's crude oil exports (BP 2001).

My second example of recent unpredictable events concerns the generation of CO_2 from the combustion of coals and hydrocarbons. Although its former dominance among the greenhouse gases has greatly diminished (see chapter 1), CO_2 remains the single largest contributor to the anthropogenic warming of the troposphere, and its emissions are greatly affected by socioeconomic discontinuities that are, inevitably, translated into changed energy demand. Neither of the two events that had the greatest effect on global CO_2 emissions of the last 20 years of the twentieth century—the precipitous but nonviolent collapse of the Soviet empire and the rise of a surprisingly more efficient China integrated into the world economy—could have been considered by even the most astute climate modeller in 1980. After all, they came as total surprises even to the people who have spent decades studying the two countries in question.

During the 1990s, energy consumption in the successor states of the Soviet Union and in the nations of the former Soviet empire fell by about one-third (BP 2001), and as a result those countries released at least 2.5 Gt C less than if they would have consumed fossil fuels at the late-1980s level, and about 3 Gt less than if their growth of energy demand would have continued at the same rate as it did during the 1980s. And because during the 1990s energy needed per unit of GDP by China's fast-growing economy was cut by half, the country's aggregate carbon emissions for the decade were about 3.3 Gt C lower. Consequently, the demise of the Soviet empire and unexpectedly strong efficiency gains in China prevented the release of nearly 6.5 Gt C during the 1990s, an equivalent of the world's average annual carbon production from fossil fuels during the decade's last two years (Marland et al. 1999). Who among the forecasters of global CO_2 generation (they are mostly atmospheric physicists) would have even dreamed about including these massive energy consumption shifts in the global climate models they were building during the early 1980s?

At this point there is no need for more examples of failed forecasts. The conclusion is pretty clear: long-range forecasters of energy affairs have missed every important shift of the past two generations. They paid hardly any attention to OPEC's unassuming rise (1960–1971) and they were perhaps stunned as much by the quintupling (1973–1974) and then the additional quadrupling (1979–1980) of the world crude oil price as they were by the cartel's sudden loss of influence (1985). They were surprised by the Iraqi occupation of Kuwait and the ensuing Gulf War (1990–1991), as well as by the recent tripling of the world oil price (1999–2000). They failed to anticipate first the drastic reduction of electricity demand throughout the Western world after 1970, and then the collapse of the oversold promise of nuclear generation that was to switch rather rapidly to fast breeders and to rely eventually on fusion.

At the same time, most forecasters have repeatedly and vastly overestimated the potential of new energy conversions, be they synthetic fuels derived from coal or nonconventional hydrocarbons, various uses of biomass, wind, geothermal, and central solar power, or the rates of commercial diffusion of fuel cells, hydrogen, and electric cars. Conversely, they have greatly underestimated both the cumulative contributions of individual actions of mundane energy conservation (ranging from better wall and roof insulation to more efficient lighting and household appliances) and the large remaining potential for reducing energy intensity of major economies, including the ones in rapid stages of modernization that were traditionally seen as the worst possible candidates for such impressive savings.

Most of these grossly inaccurate forecasts can be explained by two opposite expectations. The first one is the spell cast by unfolding trends and by the mood of the moment, the widespread belief that the patterns of the recent past and the aspirations of the day will be dominant, not just in the medium term but even over a long period of time. Its obvious corollary is the recurrent phenomenon of the herd instinct in forecasting as immediate experiences and reigning expectations color the outlook of all but a few maverick seers. The second one is the infatuation with novelties and with seemingly simple, magical solutions through technical fixes: such wishful thinking elevates the future performance of favorite techniques or influences of preferred policies far beyond their realistically possible contributions.

After I came from Europe to the United States in the late 1960s and was given access to one of the better mainframes of that time (now the laughably wimpy IBM 360/67), I was fascinated with the possibilities offered by computer modeling of complex systems. But my initial enthusiasm diminished gradually as I faced the necessity to plant many numerical assumptions into so many unquantified, and seemingly unquantifiable, fields. By the mid-1970s, after my detailed dissection of *The Limits to Growth,* I was rather skeptical about the use of such models in any decision making. Then a new crop of even more complex models helped to turn my skepticism into incredulity. Misleadingly precise estimates of future shares, rates, and totals that were offered with apparently straight faces by these models were the last push I needed to abandon complex, formal quantitative forecasting. My overwhelming feelings were: how could there be so much pretense, so much delusion, so much autosuggestion; why play such laughable games?

I am always pleased when I discover that somebody dealing with the same problem was able to express the same thoughts in a more succinct manner. As far as the forecasting of energy futures goes, Alvin Weinberg did it for me a long time ago in a fine paper on limits of energy modeling. Weinberg (1979a) pointed out the possibility that not only is the mathematics behind such models nondeterministic but that "the underlying phenomena simply cannot be encompassed by mathematics," that energy analysis is beset with propositions that are either intrinsically or practically undecidable, and that "we must maintain a kind of humility" when engaging in these intellectual exercises (Weinberg 1979a, p. 9, p. 17). So what is then the most sensible course of action? To do what really matters.

In Favor of Normative Scenarios

Forecasting is now an entrenched, highly institutionalized multibillion dollar business. An overwhelming majority of it proceeds along the lines of what Bankes (1993) calls consolidative modelling, whereby known facts are assembled into a single, and increasingly complex, package that is then used as surrogate for the actual system. Where the match between such a model and reality is sufficiently close this technique offers a powerful tool for understanding and predicting the behavior of complex systems. But the consolidative approach will not work whenever insufficient understanding or irreducible uncertainties will preclude building an effective surrogate for the studied system. In spite of the fact that the global energy system, and its national and regional subsystems, clearly belong to this unruly category many experts keep on using the unsuitable approach.

Typical forecasts offer little else but more or less linear extensions of business as usual. For example, the U.S. Department of Energy projects the global TPES rising by almost 60% by the year 2020, with the consumption in low-income countries going up nearly 2.2 times and the U.S. demand rising by nearly one-third (EIA 2001d). Results of these efforts are predictable as these models keep on producing thousands of pages of almost instantly obsolete forecasts, and as corporate offices and libraries continue to be littered with computer printouts, binders, and books whose usefulness should have been suspect even before they were commissioned.

Given the prevailing approaches and practices of energy forecasting, new embarrassments and new misses lie ahead. There will be new crops of naïve, and (not only) in retrospect incredibly shortsighted or outright ridiculous predictions. Conversely, we will be surprised repeatedly by unanticipated turns of events. Some extreme predictions, often the easiest possibility to outline, may eventually come to pass. What will remain always the most difficult challenge is to anticipate the more likely realities arising from a mix of well-understood and almost inevitable continua on one hand and of astounding discontinuities and surprises on the other. In this respect a new century will make little difference to our ability of making point forecasts: we will spend more time and money on playing the future game—but our predictions will continue to be wrong.

We will not do better just because available computing power will make it easy to handle an unprecedented number of variables and we will weave their interactions into ever more complex models. We will not do better as we try to include every conceivable factor in our assessments because many of these attributes are either unquantifiable or their quantification cannot go beyond educated guessing. We will not be more successful in coming up with single numbers, be it for absolute performances, rates, or shares, for specified periods of time decades, or even a century, ahead. And we will not get more profound insights from preparing very wide fans of alternative scenarios that try to capture every imaginable contingency and eventuality. But this does not mean that I advocate a complete abstention from looking far ahead.

Forecasts have always been done in order to better anticipate future realities and to get clearer insights into the kind and rate of requisite actions needed to prevent, or at least to minimize, unwanted outcomes and to reinforce desirable trends: simply put, to be able not just to react but to better mold our collective futures. As the scope of controls exercised by an increasingly complex society expands and as our continued existence as a high-energy civilization is putting a greater stress on the biosphere's vital structures and functions, we should move beyond passive, agnostic forecasting of possible outcomes. Instead, we should work earnestly toward holistically defined goals that reconcile the maintenance of human dignity and a decent quality of life with the protection of the biosphere's irreplaceable integrity. As the future is inherently unpredictable it is the setting up of desirable goals and formulation of effective no-regret pathways leading to them that can make the greatest difference to our quest for rational management.

Normative scenarios, outlining what should happen rather than what is likely to happen, are thus particularly useful. Exploratory modeling, a series of formal forecasting experiments that may include quantitative consolidative models, can be then used in order to assess the implications of varying assumptions and hypotheses (Bankes 1993). And given the increasing need for preventing long-lasting damage to the global commons, I strongly believe that normative scenarios are not just useful, but essential. Intent is, of course, one thing; execution is another matter. To be both realistic and useful, normative scenarios must combine the recognition of numerous limits and uncertainties in the real world with sensible aspirations for a higher quality of life and commitments to preserving the biosphere's integrity. They should also remain probing, flexible, and critical, rather than being just rigid and, in spite of their seeming boldness, unimaginative advocacy tools promoted by true believers.

And they should always try to feature no-regret prescriptions as perhaps the best way to avoid the impacts of both unpleasant and welcome surprises.

Formulation of desirable goals may seem to be a hopeless task in modern affluent societies splintered by special interests and driven by disparate visions of the future. I refuse to subscribe to such defeatist thinking and believe that basic normative scenarios are not actually so hard to formulate. As the goals should not be spelled in excessive detail—both in order to retain needed flexibility and to prevent the obvious perils of micromanagement—their definitions can remain qualitative or can be quantified in broad, nonrigid terms. But before we actually begin the construction of normative scenarios we must critically appraise all available options that we could use in order to arrive at desired socioeconomic and environmental destinations.

Given the complexity of modern energy systems this means that these appraisals must focus above all on a multitude of uncertainties concerning possible contributions to be made by coal, traditional and unconventional hydrocarbons, nuclear and renewable energies, and by rational use of fuels and electricity. Some of these concerns have been with us for generations, others are relatively new, and in either case perceptions and interpretations keep shifting with the changes in economic outlook, national and international security, as well as with the emergence of new environmental disruptions. An imminent end of the oil era is a perfect example in the first category while second thoughts about the benign character of renewable energy conversion belong to the second group of concerns.

And, to use perhaps the most prominent environmental example, any long-term appraisals of energy prospects will be strongly influenced by changing perceptions concerning the rate and degree of global climate change. Rapid and pronounced global warming would have obviously different consequences for the fate of coal-based conversion and the diffusion of fuel cells than would a limited and slowly advancing process hardly recognizable from long-term natural fluctuations. We should not be so naïve, or so arrogant, as to believe that even the best possible critical assessments of limits and uncertainties can eliminate all the extremes of exaggerated hopes and unnecessarily timid expectations—but they should at least help us minimize the likelihood of making such common errors of judgment. I will deal with these uncertainties in the next two chapters, addressing first the future of fossil fuels and then taking a systematic look at the prospect for nonfossil energies.

4

Fossil Fuel Futures

Appraising the future of fossil fuels means much more than just looking at their reserves and resources. But, inevitably, the two terms will be recurring throughout this chapter and hence their brief definitions are in order. Geologically speaking, the two nouns are not synonyms, and resources are not used to describe, as they are in common parlance, any natural (mineral or living) endowment but represent the totality of a particular mineral present in the Earth's crust. Reserves are those well-explored shares of total resources that can be extracted with available techniques at an acceptable cost; advances in exploration and extraction constantly transfer the minerals from the broader, poorly known and often just guessed at, resource category to the reserve pool (Tilton and Skinner 1987).

Simply, resources of a particular mineral commodity are naturally in place, reserves of it become available through human actions. Standard graphic representation of the two concepts in fig. 4.1 also shows commonly used subcategories. Resource exhaustion is thus not a matter of actual physical depletion but rather one of eventually unacceptable costs, and the latter designation may now mean not just in an economic sense (too expensive to recover or to deliver to a distant market) but also in environmental (causing excessive pollution or unacceptable ecosystemic destruction or degradation) and social terms (where their recovery would necessitate displacement of a large number of people, or bring serious health problems).

As I will demonstrate, the highly dynamic nature of the exploitation process means that although the fossil-fueled civilization is energized by the recovery of finite, nonrenewable resources it is dubious to offer any fixed dates for the end of this critical dependence based on specific exhaustion scenarios. Commonly used reserve/production (R/P) ratios, calculated simply by dividing known reserves of a commodity by its annual output, may be a useful warning sign of an impending exhaustion—or they may tell us little about the long-term prospects for recovering a mineral. A

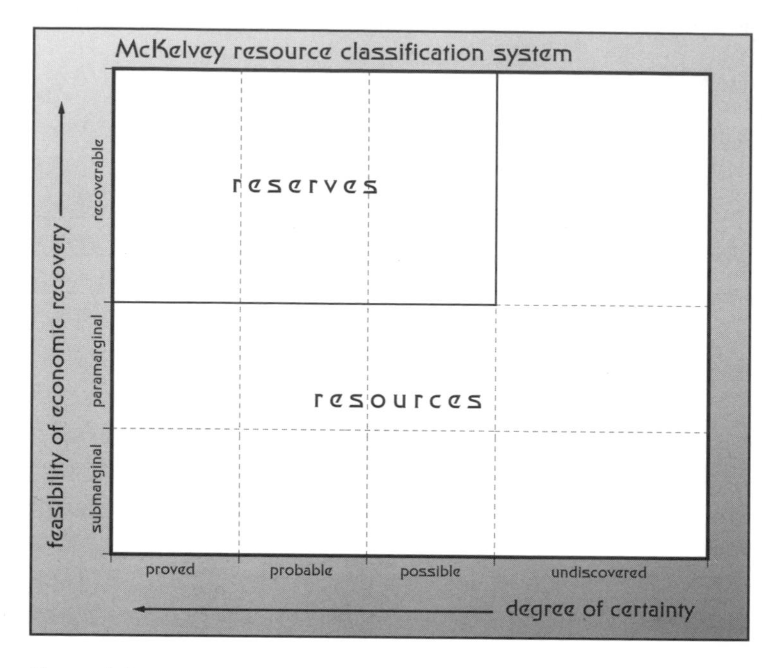

Figure 4.1
Graphical representation of the standard U.S. resource classification system (McKelvey 1973). A number of modified versions of McKelvey's box are now in use.

very low R/P ratio for a locality where a mineral is being extracted from a clearly circumscribed and perfectly known deposit will clearly signal an imminent local exhaustion. A low global R/P ratio of a commodity whose exploration is still in the comparatively early stages and whose production costs are bound to decline with technical advances is largely irrelevant in judging that mineral's future output.

In spite of enormous cumulative recoveries of fossil fuels during the past century—almost 250 Gt of coal, nearly 125 Gt of oil, and more than 60 Tm^3 of gas—R/P ratios of all of these minerals were substantially higher in the year 2000 than they were a century ago (BP 2001). Coal's R/P ratio of about 225 years and the enormity of coal resources makes the fuel's extraction and transportation costs and the environmental consequences of its combustion, rather than its availability, the foremost concerns determining its future use. The situation appears to be very different for hydrocarbons, and particularly for crude oil as the twentieth century saw many forecasts pinpointing the onsets of irreversible decline of petroleum extraction. Some of these forecasts—based on the notion that a lifetime of resource production must follow, more or less, a normal (Gaussian) curve—have been presented as unalterable realities, but I will show that such outcomes are not preordained.

They would be credible only if well-defined categories of global resources were known with a high degree of accuracy, if we were to assume no improvements in our technical capabilities, and if we had a perfect knowledge of future demand. None of these conditions apply in the real world. Ultimately recoverable resources of conventional crude oil or natural gas may not be three times as large as today's modal estimates, but they may be easily 50% or even 100% higher. Whatever the ultimate resources may be, technical advances will boost their recovery rates and innovation in exploration and extraction will be also steadily shifting the divide between conventional (liquid) and nonconventional oil resources. Finally, the pace of any resource exhaustion is obviously determined not just by the presence of a mineral in the crust but also by demand, and the need for a particular commodity may greatly weaken, for a variety of reasons, long before we approach its physical exhaustion.

Given the inherently inertial nature of complex energy systems there is a very low probability that any fundamental, large-scale shifts in the use of principal energy resources will take place during the next 10–15 years. Such dramatic moves are not impossible, but during the second half of the twentieth century their occurrence has been limited to those lucky nations whose discovery of large hydrocarbon reserves enabled them to shut down abruptly, or at least to scale down drastically, their coal extraction. The Dutch discovery of Groningen gas, and the U.K.'s North Sea oil and gas bonanza would be perhaps the best examples. Consequently, we may be surprised repeatedly, as we were during the last quarter of the twentieth century, by the resilience of long-established fuels and their conversions and by the inventiveness of markets to give them new lives by making them more economical and environmentally more acceptable.

Conversely, when looking 100 years ahead there is also a very low probability that fossil fuels will energize the world of the early twenty-second century to the same extent as they do now. The intervening decline of our reliance on coal and hydrocarbons and the rise of nonfossil energies will not take place because of the physical exhaustion of accessible fossil fuels but, as in the case of past resource transitions, because of the mounting cost of their extraction and, even more important, because of the environmental consequences of their combustion. From our vantage point it is thus the contours of the period between the years 2020–2080 that remain particularly uncertain. However, revealing analyses of both time periods are influenced by the same ephemeral distractions that range from yet another threatening OPEC meeting and a spike in gasoline prices to a new warning about the rate of global warming and to yet another Middle Eastern conflict. As I address the future of fossil fuels in a systematic fashion I will try to look beyond such temporary distractions, as well as beyond numerous advocacy claims. I will start with the assessment of currently fashionable predictions of the imminent peak of global crude oil production.

Is the Decline of Global Crude Oil Production Imminent?

Before I begin a brief historical review of oil extraction and present the most recent round of the controversy regarding oil's global future I must note that I will make an exception to my normal adherence to metric units and will repeatedly refer to barrels of oil. This antiquated unit—abbreviated as bbl (originally for blue barrel, identifying the standard container of 42 U.S. gallons), going back to the early years of the U.S. oil industry and adopted officially by the U.S. Bureau of the Census in 1872—remains the oil industry's most common measure of both reserves and production. Given the various densities of crude oils there is no one way to convert these volume units to mass equivalents. Crude oils have densities as low as 740 and as high as 1,040 kg/m^3, which means that anywhere betwen 6.04 and 8.5 barrels make a tonne, with most crude oils falling between 7–7.5 bbl/t. The average of 7.33 bbl/t has been used commonly, while BP's latest mean is 7.35 bbl/t (BP 2001).

Oil production during the last four decades of the nineteenth century (beginning with the first commercial operation in Pennsylvania in 1859) grew slowly, and it was limited to just a few major players dominated by the United States, Russia, and Romania. By 1900 annual global output reached about 150 million barrels (Mb) of oil, the amount that we now extract worldwide in just two days. By 2000, annual flows were above 26 billion barrels (Gb) of oil drawn from reserves two orders of magnitude larger than in 1900, and sold at an average (inflation-adjusted) price lower than 100 years ago (fig. 4.2; for price trends see figs. 2.8 and 3.15). By any standard, these realities are an admirable testimony to both technical and organizational acumen of modern civilization. How long will this comfortable situation continue? How soon may we expect the permanent downturn in the global output of crude oil? How much will all of this affect affluent populations consuming most of the world's crude oil and the still growing populations of low-income countries where average consumption remains very low but economic aspirations are enormous? And are we now, after generations of contradictory opinions, finally closer to answering these fascinating questions?

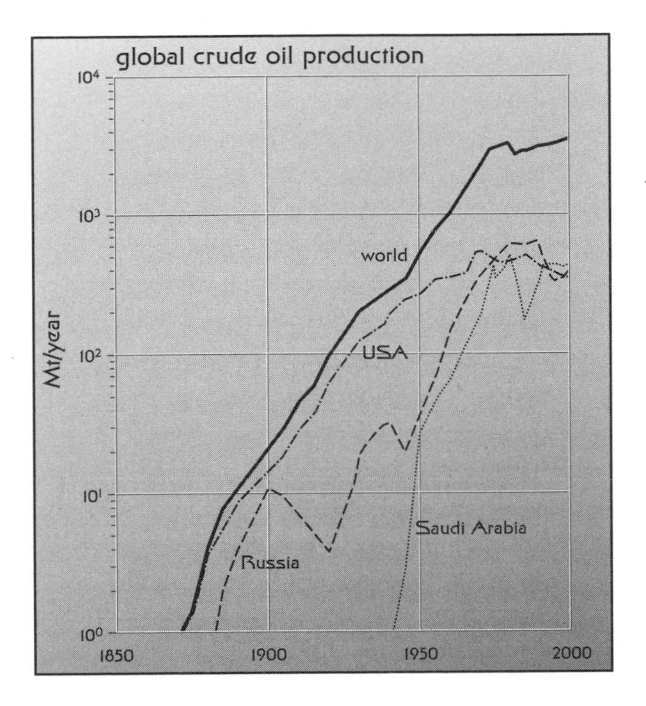

Figure 4.2
Global oil production, 1850–2000. Based on Smil (1994a) and additional data from BP (2001) and UNO (2001).

Only the last question has an easy answer: not really, as extreme opinions are not easily reconciled. There is, on one hand, the anticipation of a long spell of comfortable low-priced oil supply that precedes a fairly smooth transition to other energy sources. On the other hand there is a firm belief in an imminent end of cheap oil brought by steeply declining extraction that will follow the rapidly approaching peak of global crude oil production. What is no less remarkable is the repetitiveness of these resource debates: individuals come and go, but irreconcilable conclusions remain. Whom the policymakers tend to believe more at any given time depends mostly on the immediate impacts of short-term price trends.

Extreme arguments about oil resources and their future extraction belong to two much broader patterns of assessing the planet's resource futures—and hence civilization's long-term prospects. The optimistic position has been staked out by those who had studied the history of resource use and became fervent believers in the commanding role of prices, unceasing human inventiveness, and the surpassing power of technical fixes. Proponents of this way of thinking—a group comprised mostly of economists (representative writings are: Hirsch 1976; Simon 1981, 1996; Goeller and Zucker 1984; De Gregori 1987; Tilton and Skinner 1987; Adelman 1997)—do not believe that the crustal presence of any particular natural resource is an inexorable determinant of the fate of civilization.

For their way of thinking the obvious fact that the Earth's crust contains finite quantities of minerals is irrelevant because our effort to extract these resources will cease long before we approach their global physical exhaustion. As the cost of creating new reserves (in-ground inventories ready for commercial exploitation) through exploration and development becomes too high, an industry disappears. An unknown amount of a resource left in the crust ceases to be a sought-after commodity as substitutes take its place. As far as crude oil is concerned, the cost of translating its resources into reserves does not seem to be giving any signals of imminent exhaustion. This means that we should see continued additions to global oil reserves for many years to come.

The real challenge is then to avoid the situation where the extraction of the remaining resource would be so expensive that it would disrupt economies and social structures at a time when no acceptable substitutes would be ready on the market. Industrial civilization has been able to avoid repeatedly this undesirable state of affairs by raising utilization efficiencies, by developing new techniques for economic exploitation of lower-grade resources and, eventually, by managing to have effective substitutes ready to fill the place of resources that have become either too costly or otherwise unacceptable (mainly because of the environmental impacts of their extraction and use). Consequently, the fear of physical exhaustion of a particular resource has been repeatedly exposed as groundless. "Adaptive cornucopian" would be perhaps the best description of this worldview.

Julian Simon was probably the best known cornucopian among modern economists (Simon 1981; Simon and Kahn 1984; Simon 1996). He liked to use the recent history of oil prices to demonstrate the effectiveness of market adjustments. He also used U.S. and global R/P ratios for oil to illustrate the uselessness of known reserves as a measure of future supply, noting how these reserves always stay just a step ahead of demand—and had no doubt that "never" was the only correct response when answering the question "When will we run out of oil?" (Simon 1996). Morris Adelman, who spent most of his career as a mineral economist at the MIT, has been the most eloquent advocate of this view among the academics studying oil industry. "Finite resources," in Adelman's (1992, pp. 7–8) succinct verdict, "is an empty slogan; only marginal cost matters."

Any talk about infinite supplies and unceasing inventiveness is an anathema to those who are preoccupied with resource exhaustion in general, and with inevitable decline of conventional oil extraction in particular. They worry about the imminent end of the oil era with the irreversible decline of global oil extraction beginning before the end of the first decade of the twenty-first century. Knowing this, they cannot foresee how the anticipated scarcity of the fuel that has been so instrumental in creating modern civilization could be managed without profound worldwide economic, social, and environmental consequences. People belonging to this rather heterogeneous group range from unabashed environmental catastrophists to cautious consulting geologists.

Although most of them do not know it, they are all intellectual heirs of an old British tradition whose most famous representative was William Jevons (1865) someone whose spectacularly incorrect forecasts were already introduced in chapter 3. Not surprisingly, worries about the finiteness of coal supply arose for the first time in the late eighteenth-century Britain, but estimates published before 1830 were comforting, being as high as 2000 years before exhaustion (Sieferle 2001). But by the 1860s the consensus shrank to a few centuries and at a time when a steadily rising British coal production dominated the fuel's global extraction, Jevons worried about the coming exhaustion of the country's coal resources that would usher the decline of imperial power. Nearly a century and a half later the United Kingdom has plenty of coal left in the ground and can produce it cheaper than any other nation in the EU (UK Coal 2001)—but, as I already illustrated in chapter 1 (see fig. 1.6) its coal extraction has shrunk so much that it provided less than 10% of the country's TPES in the year 2000.

Darker versions of catastrophist prophecies have maintained that modern civilization has already moved very close to, or perhaps even beyond, the point of no return as it rushes toward exhaustion of nonrenewable resources, irreparable deterioration of the global environment, and deepening social crises (Meadows et al. 1972; Ehrlich and Holdren 1988; Ehrlich 1990; Meadows and Meadows 1992). Conservative geologists buttress such opinions by arguing that the extraction of conventional oil will begin to decline sooner than most people think and that such a permanent shift will lead to a radical increase in oil prices.

These conclusions translate not just to forecasts of a rapidly approaching third round of a global oil crisis (Criqui 1991) but also to more worrisome scenarios about an early end of the oil era that see the time of permanent extraction downturn as "the inevitable doomsday" followed by "economic implosion" that will make "many of the world's developed societies look more like today's Russia than the United States." (Ivanhoe 1995, p. 5). Are these conclusions based on incontrovertible evidence—or should they be dismissed as just the latest installment in a long series of similarly dire forecasts? And even if it comes soon, must the permanent decline of global crude oil production have crippling economic consequences? I will try to answer these critical questions after taking a closer look at the arguments offered by the opposing camps of this fascinating debate.

Cornucopians are not particularly concerned about accurate definitions of technical terms used in categorization of natural resources or about magnitudes and probabilities of resource estimates: they are satisfied just to use either long-term price trends or simple R/P ratios to demonstrate the continuing abundance of supply and its fundamentally secure prospects. Generally declining long-term oil price trends were already described in chapter 2. The acceptable record for the global R/P ratio begins only after the Second World War when the oil industry started to publish annual worldwide reserve surveys. After a rapid doubling from 20 to 40 years in less than a decade, followed by a fall below 30 years between the years 1974 and 1980, the ratio has rebounded and it is now just above 40 years, higher than it stood at any time since 1945 (fig. 4.3). Clearly, neither the price nor the R/P history reveal any discouraging trends and cornucopians argue that the combination of abundant resources and continuing innovation will keep it that way.

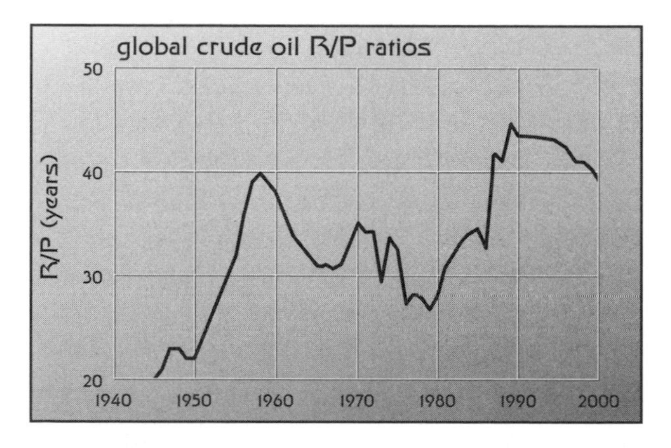

Figure 4.3
Global R/P ratio for crude oil, 1945–2000. Based on data published annually by the *Oil & Gas Journal.*

Their enthusiasm about the potential for new discoveries, often shared by some of the industry's managers, is frequently exaggerated. Expectations for offshore waters along the U.S. East Coast (Baltimore Canyon) whose enormous promise was touted during the 1970s, and in the South China Sea, whose vastly overstated potential lured in virtually every major oil company during the 1980s (Smil 1988), are perhaps the two most spectacular examples of this excess. And the Caspian Sea, whose total oil reserves have been put speculatively as high as two-thirds of the Saudi total, may turn out to be the latest disappointment (EIA 2001g).

But these failed expectations make little difference to the cornucopian argument. Even if the Baltimore Canyon or the Pearl River Delta or the Caspian Sea have not turned out to be as oil-rich as once foreseen, the cumulative worldwide oil discoveries still keep adding new reserves well in excess of consumption. There may be deficits for particular years, but the long-term trend is unmistakable: using the *Oil & Gas Journal* reserve and production estimates, the world extracted roughly 680 Gb of oil during the three decades between 1970 and 2000—but about 980 Gb of new reserves were added, leaving the global R/P ratio above 40 years, about five years above the 1970 rate and near its all-time high.

Adelman (1992) pointed out that in 1984, when the Middle East's proved reserves stood at 398 Gb, the U.S. Geological Survey estimated that there is no more than a 5% probability of discovering an additional 199 Gb (Masters et al. 1987)—yet by the end of 1989 the gross increase in proved reserves was 289 Gb. As I will note in the next section, a different interpretation of this claim is used to prove the *very opposite* point by advocates of early oil exhaustion!

Studies by Goeller and Marchetti have been widely seen to provide particularly strong endorsements of a relaxed view of resource availability and substitutions. Goeller based his arguments about infinite resources on detailed examinations of past and future substitutions of both major and rare metals (Goeller and Weinberg 1976; Goeller and Zucker 1984). But as the effective fuel substitutions of the last 150 years demonstrate, these conclusions are no less valid for the sequence of primary energies. Marchetti's argument about clockwork-like substitutions of primary energy sources has been already reviewed (in chapter 3) and found less than perfect. Economic and technical imperatives will, indeed, work to displace oil by natural gas, renewables, or even nuclear energy—but the timing of these transitions is far from as orderly as Marchetti would like us to believe.

Any review of the recent flood of writings on the imminent peak of global oil production should start with the writings of the most prominent contributors to this fashionable genre, with the forecasting work done by Colin Campbell and Jean Laherrère (Campbell 1991, 1996, 1997, Campbell and Laherrère 1998; Laherrère 1995, 1996, 1997, 2001). Their comprehensive arguments are built on questioning the reliability of published reserve figures, on correcting the recent rate of discoveries by backdating the reserve revisions, and fitting a fixed total of reserves and likely future discoveries into a symmetrical exhaustion curve. This curve was introduced as a standard forecasting tool by M. King Hubbert more than four decades ago (Hubbert 1956) and it was popularized by Hubbert's predictions of the permanent decline of U.S. oil extraction published during the 1960s (fig. 4.4; Hubbert 1969). Other researchers that have used Hubbert's approach to forecast an early peak of global oil production—and to conclude that a world oil shortage, and even the end of industrial civilization, are inevitable—include Ivanhoe (1995, 1997), Duncan (2000), Duncan and Youngquist (1999), and Deffeyes (2001). Most of their recent, and invariably pessimistic writings, are accessible through a Web site devoted to the supposedly imminent global oil crisis (The Coming Global Oil Crisis 2001).

Main arguments of all of these proponents of an early end-of-oil era rest on the following key points. Exploratory drilling has already discovered about 90% of oil that was present in the Earth's crust before the fuel's commercial extraction began during the latter half of the nineteenth century. This means that both the recent revisions of some national oil reserve totals or claims of new spectacular discoveries

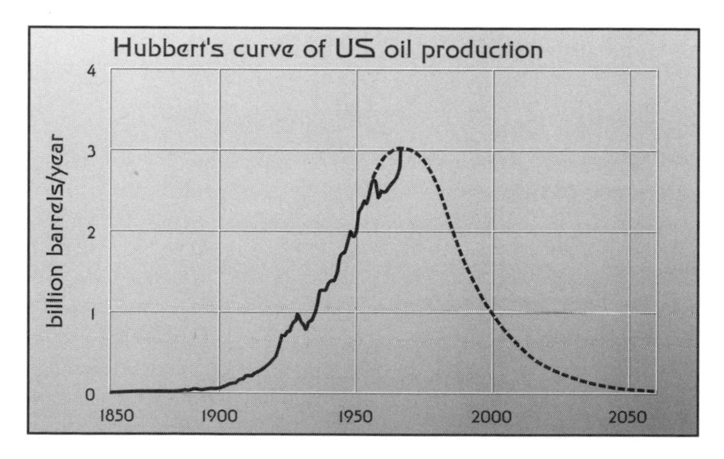

Figure 4.4
Hubbert's curve of the U.S. oil production. Based on Hubbert (1969).

are at best exaggerated, at worst outright fraudulent. Since the late 1970s oil producers have been extracting (every year) more oil than they have been discovering, and some four-fifths of global production now come from fields found before 1973 and most of these have steeply declining rates of extraction. Moreover, neither new exploratory and extraction techniques, nor the recovery of unconventional oil can prevent this decline. Deffeyes (2001, p. 149) goes even further, concluding that

> No initiative put in place today can have a substantial effect on the peak production year. No Caspian Sea exploration, no drilling in the South China Sea, no SUV replacements, no renewable energy projects can be brought on at a sufficient rate to avoid a bidding war for the remaining oil.

As with any analyses of a complex phenomenon, authors of these forecasts make a number of correct observations and conclusions. To begin with, there is the unfortunate absence of rigorous international standards in reporting oil reserves: while the United States includes only proved reserves, other countries report, with widely differing rigor, mixtures of proved and probable totals. Moreover, releases of many of these official totals are politically motivated, with national figures that either do not change at all from year to year or take sudden suspicious jumps.

Consequently, not only are the totals reported by the two most often cited annual surveys—compiled by the *Oil & Gas Journal* (and given an even wider publicity through their inclusion in *BP Statistical Review of World Energy*) and *World Oil*, and often listed with misleading accuracy to the third decimal value—not identical but their impressively ascending secular trend (fig. 4.5) conveys misleading information about the future trend of oil reserves. This becomes clear when these totals are compared with present estimates of what Laherrère calls "technical" reserves, means of proved and probable reserves listed in confidential data bases of oil companies backdated to the actual year of their discovery: this measure shows an unmistakable downward trend since 1980 (fig. 4.5).

After sifting through what is the world's most extensive oil field data base, Campbell and Laherrère concluded that in 1996 the remaining reserves of conventional oil are only 850 Gb—almost 17% less than the *Oil & Gas Journal* total of 1,019 Gb, and nearly 27% below the *World Oil* figure. The single largest reason for this discrepancy is well known to anybody interested in global oil prospects: the biggest ever annual jump in world oil reserves occurred not as a result of discovering new fields but almost completely as an accounting upgrade performed during the single year, in 1987, by a group of OPEC members (fig. 4.5; *Oil and Gas Journal* Special Report 1987). While upward revaluations of existing reserves are to be expected,

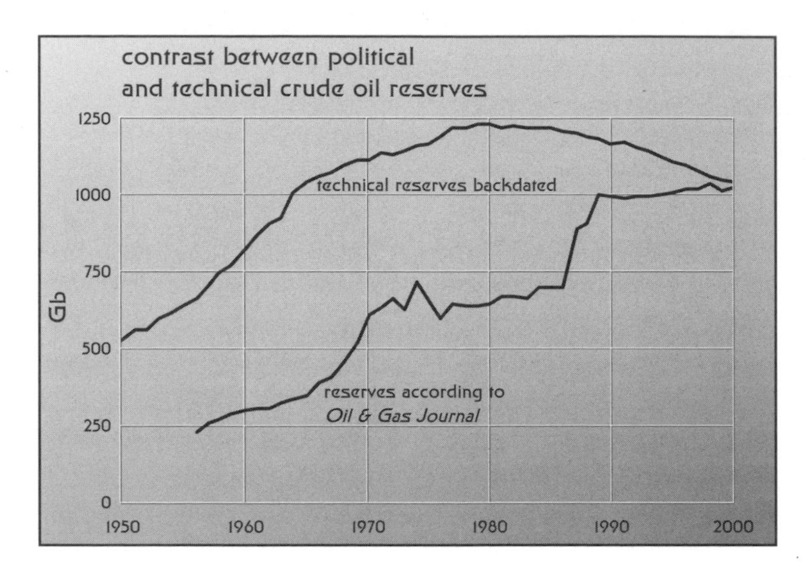

Figure 4.5
Laherrère's (2001) contrast between what he calls political and technical oil reserves for the period 1950–2000.

Campbell and Laherrère concluded that both the magnitude of this particular increase (27% jump compared to 1986), and its timing in the wake of weakened world oil prices, make it a prime example of politically generated reserves, and that the questions about their validity could be answered only if the relevant state oil companies would open their books for detailed inspection.

No less than 165 Gb of the 190 Gb increase was due to revised estimates in six OPEC countries, five of them in the Persian Gulf region. Most notably, the two countries that were at war at that time reported huge reserve additions: Iraq's estimate was up by 112%, Iran's by 90%. During the late 1980s six of OPEC's 11 nations revised their reserves by a total of 287 Gb, the volume equal to about one-third of all oil produced worldwide during the past 140 years. Another source of exaggerated reporting is due to the fact that an increasing number of countries have been claiming unchanged reserve totals for a number of consecutive years.

At the beginning of the twenty-first century about three-quarters of the world's known recoverable oil were in 370 giant fields (each containing at least 500 Mb of oil) whose discoveries peaked in the early 1960s, and whose reserves became rather reliably known within a short time after their discovery. This is why Campbell and Laherrère see no possibility of large-scale additions to the known oil pool. They conclude that some 92% of all oil we will eventually recover has been already discovered. As a result, "creaming" curves that plot cumulative discoveries against a cumulative number of new wildcat wells show typical decreasing slopes of diminishing returns, be it for individual giant fields or for entire countries, regions, or continents (fig. 4.6). They also question the commonly held opinion that additions to reserves in existing oil fields can appreciably delay the onset of declining output. They point out that the experience with U.S. onshore wells, whose initially reported reserves are nearly always revised substantially upwards over the lifetime of a field, result from the strict reserve reporting rules imposed by the Securities and Exchange Commission, which consider only proved developed reserves.

As a result, annual U.S. reserve growth in old fields exceeds new discoveries—but this unique experience is unlikely to be matched around the world where the totality of reserves is appraised as best as possible at the time of discovery and where subsequent revisions may be upward as well as downward. Consequently, it may be unrealistic to expect that large revisions common for U.S. onshore fields will be applicable worldwide—but we do not have any detailed, comprehensive evaluation of world reserve revisions in fields that would allow us to estimate more reliably the most

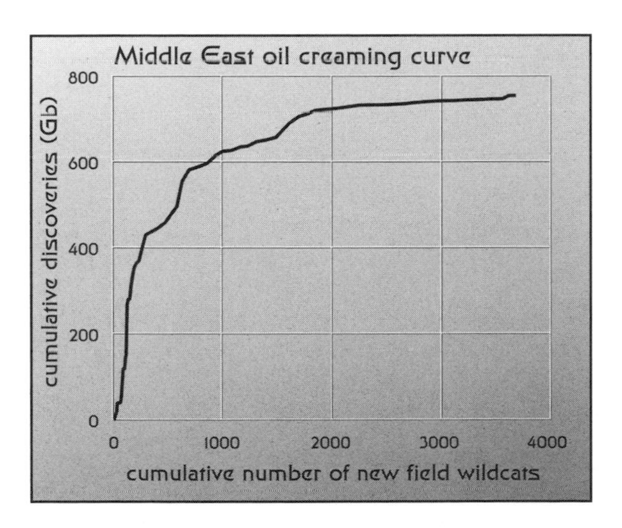

Figure 4.6
Laherrère's (2001) creaming curve for the Middle Eastern crude oil and condensate.

likely revision factor. Moreover, in order to avoid overstating current discovery rates, any reserve additions should be backdated to the time of the field discovery. Without this backdating it appears that reported reserves are still rising—but, as already noted, with it the reserves have been declining since 1980, together with new field discoveries.

Campbell and Laherrère also largely discount future contributions to be made by enhanced recovery in existing fields. They conclude that the recovery factors are not primarily functions of extraction techniques but of reservoir makeup and that apparent recovery improvements may simply reflect initial underestimate or understatement of the amount of oil-in-place. And while they acknowledge the magnitude of nonconventional oil resources they think the industry would be hard-pressed, both in terms of time and capital, to make up for the declining conventional extraction. That is why they do not see oil shales, tar sands, and heavy oil as realistic substitutes for conventional extraction. As for rising oil prices, they will have an effect on the decision to drill in marginal areas but they will not increase the total volume of oil to be recovered: unlike low-quality ores whose extraction becomes economical with higher prices, nine-tenths of all oil have been already discovered and working in marginal areas will only decrease the success ratio and increase the net energy cost.

With 850 Gb in existing reserves, and with no more than 150 Gb of oil to be discovered, we would have no more than 1,000 Gb to produce in the future, only about 20% more than we have burned already. Because the curve of the long-term course of extracting a finite resource should have a fairly symmetrical bell shape, global oil extraction would begin to fall once the cumulated production passes the midpoint of ultimately recoverable resources, that is about 900 Gb according to Campbell and Laherrère. They see the inevitable peak sometime during the twenty-first century's first decade (fig. 4.7) and other forecasts concur. Ivanhoe (1995, 1997) predicted the peak, even with restricted production, as early as the year 2000. Deffeyes (2001) believes 2003 is the most likely year but settles on a range between 2004–2008. Duncan and Youngquist (1999) prefer the year 2007. Hatfield (1997), using the ultimate reserves of 1,550 Gb, gives us a few more years, predicting that the maximum production will come between the years 2010 and 2015. Afterwards comes the permanent decline of conventional oil output, world oil shortages (Deffeyes' belief) or, in Duncan's (2000) darker vision, unemployment, breadlines, homelessness, and the end of industrial civilization.

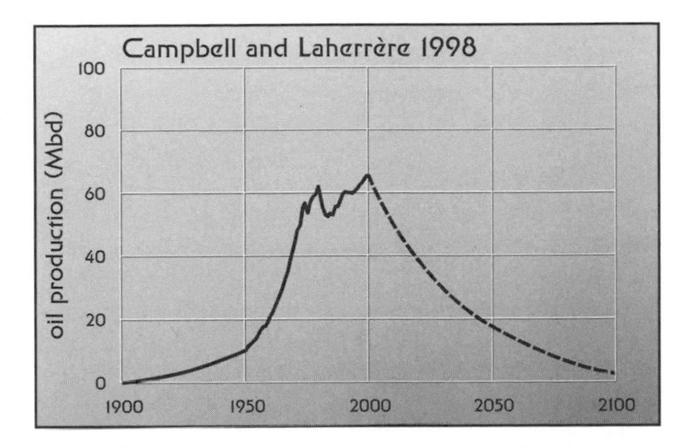

Figure 4.7
Campbell and Laherrère's (1998) forecast of the global oil extraction peaking in the year 2007.

Different Perspectives on the Oil Era

A new conventional wisdom of an imminent global oil production peak has been taken up enthusiastically by media always eager to report new bad news, and we have been repeatedly told that "the oil era is over." The incongruity of such pronouncements is evidently too obvious for their purveyors to notice: even if one would subscribe to the inevitability of the oil extraction peak coming before the year 2010, the Hubbertian reality is that half of all oil is yet to be extracted after that date, and if the production curve is to be anywhere near its required normal (Gaussian) shape, then we still have more than a century of, albeit declining, oil production ahead of us (see fig. 4.7).

If there were no doubt about these forecasts it would be almost certainly too late to begin determined and wide-ranging preparations designed to meet this unprecedented challenge. If the proponents of early exhaustion are correct, this would be the first time in history when the world would be facing a permanently declining supply of the dominant fuel: transitions from wood to coal and from coal to oil were not brought by global exhaustion of the substituted resource, but by lower prices and better qualities of new fuels. But explicit forecasts of peak years of oil extraction are not new and the challenge is to decide if we are seeing just the latest version of an old story or if, this time, the situation is really fundamentally different.

I could have placed a large part of this section in the preceding chapter as yet another illustration of failed energy forecasts: categorical declarations of an early end of the oil era—ushered by an imminent and fairly precipitous decline of global oil extraction—are just the latest additions to a long list of failed predictions concerning the future of oil (Smil 1998b). Their authors have continued to overlook the fundamental fact that the timing of oil's global demise depends not only on the unknown quantity of ultimately recoverable crude oil resources (which has been, so far, repeatedly underestimated) but also on the future demand whose growth they have usually exaggerated and that is determined by a complex interplay of energy substitutions, technical advances, government policies, and environmental considerations.

Consequently, all past efforts to pinpoint the peak years of global oil output and its subsequent decline have been unsuccessful, and I will argue that the latest wave of these timings will not fare any better. Pratt (1944) and Moody (1978) give some instructive examples of mistaken predictions of oil exhaustion in the United States predating Hubbert's work. Hubbert's (1969) original timing for the global peak of conventional oil extraction was between the years 1993 and 2000 (fig. 4.8). I will cite some of the most notable examples of the already failed forecasts (some of them turned out to be wrong in just a few years!). The Workshop on Alternative Energy Strategies, whose exaggerated forecasts of global energy needs were already noted in chapter 3, concluded that the global oil supply would fail to meet the growing

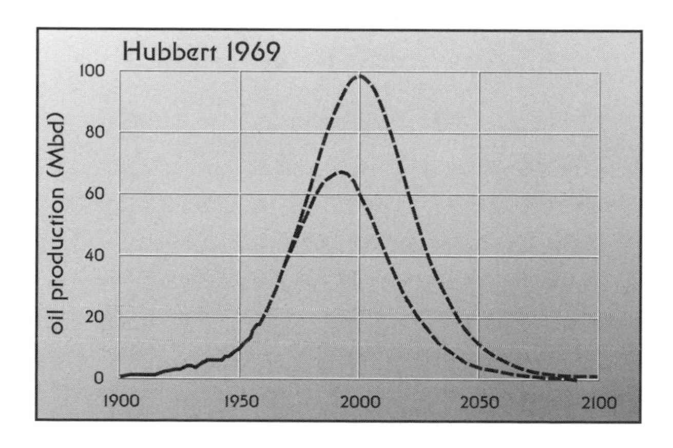

Figure 4.8
Hubbert's (1969) forecast of the global oil extraction for two different production totals.

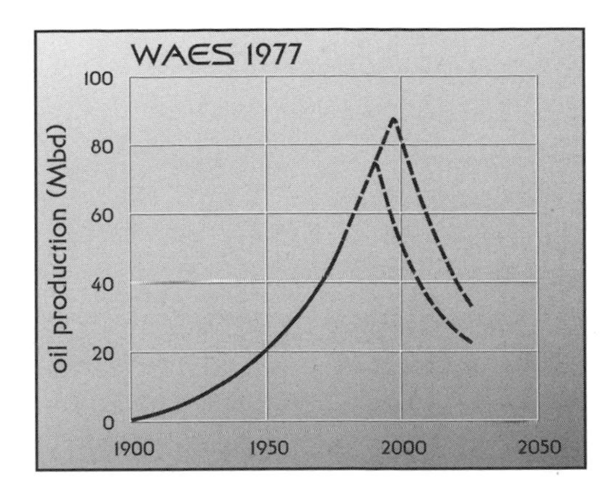

Figure 4.9
Global crude oil exhaustion curves generated by the Workshop on Alternative Energy Strategies (WAES 1977).

demand before the year 2000 (Flower 1978; Workshop on Alternative Energy Strategies 1977). Oil exhaustion curves generated by the project showed the global output peaking as early as 1990—and no later than in the year 2004, with the most likely timing of the peak production put between 1994 and 1997 (fig. 4.9).

A year later the U.S. Central Intelligence Agency offered an even more panicky projection issued at the height of the Iranian revolution. The Agency concluded that "the world energy problem reflects the limited nature of world oil resources," and that, with consumption greatly exceeding supplies, the global "output must fall within a decade ahead" (National Foreign Assessment Center 1979, p. iii). As a result "the world can no longer count on increases in oil production to meet its energy needs" and it "does not have years in which to make a smooth transition to alternative energy sources." The last sentence is truly astonishing as it implied the necessity of doing something utterly impossible: converting the world's primary energy supply to a different source in a matter of months! A generation later I still marvel at the level of ignorance behind such a statement.

In the same year BP's study of crude oil futures outside of the (then) Soviet bloc was no less panicky, predicting the world production peak in 1985 and the total output in the year 2000 nearly 25% below that maximum (BP 1979; fig. 4.10). In reality, the global oil output in the year 2000 was nearly 25% above the 1985 level! More recently, Masters, Root, and Attanasi (1990) predicted the peak of non-OPEC oil production at just short of 40 Mbd before the year 1995—while the actual non-

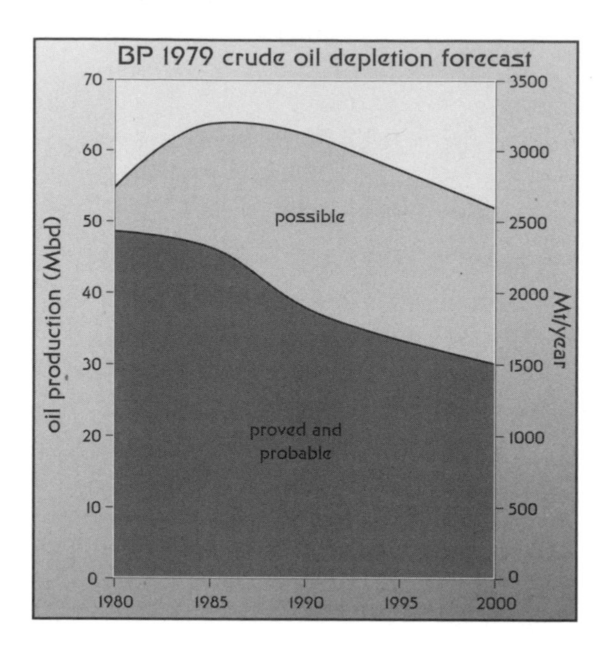

Figure 4.10
In 1979 British Petroleum forecast the peak of global crude oil production to take place in just six years. Based on BP (1979).

OPEC output in the year 2000 was above 42 Mbd. Similarly, in 1990 the UNO concluded that non-OPEC production has already peaked and will be declining (UNO 1990), but 10 years later non-OPEC extraction was nearly 5% above the 1990 total.

Failed regional or field forecasts are equally easy to find. Production from the North Sea is an excellent example: it was expected to enter permanent decline in the early 1980s—yet in the year 2000 its extraction reached a new record of more than 6 Mbd, nearly three times the 1980 level (EIA 2001h)! And, in spite of more than 20 years of repeated forecasts of sharp production declines, Daqing, China's largest oilfield in Heilongjiang (discovered in 1961) still produces about 55 Mt/year, or as much as in the late 1980s and more than two decades ago (Fridley 2001).

Not surprisingly, the long history of overly confident but wrong past pronouncements make any new forecasts of imminent production downturn ready targets of cornucopian scorn. Advocates of imminent oil production decline acknowledge this skepticism but argue that this time the situation is *totally* different from previous experiences, including the forecasts of the 1970s. At that time the world could rely on additional supply coming from the new oil provinces of Alaska, the North Sea, and other non-OPEC oilfields that had already been found. "This time none are even in sight and very possibly do not exist" (Campbell 1996, p. 12). Or, in Laherrère's (1997, p. 17) conclusion, "Today we consume three times more than we discover. No technological breakthrough is foreseen!" This, all advocates of the imminent oil peak output agree, does not leave much room for radically different interpretations regarding the future of oil.

But how can these advocates be so sure that this time they have been really able to avoid the two key recurrent mistakes that explain failed forecasts of their predecessors: underestimating the amount of the ultimately recoverable oil, and overestimating the role of oil in future energy supply? They claim that their resource accounts are superior to any other published estimates. Laherrère (2001) judges the latest U.S. Geological Survey's global assessment of undiscovered oil (USGS 2000) in a particularly harsh way. The previous USGS study (Masters, Attanasi, and Root 1994) used 2,900 Gb as the high estimate of ultimate oil resources, which would mean that we could still extract about 2.5 times as much oil as we have produced since the beginning of the oil era (just over 800 Gb by the end of 1997). But the latest study concluded that the *mean* grand total of undiscovered conventional oil, reserve growth in discovered fields, and cumulative production up to the year is 3,012 Gb, 20% higher than the previous assessment, and 72% above Campbell and Laherrère's total of 1,750 Gb.

Laherrère sees it as a product of "lonely academic U.S.G.S. geologists" working without any access to well and seismic data kept in confidence by oil companies, with "main assumptions filled up by one geologist on one sheet of paper giving number and size of undiscovered fields." His final verdict: the study is "just pure guesses and wishful thinking" and "no scientific credence can be given to work of this sort." Actually, Laherrère is dismissive of all "academic" and "theoretical" writers; only retired "geologists free to speak when retired, with large experience and access to confidential technical data" (i.e., people such as himself, Campbell, Perrodon, Ivanhoe, and Youngquist) know the truth (Laherrère 2001, p. 62). But the latest U.S.G.S. assessment is just one of several recent estimates of ultimately recoverable oil that are substantially higher than Campbell and Laherrère's total.

In 1984 Odell thought the total of 2,000 Gb to be ultrapessimistic and impossibly low given the large areas yet to be explored for oil (Odell 1984). In 1992, while reviewing the post-1950 additions to reserves, he concluded that "the world is running into oil, not out of it" (Odell 1992, p. 285). In *The Future of Oil* (Odell and Rosing 1983) he assigned a 50% probability to a scenario consisting of a 1.9%

growth consumption growth rate and 5.8 Gb of conventional and nonconventional oil that would see the peak global oil extraction coming only in the year 2046 at roughly 3.5 times the 1980 flow. With 3,000 Gb of ultimately recoverable conventional oil, Odell predicted the peak recoveries between 2015–2020, and the global flow in the year 2045 above 20 Gb, still as large as in the mid-1980s. And Laherrère (1996) himself conceded that adding the median reserve estimates of natural gas liquids (200 Gb) and nonconventional oil (700 Gb) would result in up to 1,900 Gb of oil that is yet to be produced, double the amount of his conservative estimate for conventional crude oil alone.

As already noted, advocates of the early end-of-the-oil era also make much of the fact that the annual means of scores of estimates of ultimately recoverable oil that have been published since the 1940s were rising until the late 1960s but that they have been flat or falling since the early 1970s, with most of the values clustering around 2,000 Gb, or 300 Gtoe. But that is a biased interpretation. A complete set of such estimates shows that all but four totals published since 1980 are above 300 Gtoe and that the midpoint of all recent assessments is about 3,000 Gb, or just over 400 Gtoe (fig. 4.11). A complete production curve based on this ultimately

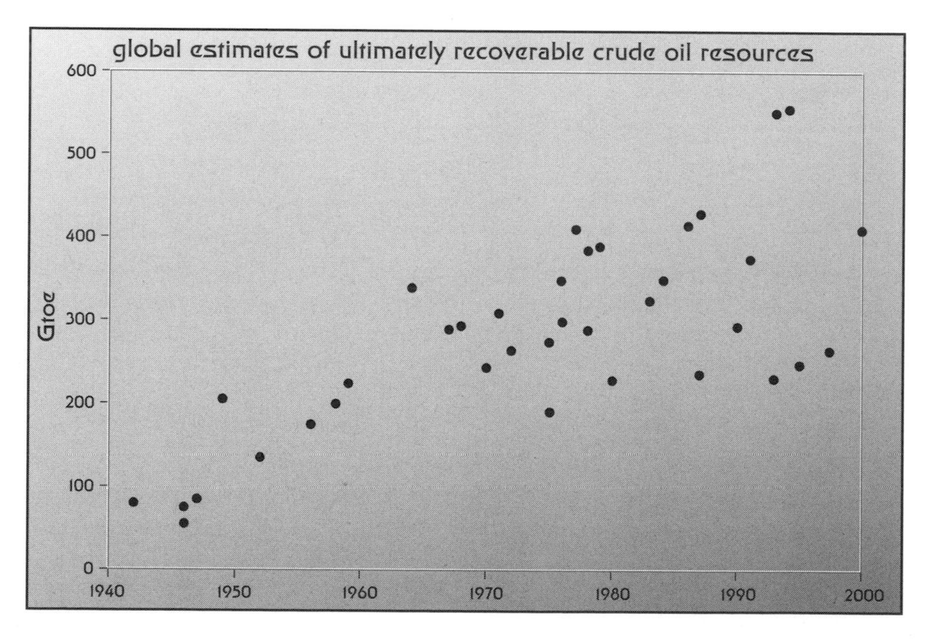

Figure 4.11
Global estimates of ultimately recoverable crude oil, 1940–2000. Based on Smil (1987) and Laherrère (2000).

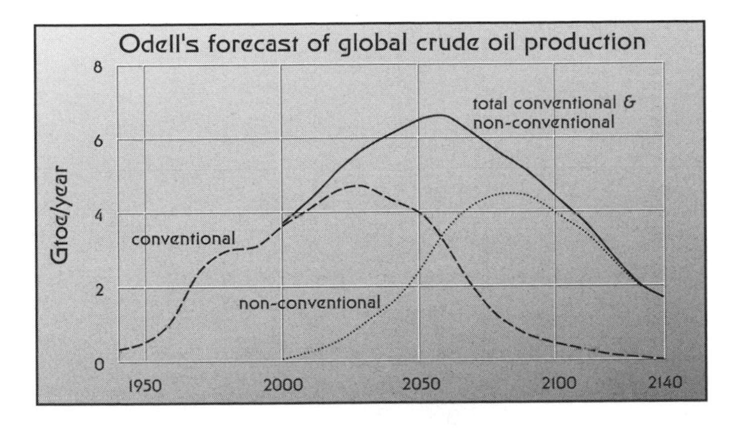

Figure 4.12
Odell's (1999) forecast of global recovery curves for conventional and nonconventional crude oil. Extraction of the latter resource could delay the peak recovery past the year 2060.

recoverable total shows a peak of conventional oil extraction at 4.33 Gtoe in the year 2030 and continued production well into the twenty-second century. Gradual increase of nonconventional oil extraction could raise this peak to 6.5 Gtoe and delay it to the year 2060 (fig. 4.12).

And Campbell's argument about backdating currently known reserves to the year of their discovery is refuted by Odell (1999) who points out that this procedure makes the past look more attractive than it actually appeared to those who had to take the relevant economic decision while making the recently discovered reserves less attractive as these can be clearly expected to appreciate substantially during the coming decades. As a result, Odell offers a more appropriate graphic representation of the historic appreciation of proven reserves of conventional oil since 1945 (fig. 4.13).

Before leaving the evaluations of the world's oil resources I must mention an intriguing Russian–Ukrainian theory of the abyssal abiotic origin of hydrocarbons. Because of its provenance this view has received little serious attention in the West but it has been based on extensive field experience as well as on theoretical considerations that have been subjected to decades of criticism. Beginnings of the theory go to the early 1950s and since that time it has been discussed in more than 4,000 papers, book chapters, and books (Kudryavtsev 1959; Simakov 1986; Kenney 1996). More importantly, the theory has guided extensive exploratory drilling that resulted in oil production from crystalline basement rock scores of hydrocarbon fields in the Caspian Sea region, western Siberia, and the Dnieper–Donets Basin.

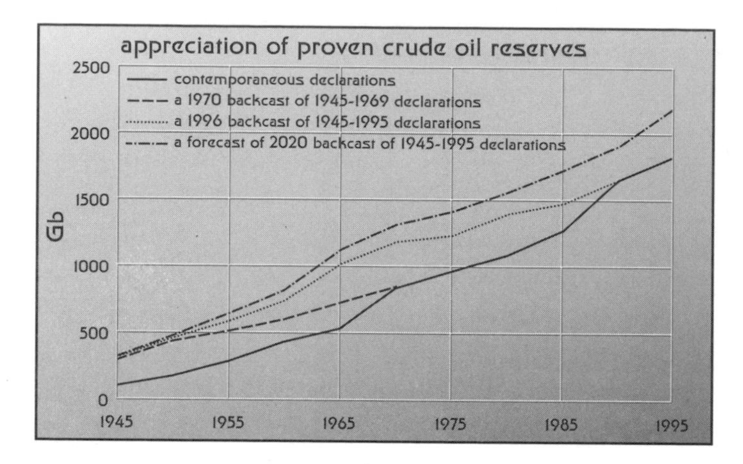

Figure 4.13
Appreciation of proven reserves of conventional oil: Campbell versus Odell. Based on Odell (1999).

Instead of considering highly oxidized low-energy organic molecules as the precursors of all highly reduced high-energy hydrocarbons, as does the standard account of the origins of oil and gas, Russian and Ukrainian scientists believe that such an evolution would violate the second law of thermodyanmics and that the formation of such highly reduced molecules calls for high pressures encountered only in the Earth's mantle. They also concluded that giant oilfields are more logically explained by inorganic theory rather than by postulating extraordinary accumulations of organic material because calculations of potential hydrocarbon contents in sediments do not show organic materials sufficient to supply the volumes of petroleum found in such enormous structures (Porfir'yev 1959, 1974).

The theory would have us looking for hydrocarbons in such nonsedimentary regions as crystalline basements, volcanic and impact structures and rifts, as well as in strata deeper below the already producing reservoirs. Moreover, if the origin of hydrocarbons is abyssal and abiotic then there is a possibility of existing producing reservoirs being gradually repleted, a phenomenon that would radically shift many oil and gas deposits from the category of fossil, nonrenewable energy resources to renewables (Mahfoud and Beck 1995; Gurney 1997). The theory has been dismissed by all but a few Western geologists. Isotopic composition of hydrocarbons—their content of heavier ^{13}C matches that of terrestrial and marine plants and it is lower than the isotope's presence in abiotically generated CH_4 (Hunt 1979)—is cited as the best argument in support of the standard view of their biogenic origin.

This is not a place to adjudicate controversies of this kind but I would just ask all of those who would simply dismiss the theory of abiotic hydrocarbon origins as entirely impossible at least to consider the possibility of such a development for at least a part of the world's hydrocarbon resources. Two reasons favor this consideration. The first one is the undoubted success of oil production from nonsedimentary formations (explainable, the standard view says, by migration from adjacent strata). The second is the fact that most geologists resisted for several generations the idea that is now a key paradigm of their science, the theory of plate tectonics championed by Alfred Wegener (Oreskes 1999). I will return to the abiotic origins of hydrocarbons in the very next section when I will examine the prospects for global production of natural gas during the twenty-first century.

Another way to look at low estimates of ultimately recoverable oil is to approach them from an economic point of view. Adelman's comments are surely best suited to this task. He believes that the estimates of discovered and undiscovered recoverable reserves do not tell us what is in the ground. "They are implicit unarticulated cost/price forecasts: how much it will be profitable to produce" (Adelman 1990, p. 2)—but they do not measure "any stock existing in nature." Quoting a former Exxon chief geologist, Adelman sees these figures as mere *ordinals,* indicating which areas are promising but implying nothing about current and future scarcity. He points out that the past experience makes it clear that increasing knowledge has, so far, more than kept its pace with decreasing returns, resulting in declining or flat prices for most mineral resources. Adelman (1997, p. 40) concludes that "the burden of proof, if it rests on anyone, rests on those who think the process will reverse in the foreseeable future or has already reversed." Campbell and Laherrère clearly think they have answered this challenge.

But one does not have to share Adelman's dismissal of ultimate resource estimates in order to note that all of the published forecasts of an early onset of a permanent decline end of oil extraction fail to consider prices—or to believe that even with today's low prices the highly conservative valuations of ultimately recoverable oil are almost certainly too pessimistic. Granting the argument about an extremely low probability of discovering additional giant fields is much easier than believing that higher prices will not lead to more intensive exploration that will locate a large number of smaller finds with substantial total reserves.

These assumptions rest on what is still a relatively low density of exploratory drilling outside the North American continent, and particularly in Africa and Asia—the fact that was first convincingly pointed out a generation ago by Bernardo Grossling (1976)—and on oil discoveries around the world that have been made since 1985 in spite of the world oil prices being lower in constant money terms than a generation ago. Not surprisingly, assumptions of relatively high ultimate recoveries are behind some of the recent longer-term forecasts of oil production. World Energy Council's high growth scenarios projected annual oil consumption of almost 34 Gb in the year 2020, and the Council's latest forecast for the year 2050 suggests that the world may be now only one-third of the way through the oil age and that the global oil consumption according to its middle-course scenario may be close to 30 Gb even in the year 2050 (IIASA and WEC 1995). And the U.S. EIA (2001d) does not foresee the peak global production rate before 2020.

I also believe that recent technical innovations will make a substantial difference to the total amount of oil that will be eventually recovered from existing fields and from those yet to be discovered. Campbell and Laherrère's virtual dismissal of these contributions underestimates the ingenuity with which we can create new reserves from the unknown volume of oil remaining in the crust. Cumulative effects of technical innovation on oil recovery have been already impressive and there are obvious prospects for further gains (Anderson 1998). Perhaps the two best recent examples are advances in hydrocarbon field imaging and in directional drilling whose recent remarkable accomplishments I already noted in the first chapter. Combination of advanced reservoir imaging and directional drilling raises potential oil recovery from the traditional rates of 30–35% to at least 65%, and in the near future to as much as 75% of oil-in-place: it is most unlikely that this large increment would have been anticipated at the time of original valuation of all non-U.S. reserves and hence we will see substantial appreciation of non-American reserve totals as well.

And any long-term outlook must consider the enormous resources of nonconventional oil, amounting globally to more than 400 Gtoe, with about two-thirds of that total in oil shales, roughly one-fifth in heavy oils and the rest in tar sands (Rogner 2000). Much of the past enthusiasm regarding the recovery of these resources, particularly the oil extraction from shales, has been clearly misplaced but recent technical innovations also mean that the exploitation of these nonconventional resources is not a matter of the distant future: commercial operations are rewarding even with today's low oil prices and there are encouraging prospects for further advances (George 1998).

As already noted in chapter 3, Canada's Suncor Energy has been producing a modest, but rising, volume of crude oil from northern Alberta's huge Athabasca oil sand deposits since 1967. Annual extraction rate during the late 1990s was equivalent to about 7 Mt of crude oil and an expansion currently underway will nearly quadruple the output by 2010–2012 (Suncor Energy 2001). A lesser known example of nonconventional oil is the production of Orimulsion by Petróleos de Venezuela (PDVSA 2001). This liquid fuel is 70% natural bitumen from the enormous deposits in the Orinoco Belt, 30% water, and a small amount of an additive that stabilizes the emulsion. Orimulsion makes an environmentally acceptable and cost-competitive fuel for electricity generation and industrial boilers and almost 4 Mt/year are now being exported in double-hulled tankers to nearly a dozen countries, including the United States, Japan, Canada, the United Kingdom, and Germany.

In view of these developments the advocates of an early-end-of-the-oil era would be more credible if they would go along with Houthakker (1997), who concluded that a more realistic interpretation of recent concerns is to believe that sometime between the years 2000 and 2020 prices of conventional oil may rise to a level where increasing amounts of oil can be profitably extracted from nonconventional sources. In any case, any predictions of long-term rates of oil extraction can be meaningful only if they take into account plausible changes in prices—but this key factor is almost uniformly, and inexplicably, missing from all accounts of any coming oil depletion. Rising oil prices would obviously accelerate transitions to other energy sources, a lesson painfully learned by oil producers during and after the OPEC's'price overshoot during the years 1979–1983. But even steady or falling oil prices may not prevent substantial declines of demand for oil.

As the coal precedent shows, neither the abundant resources nor the competitive prices were decisive determinants of the fuel's future. Coal extraction is now completely determined by limited demand and it has virtually nothing to do with the vast amount of remaining fuel in the ground and production in the remaining major coal-producing countries is either stagnating or declining in spite of fairly low prices. Analogously, reduced demand could significantly prolong the life span of oil recovery, whatever the ultimate resource may be. Post-1974 trends in global oil extraction have been a clear shift in that direction. There has been an abrupt shift from the average growth rate of 7% a year between 1950 and 1974 to less than 1.4% between 1974 and 2000. The 1979 production peak was not equalled again until 1994, and the aggregate crude oil consumption during the last 30 years of the twentieth century was about 93 Gtoe instead of about 250 Gtoe forecast in the early 1970s by Warman (1972).

Even if the world's ultimately recoverable oil resources (both liquid and nontraditional) were known with a very high degree of certainty, it will be the poorly known future demand that will also shape the global production curve—and it is not difficult to present realistic scenarios based on moderate, and eventually declining, rates of this demand in decades to come. These would bring declining oil extraction not because of any imminent exhaustion of the fuel but because the fuel would not be simply needed in previously anticipated quantities. Expectations of strong growth in global oil demand are based on four well-established trends. The first one is the continuing growth of Asian, African, and Latin American populations whose well-being will require, even when coupled with only modest improvements in the average standard of living, relatively rapid rates of economic expansion.

The second factor, a corollary of the first trend, is a massive transition from agricultural subsistence to incipient affluence in increasingly urbanized societies; this shift, perhaps best illustrated by recent developments in China, is always connected with substantial increases of per capita energy consumption. The third factor is a powerful lure of the private car; this subset of the second category must be singled out both because of its truly global nature and because of its special impact on crude oil demand. Finally, I must mention profound infrastructural changes in the United States, the world's largest oil market, where ex-urban (no more merely suburban) commuters are driving greater numbers of improbably sized, gasoline-guzzling vehicles and setting a most undesirable example to affluent populations around the world.

But powerful countervailing trends must be also taken into account. The latest population projections indicate an unexpectedly rapid slowdown in global population growth. Its relative rate has been decreasing since the late 1960s, and after going down to less than 1.5% during the 1990s it means that the growth is now declining even in absolute terms. As a result, in 1990 we were anticipating as many as 3.1 billion additional people in poor countries by the year 2025 (UNO 1991), but in 2001 the most likely forecast was for an increase of 2.5 billion (UNO 2002), a 20% difference. As shown in chapter 2, the declining energy intensity has already made an enormous difference to the world's demand for primary energy, and the oil intensity of economies has declined even faster. In spite of low post-1985 oil prices, oil intensity has decreased even in the relatively wasteful United States. After holding steady between 1950 and 1975, it fell by 50% between 1975 and 2000 (EIA 2001a). The decline in the dependence of China's economy on oil was even more precipitous. When using official GDP figures it amounted to a drop of more than 70% between 1980 and 2000 and even with adjusted GDP totals it would be just over 60% (Fridley 2001).

The gross world economic product is not a precise figure but using its best available aggregates in global oil intensity calculations points to a 30–35% reduction of that rate between 1975 and 2000. As ample opportunities remain for further efficiency gains in all economic sectors there is no reason why oil intensities should not continue falling. A significant share of this reduction will come from more efficient transportation and, eventually, from a large-scale adoption of cars not powered by internal combustion engines. Today's transportation sector is generally more than 90% dependent on liquid fuels but introduction of much more efficient cars can reduce its absolute demand quite impressively. The wastefulness of America's all-terrain trucks and large vans was made possible by low oil prices—but large cuts in consumption do not have to wait for futuristic hypercars promoted by the Rocky Mountain Institute (RMI; 2001).

Average fuel consumption of U.S. private vehicles could be cut by one-third in just a matter of years by limiting the sales of gasoline-guzzling SUVs (classified as light trucks and hence exempt from minimum CAFE requirements for passenger cars) for urban use and by raising the new CAFE rate (unchanged since 1987) from 27.5 mpg to 35 mpg; that is, to the level of fuel-efficient Japanese sedans. And there are no technical obstacles to cutting U.S. gasoline consumption by half within a decade and then to go even further. Combination of these trends means that we can realistically envisage the continuing transformation of oil from a general fuel to a few niche uses, a shift similar to the one already accomplished in the coal market (see chapter 1).

During the 1950s refined oil products were also widely used for household and institutional heating, in industrial production, and for electricity generation. By the century's end natural gas replaced coal in most of the heating applications, and oil generated only about half as much electricity in OECD countries as it did in the early 1970s, and during the same period of time its global share of electricity generation fell from just over 20% to less than 10% (IEA 2001). In many affluent countries refined liquids are already used almost exclusively in transportation, a sector where major efficiency gains lie ahead.

Finally, future oil consumption may be restricted because of the necessity to reduce the emissions of CO_2. (I will address this major environmental uncertainty in some detail in the last chapter of this book.) This concern would obviously favor a faster substitution of oil by natural gas. An important advance whose eventual perfection could facilitate this transition is the recent discovery of platinum catalysts for the high-yield (greater than 70%) oxidation of methane to a methanol derivative

(Periana et al. 1998). This discovery points the way toward an eventual large-scale conversion of natural gas to methanol near remote wellheads. The conversion would solve two key problems by producing a liquid much easier to transport than methane and highly suitable for internal combustion engines used in transportation. As it stands, the end product of the catalytic reaction is methyl bisulfate rather than methanol and the necessary conversion to methanol would probably not make the process economical on a large scale, but more benign solvents and better catalysts may result in direct methanol production at high yields (Service 1998).

Combining all of these factors that will, or may, limit future demand for oil makes it possible to question what is perhaps the most widely shared opinion about the future of oil: the inevitability of OPEC's, or more precisely of the Persian Gulf's, reemergence as the arbiter of world oil pricing and as the supplier of most of the new demand during the coming generation. OPEC now expects that virtually all incremental oil demand during the twenty-first century will be met by the increased extraction of its member countries, and that its share of the global oil market will rise from about 40% in 1995 to more than 46% by 2010 and to 52% by 2020. Standard expectations are that once OPEC's share of the global oil production passes 50% once again, as it did in 1973, the cartel will reassert its pricing power and the world will be at the mercy of extortionary OPEC producers, or, a development even more frightening to contemplate, of radical Muslim regimes.

The latter fear was explicitly voiced in Richard Duncan's (1997) letter to President Clinton. Duncan argued that perhaps as early as 2010 the Muslim countries would control nearly 100% of the world's crude oil exports and that they, instigated by Arafat's call to liberate Jerusalem, could set up a new Alliance of Muslim Petroleum Exporting Nations whose mere appearance could bring the world's stock markets down by 50% in a day or, much worse, lead to a jihad. Undoubtedly, many catastrophists would see such a scenario to be much more likely after September 11, 2001.

And, given the known size of the Middle Eastern reserves (fig. 4.14) and the inexpensive production from the region's supergiant and giant fields, even a dispassionate analysis must see at least a renewed OPEC domination of the global oil market as highly likely—but not inevitable. Adelman (1992, p. 21) summed up this possibility by concluding that "Middle East oil will remain very important in the world market in the next century, dependence on the Middle East will not greatly increase, and may decrease." And to make clear that these opinions are not limited to a few (mostly American) oil economists I will cite recent conclusions of two knowledgeable OPEC

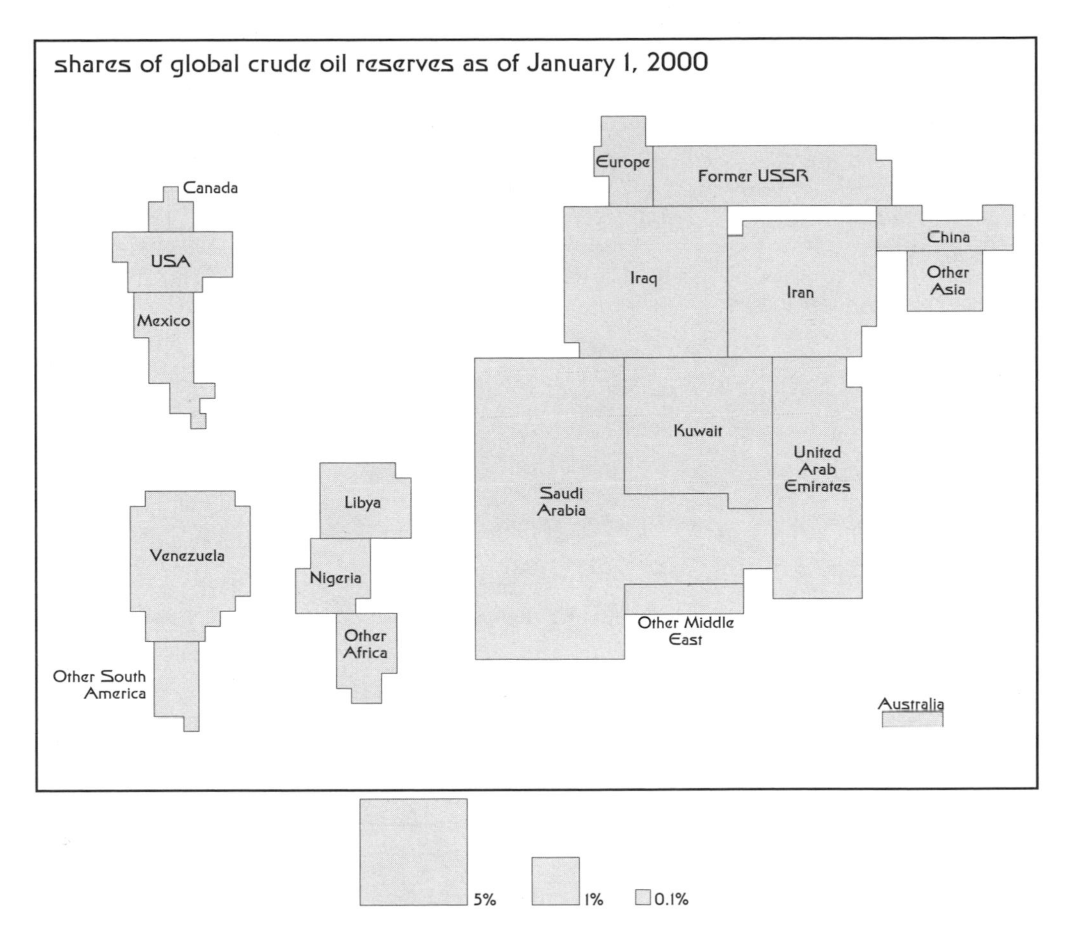

Figure 4.14
Global distribution of crude oil reserves shown in a map with nations and continents proportional to the size of their reserves. Areas of individual shapes correspond to reserves at the end of the year 2000 listed in BP (2001).

insiders whose views leave no room for scenarios of collapsing global economy and oil-financed jihad.

Nagy Eltony, who is with the Kuwait Institute for Scientific Research, warned the Gulf Cooperation Council countries not to rely solely on the assumption of a growing dependence on Gulf oil, arguing that non-OPEC supplies will remain strong and that the combination of technical advances, declining energy intensities, and concerns about the environment may substantially lower the future demand for the

Middle Eastern oil and the region's nations "may run the ultimate risk of being left with huge oil reserves that no one wants" (Eltony 1996, p. 63). And Sheikh Ahmed Zaki Yamani, the Saudi oil minister from 1962 to 1986 and one of the architects of OPEC, strongly believes that high crude oil prices will only hasten the day when the organization "will be left staring at untouched fuel reserves, marking the end of oil era" because new efficient techniques "will have cut deep into demand for transport fuels" and much of the Middle Eastern oil "will stay in the ground forever" (Yamani 2000, p. 1).

Adelman, Eltony, and Yamani may be to be too optimistic (or too pessimistic, depending which side one takes) but their conclusions could be easily dismissed only in a world where the possession of a desirable resource would give an automatic right of dominance. Technical and social adjustments and, perhaps most important, the eventual need to safeguard a tolerable environment, make the probability of a less-than-expected dependence on the Gulf during the next 10–20 years far from negligible. Before resolutely dissenting with this conclusion, recall that OPEC's share fell from 52% in 1973 to 29% by 1984, then review the failures of forecasts of primary energy needs in chapter 3 and the predictions of imminent oil extraction peaks detailed earlier in this chapter.

A more realistic appraisal of global oil futures can be summed up in the following paragraphs. There is a large mass of oil, in conventional and nonconventional deposits, in the Earth's crust and although oil's contribution to total hydrocarbon supply will be declining (slowly until 2020, faster afterward), substantial amounts of crude oil will remain on the world market throughout the twenty-first century. As the extraction of oil deposits will become dearer we will become much more efficient, and much more selective, in the use of that flexible fuel. There is *nothing inevitable* about any particular estimate of an imminent peak of global oil extraction: this event may, indeed, happen during the first decade of the new century, but it may come at any time from a few years to a few decades in the future.

Timing of this event depends on the unknown (or poorly known) quantity of ultimately recoverable resources: we can estimate this amount with much less confidence than claimed by some geologists. There is more conventional oil to be discovered than the maximum of 150 Gb estimated by Campbell and Laherrère—and there is much more oil to be recovered from existing fields and from new discoveries than 30%–40% we used to count on, as well as more from nonconventional sources. All standard predictions of imminent oil production peak assume extraction at full global production capacity but its real rate is uncertain, as the producing countries benefit from misrepresenting it. Timing of peak oil production is obviously dependent not only on supply but also on demand, and changes in demand result from a complex interplay of energy substitutions, technical advances, government policies, and environmental considerations.

Combination of slightly larger ultimate oil resources and unrestrained demand would postpone the timing of the peak production by only a few years—but a combination of appreciably larger oil resources, relatively modest growth rates of future oil consumption, and rapid rates of substitution by other sources of primary energy and by more efficient energy converters would extend the duration of the era of oil much further into the future. An extreme interpretation of this development is that efficiency and substitution gains may outpace the rate of oil depletion and crude oil, much as coal or uranium "could become uncompetitive even at fairly low prices long before it will become unavailable at any price" (Lovins 1998, p. 49).

Hubbertian curves charting the world's ultimate oil production are thus constrained by a number of factors—but there is a considerable range of plausible outcomes within these constraints (fig. 4.15). Post-1975 departure of the global crude oil output from an expected trend of declining importance in the global primary energy supply (see fig. 3.24) allows only two logical conclusions. The fact that oil's share in the world's primary energy consumption has not declined as precipitously as predicted by a long-term substitution model is either a strong indicator that there

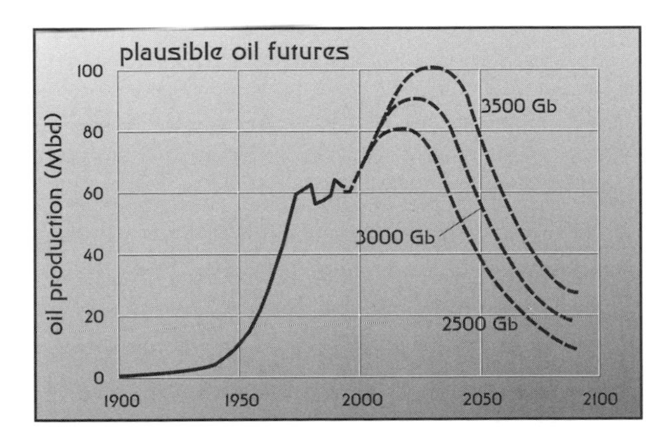

Figure 4.15
Plausible oil futures constrained by different estimates of ultimately recoverable oil show that crude oil may remain a leading source of the world's primary energy well into the twenty-first century.

is more oil to be tapped than anticipated—or the result of our unwillingness to part with this excellent fuel. But if the latter is true, then this higher-than-expected contribution would translate into steeper collapse of future oil production, greatly violating the symmetry of the Hubbertian exhaustion curve, and thus undermining the validity of the key forecasting tool used by the proponents of an early-end-of-oil era.

Finally, whatever the actual course of future oil extraction may be, there is no reason—historical, economical, or technical—to interpret the demise of today's cheap oil as a harbinger of unmanageable civilizational difficulties. Energy transitions have been among the most important processes of technical evolution: they were driving our inventiveness, shaping the modern industrial, and postindustrial, civilization, and leaving their deep imprints on the structure and productivity of economies as well as on the organization and welfare of societies (Smil 1994a). Of course, these transitions inevitably produce enormous problems for the providers of energies that are being replaced, and they necessitate scrapping or reorganization of many old infrastructures and require the introduction of entirely new links, procedures, and practices. Sectoral and regional socioeconomic dislocations are thus common (and, as witnessed by economic depression in former major coal-mining regions, they could be quite severe), infrastructural transformations are often costly and protracted and their diffusion may be very even. And societies take generations to adjust to new sources of energy and to new modes of its conversion.

But historical perspectives show that every one of these transitions—from biomass fuels to coal, from coal to oil, from oil to natural gas, from direct use of fuels to electricity—has brought tremendous benefits for society as a whole. So far, every one of these transitions has been accomplished not only without damaging global economic performance, but with elevating economies and societies to new levels of productivity and affluence, and with improving quality of the environment. So even if we were to experience an early global decline of conventional oil production we should see this trend as an opportunity rather than as a catastrophe. Energy transitions have always presented astounding benefits for producers of new supplies, and historical experience also makes it clear that from a consumer's perspective there is nothing to be feared about the end of an energy era.

Moreover, the ongoing shift from crude oil to natural gas is much less traumatic for most of the major oil producers than the transition from coal to hydrocarbons had been for large coal producers. The majority of today's leading producers of conventional oil are also the countries with large resources of natural gas. As with any such transformation, actual rates of this transition are being determined by a complex interplay of economic, technical, political, and environmental concerns and major surprises may have no smaller role to play than will the expected trends. The obvious question then is: how far can natural gas go?

How Far Can Natural Gas Go?

As we have seen in chapter 3, this question would have to be answered by any true believer in Marchetti's "clockwork" substitution model as: nearly all the way toward displacing crude oil. According to Marchetti's original model natural gas was to become the world's dominant fossil fuel by the late 1990s, it was to supply 50% of all commercial primary energy by the year 2000 and its peak production, reached shortly around the year 2030, was to supply a bit more than 70% of the global primary energy consumption, compared to no more than 10% coming from crude oil (Marchetti 1987; fig. 4.16). As I already explained in the previous chapter, by the year 2000 natural gas production was far below the share expected by the substitution model, providing only about 25% of the world's primary commercial energy supply, or half of the projected value of about 50% (see also fig. 3.24). But while this reality proves the indefensibility of any clockwork model of energy substitutions it would be a big mistake to use it in order to justify sharp downgrading of global prospects for natural gas.

The principal reason for this conclusion is that a delayed substitution of liquids by gases should not have been surprising. Transition from solid fuels (coal and wood) to liquids has been made much easier by the general superiority of refined oil products: they have higher energy density (about 1.5 times higher than the best bituminous coals, commonly twice as high as ordinary steam coals), they are (except for the poorest fuel oils) cleaner as well as more flexible fuels, and they are easier both to store and to transport. In fact, Mabro (1992) argued that if the price of every fuel were to be determined purely by market forces, the low price of oil, due to its low production cost from major oilfields, would provide an effective barrier against investment in coal, gas, and nuclear industries and that oil would dominate the world's energy supply leaving only a limited room for other energies.

In contrast, natural gas has a much lower energy density than crude oil (typically around 34 MJ/m^3 compared to 34 GJ/m^3 for oil, or a 1,000-fold difference) and hence its use as transportation fuel is limited. And while it is (or can be processed to be) a cleaner fuel than any refined fuel it is more costly to transport in continental

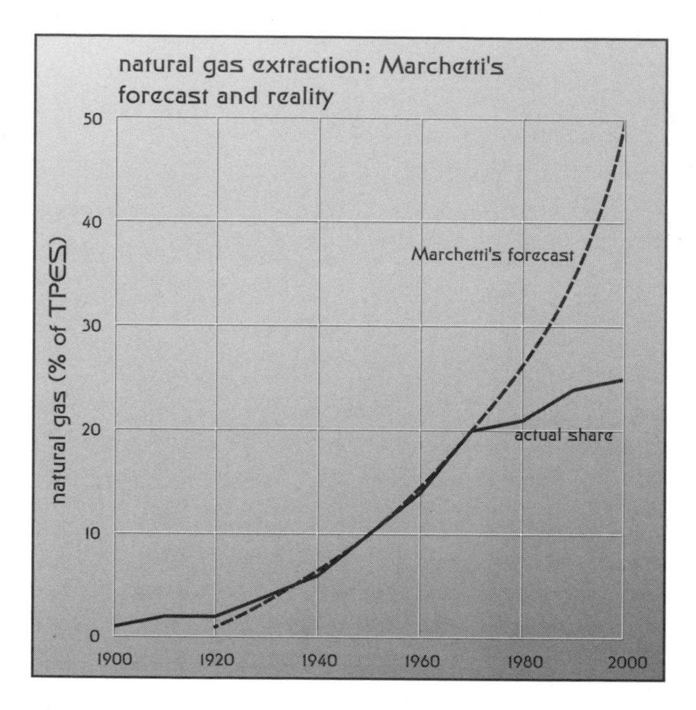

Figure 4.16
Less than 15 years after its publication, Marchetti's forecast of the global TPES share coming from the natural gas extraction turned out to exaggerate the real contribution by a factor of two. Based on Marchetti (1987), UNO (1980–2000), and BP (2001).

pipelines (and even more so by LNG tankers for transoceanic trade) and to store. Consequently, it makes sense for the world energy market to stay with the more convenient liquid fuels as long as practicable. And in a world now attuned to potential problems arising from rapid global warming there is yet another complication accompanying the rising combustion of natural gas: its contradictory role as the least carbonaceous of all fossil fuels, and hence a key to further decarbonization of the world's energy supply, on one hand, and the fact that methane, the principal or even the sole constituent of the fuel, is a very potent greenhouse gas.

While the carbon content of the best coals is more than 85% and it ranges between 84–87% for crude oil, CH_4 has, obviously, only 75% carbon and hence its combustion generates less CO_2 per unit of released energy than does the burning of coal or refined oil products. Typical emission rates are about 25 kg C/GJ of bituminous coal, 19 kg C/GH of refined fuels, and less than 14 kg C/GJ of natural gas. At the same time, CH_4 is a major greenhouse gas, and its global warming potential (GWP, compared on the molar basis) is substantially higher than that of CO_2. Over a 100-year period methane's GWP is 21 times that of CO_2, and between 1750 and 2000 the gas has accounted for about 20% of the cumulative global GWP, compared to 60% for CO_2 (IPCC 2001). Total anthropogenic emissions of CH_4 cannot be estimated with a high accuracy, but during the late 1990s they amounted to nearly 400 Mt, with livestock and rice fields being the largest sources (Stern and Kaufmann 1998). The natural gas industry (flaring, transportation) has been responsible for nearly 10% of the total.

Properly constructed and well-maintained pipelines should not lose more than 1–2% of the gas they carry, but long and ageing gas pipelines in the states of the former Soviet Union have been excessively leaky. Their losses, established most accurately by inlet–outlet difference, amounted to 47–67 Gm^3 during the early 1990s, or to at least 6 and as much as 9% of total extraction (Reshetnikov, Paramonova, and Shashkov 2000). For comparison, a loss of 50–60 Gm^3 is larger than the annual natural gas consumption in Saudi Arabia and equal to recent annual imports to Italy. Curtailing pipeline losses would thus make an appreciable difference to the fuel's availability and reduce its rising contribution to the GWP—while rising extraction and long-distance exports will inevitably add to CH_4 emissions.

Standard accounts of natural gas reserves (BP 2001) added up to about 150 Tm^3 by the end of the year 2000, an equivalent of about 980 Gb (or just over 130 Gt) of crude oil. Substantial increase of the world's natural gas reserves—they nearly tripled between 1975 and 2000, and rose by more than 25% during the 1990s—means that in spite of the doubled extraction since 1970, the global R/P ratio is now above 60 years, compared to just over 40 years in the early 1970s (BP 2001). Principal concentrations of gas reserves are in Russia (about one-third of the total), Iran (about 15%), Qatar (over 7%), and Saudi Arabia and UAE (4% each). The Middle East claims about 35% of all known reserves, a share much smaller than the one for crude oil (about 65%). North America, with about 5% of the world's reserves has more discovered gas than Europe but both continents are large consumers of this fuel with the demand driven by heating requirements and by more electricity-generating capacity fueled by natural gas. Notable national R/P ratios range from less than 10 years for the United States to just over 80 years for Russia and to more than a century for every major Middle Eastern natural gas producer.

As for the fuel's future there are, very much as with crude oil, two seemingly irreconcilable schools of thought regarding the ultimately recoverable amount of this clean and convenient fuel. On one hand, the pessimists argue that the ultimate resources of natural gas are actually smaller than those of crude oil, while the proponents of a gas-rich future believe that the Earth's crust holds much more gas, albeit at depth and in formations falling currently under the category of nonconventional resources. Once again, I will start with the gloomy arguments. Laherrère (2000) is only marginally less pessimistic about natural gas than he is about oil. He estimates the ultimately recoverable gas at 1,680 Gb of oil equivalent, or about 96% of his total for ultimately recoverable crude oil. But as only about 25% of all gas has been already produced, compared to almost 45% of all oil, the remaining reserves and undiscovered deposits of natural gas add up, according to his accounting, to about 1,280 Gb of oil equivalent, or roughly 30% more than all of the remaining conventional oil.

Turning, once more, to the distinction between "political" and "technical" estimates of remaining reserves Laherrère (2001) plotted the steady, albeit slowing, rise of the former but saw no change in the latter category since 1980 (fig. 4.17). In

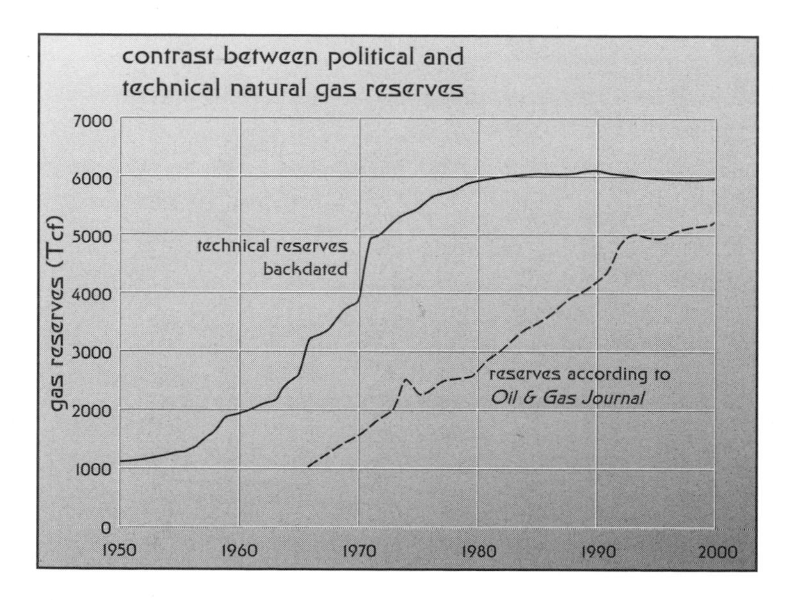

Figure 4.17
Laherrère's comparison of the world's remaining natural gas reserves contrasts the rising total from what he terms political sources (annual data from *Oil & Gas Journal* shown here) with his technical estimate that has remained virtually constant since 1980. Based on Laherrère (2000).

addition, he concludes that the estimates of nonconventional gas have been declining, and he generally dismissed the idea that gas hydrates (for more on hydrates see later in this section) could be produced economically. Limited resources would thus make it impossible to contemplate natural gas as a long-term substitute for oil: if natural gas, in addition to its late 1990s consumption rate, would displace all of the recent crude oil consumption and if the world's primary energy use were to continue at the recent pace, then the global gas supplies would be exhausted in just 35 years and the world output would peak before the year 2020.

Consequently, Laherrère (2001) rejected even the lowest variant of the 40 recent IPCC scenarios of natural gas extraction during the twenty-first century as unrealistic (fig. 4.18). Most of these scenarios assume annual natural gas consumption well above 200 EJ between the years 2030–2100 but Laherrère's projections stay well below that rate even before 2040 and fall to less than half of it afterward. If they were correct, there would be no way for the natural gas to provide eventually close to half of all global primary energy consumption. This would mean that some of

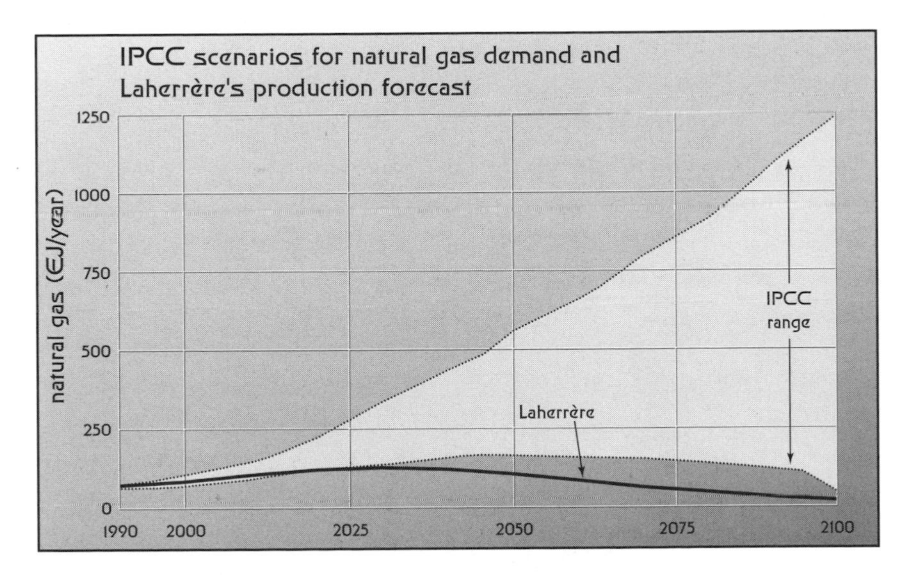

Figure 4.18
Laherrère's forecast of global natural gas extraction, based on his technical estimates of remaining reserves, is lower than all of the scenarios used by the latest IPCC assessment in modeling future CO_2 emissions. By the year 2050 it is less than half of the IPCC's median estimate. Based on Laherère (2000) and SRES (2001).

the currently contemplated gas megaprojects, most notably the large-diameter pipelines from the Middle East to Europe and India and from Siberia to China, the Koreas and Japan, would have to be either substantially scaled-down or abandoned, and that aggressive gas-based decarbonization of the world fossil fuel supply would remain unfulfilled.

Again, as with crude oil, the latest U.S.G.S. (2000) assessment of global resources of natural gas is much more optimistic, putting the total at just over 430 Gm^3, or an equivalent of almost 2,600 Gb (345 Gt) of crude oil and about 53% above Campbell and Laherrère's figure. Breakdown of the U.S.G.S. estimates shows that only about 11% of ultimately recoverable gas has been produced, that remaining reserves account for some 31%, their eventual growth should amount to nearly 24%, and that the undiscovered potential is almost exactly 33% of the total. Other recent estimates of ultimately recoverable natural gas, seen by their authors as rather conservative, range between 382–488 Gt of oil equivalent (Rogner 1997). Consequently, Odell (1999) forecasts global natural gas extraction peaking at about 5.5 Gtoe by the year 2050 and declining to the year 2000 level only at the very beginning of the twenty-second century (fig. 4.19).

All of these figures refer to conventional natural gas, that is fuel that escaped from its parental rocks and accumulated in unpermeable reservoirs, either with oil

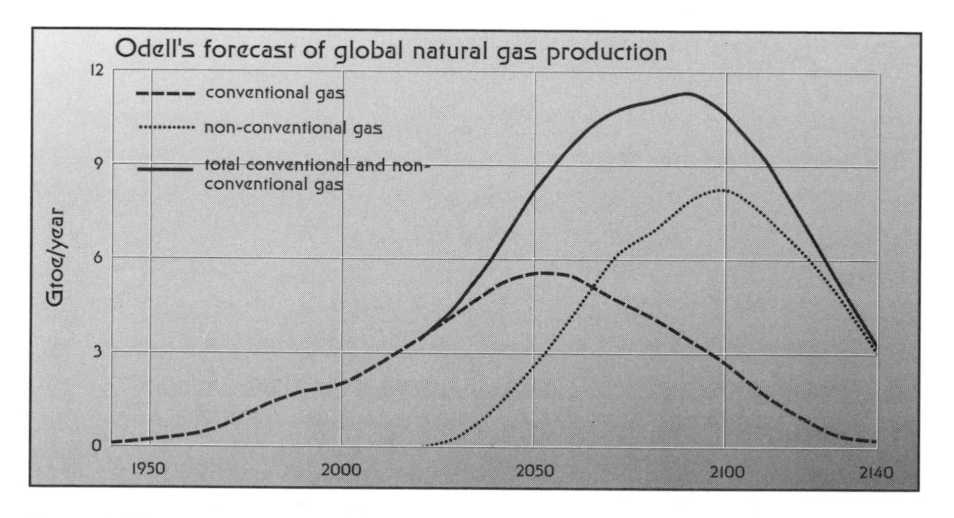

Figure 4.19
Odell's (1999) forecast of global extraction of conventional and nonconventional natural gas puts the combined production peak as late as 2090.

(associated gas, the dominant form of the fuel until the 1980s) or alone. Nonconventional gas embraces resources that are already being recovered, above all methane in coalbeds, as well as much larger deposits in tight reservoirs, high pressure aquifers, and in methane hydrates, whose eventual recovery still awaits needed technical advances (Rogner 2000). Gas in coalbeds is absorbed into coal's structure while the gas in tight reservoirs is held in impermeable rocks whose leakage rate is slower than their filling rate and that would have to be fractured inexpensively in order to allow economic extraction. Gas in high pressure aquifers is dissolved in subsurface brines under high temperature and pressure, with the total resource estimate for the Gulf of Mexico alone being larger than Campbell and Laherrère's total of the world's remaining conventional gas reserves. Global resources of geopressured gas were estimated to be about 110 times the current proven reserves of the fuel (Rogner 2000).

The second largest source of nonconventional natural gas is methane trapped inside rigid lattice cages formed by frozen water molecules (Holder, Kamath, and Godbole 1984; Kvenvolden 1993; Lowrie and Max 1999). These gas hydrates, or clathrates, appear as white, ice-like compounds, and the most common arrangement of frozen lattices allows for the inclusion of CH_4 and C_2H_6 or other similar-sized gas molecules (see the insert in fig. 4.20). As the phase-boundary diagram (fig. 4.20) shows, the upper-depth limit for the existence of methane hydrates is under moderate pressures—close to 100 m in continental polar regions and about 300 m in oceanic sediments, while the lower limit in warmer oceans is about 2,000 m—and at temperatures not far from the freezing point. Fully saturated gas hydrates have 1 methane molecule for every 5.75 molecules of water, which means that 1 m^3 of hydrate can contain as much as 164 m^3 of methane (Kvenvolden 1993). Gas hydrates are found at every latitude but the pressure, temperature, and gas volume requirements limit their occurrence to two kinds of environments, to both continental and offshore zones of polar permafrost sediments, and to deep oceans (fig. 4.21).

Most of the gas present in hydrates has been released from anoxic decomposition of organic sediments by methanogenic bacteria. Hydrate deposits have been identified on margins of every continent where they are often underlain by similarly thick zones of free gas but there is no reliable estimate of the total amount of methane these resources contain. Some localized accumulations appear to be very large: the Blake Ridge formation in the western Atlantic offshore Carolinas, emplaced beneath a ridge created by a rapidly depositing sediment, contains at least 15 Gt of carbon in hydrates and at least that much in bubbles of free gas in the sediment underlying

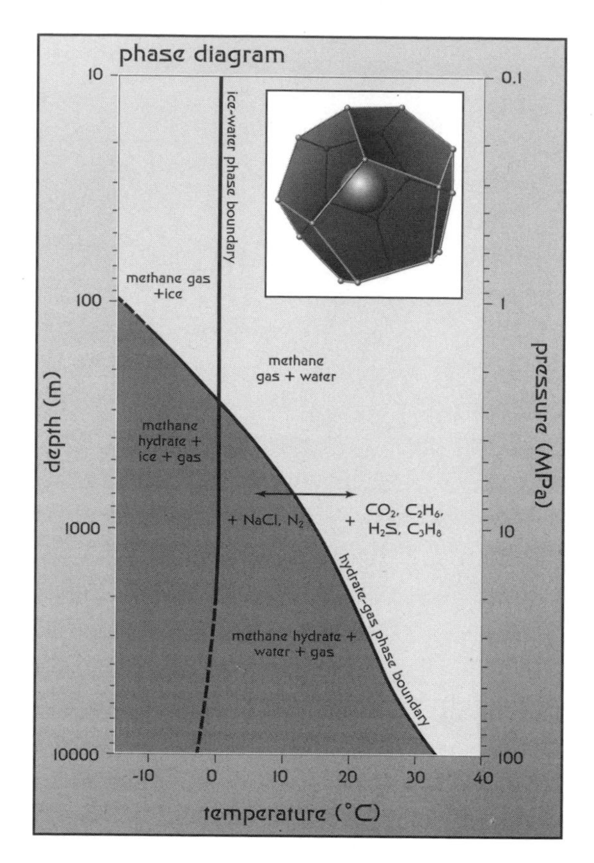

Figure 4.20
Phase diagram delimiting the natural occurrence of hydrates. The boundary between free CH_4 and gas hydrate is for pure H_2O and pure CH_4. Addition of NaCl to water shifts the boundary to the left; presence of other gases (CO_2, H_2S) moves it to the right. The inset shows a CH_4 molecule encased in an ice crystal. Based on Kvenvolden (1993) and Suess et al. (1999).

the hydrate zone (Dickens et al. 1997). The total is about six times the proven conventional U.S. reserves of natural gas, and hydrates in all coastal U.S. waters may contain as much as 1,000 times the volume of the U.S. conventional gas reserves (Lowrie and Max 1999). The U.S.G.S. estimate puts the global mass of organic carbon locked in gas hydrates at 10 Tt, or roughly twice as much as the element's total in all fossil fuels (Dillon 1992).

Needless to say, the world's natural gas industry would be radically transformed even if we were to recover just a very small share of all hydrates in shallow sediments.

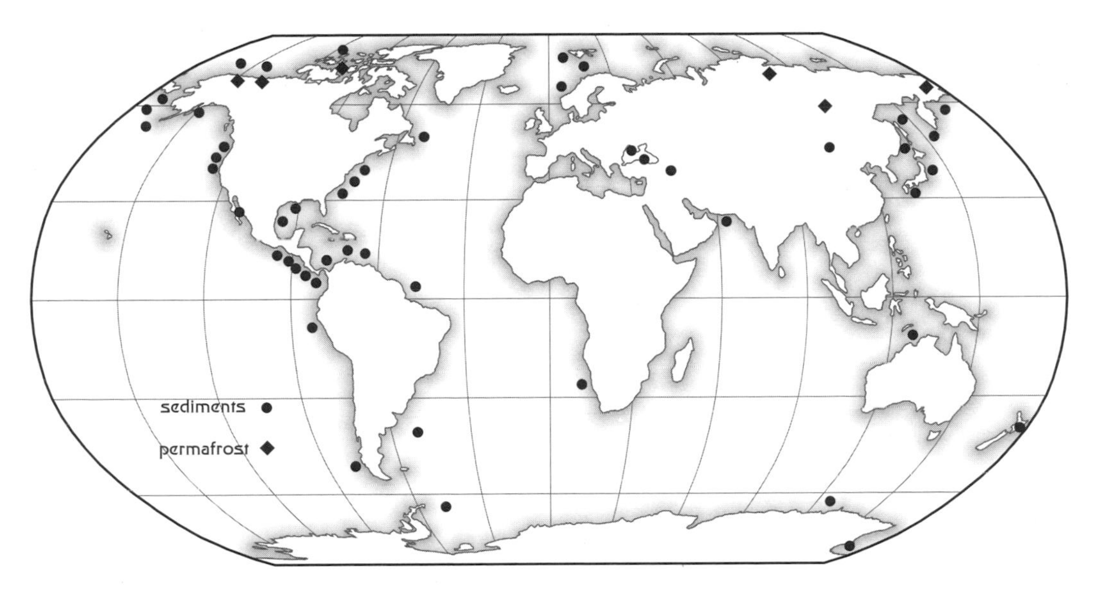

Figure 4.21
Global distribution of continental and suboceanic hydrates. Based on maps in Kvenvolden (1993) and Suess et al. (1999).

Tapping just 1% of the resource would yield more methane than is currently stored in the known reserves of natural gas. But the task will not be easy because gas from the solid hydrates is often quite dilute (for example, the Blake Ridge formation covers some 26,000 km^2), it cannot migrate into reservoir traps and a new way of recovery would have to be found before this immense resource could be tapped. Not surprisingly, pessimists consider hydrate recovery to be an illusion, while many enthusiasists are thinking about plausible means of extraction (Hydrate.org 2001; Kleinberg and Brewer 2001). Depressurization of hydrate deposits is the first obvious choice for extraction and its first applications will undoubtedly come not in giant but diffuse offshore deposits but in smaller and richer reservoirs, such as those identified beneath the permafrost of Alaska's Prudhoe Bay where hydrates make up one-third of the sediment volume.

In any case, if depressurization is not done carefully, the resulting melting of the ice and gas escape could trigger a sudden destabilization of a deposit. Such an event may be acceptable in a small formation but it could be catastrophic in large structures, leading to massive undersea landslides and to voluminous releases of methane into the atmosphere. Such naturally occurring undersea slides have been implicated in massive releases of CH_4 that contributed to a rapid global warming 55 Ma ago and even smaller-scale reprisal of such events could accelerate climate change. Obviously, melting of hydrates could be also achieved by injecting steam, hot liquids, or chemicals into hydrate formations. At this point there is no sensible way to assess future costs of recovery, but whatever the eventual cost, the economics of hydrate exploitation would be more appealing for such energy-poor countries as Japan and India. Not surprisingly, both of these countries have hydrate recovery research programs, as does the United States.

Given the post-1950 advances in hydrocarbon exploration and extraction (recall the enormous progress in three-dimensional and four-dimensional imaging of reservoirs, and in offshore, deep, directional, and horizontal drilling noted in chapter 1), I believe that only substantial new discoveries of conventional gas could prevent significant, and increasing, commercial extraction of nonconventional gas, including some hydrates, before the middle of the twenty-first century. Odell (1999), too, foresees unconventional gas production rising rather rapidly after the year 2020 and peaking around the year 2100 at more than 8 Gtoe. New nonconventional discoveries cannot be excluded because, as already noted in the previous section, there is yet another fascinating matter related to natural gas resources, namely the question of the fuel's abiogenic origin.

Unlike the almost completely overlooked Russian–Ukrainian theory about the inorganic origins of crude oil, the possibility of abiogenic gas has received fairly wide attention in the West thanks to the vigorous and untiring promotion of the idea by Thomas Gold (1993, 1996). Gold's reasoning is as follows. We know that abiogenic hydrocarbons are present on other planetary bodies, on comets and in galactic space. On Earth hydrogen and carbon under the high temperatures and pressures of the outer mantle can readily form hydrocarbon molecules, and the stable ones among them will ascend into the outer crust. All of those unmistakably biogenic molecules in hydrocarbons can be easily produced by the metabolism of underground bacteria present in the topmost layers of both continental and oceanic crust. Deep abiogenic origin of hydrocarbons would also explain why these compounds almost always contain elevated levels of chemically inert helium.

Not surprisingly, Gold's reasoning, much like the Russian–Ukrainian theory, has been rejected by most Western petroleum geologists who call it a little more than wishful thinking, and argue that any production from weathered or fractured crystalline rocks can be easily explained by migration from a flanking or overlying conventional source rock and that all fossil fuels are of biogenic origin. I am not arguing for an uncritical embrace of Gold's ideas but, once again, I plead for a greater tolerance of unorthodox ideas—and I must point out that Gold's (1992, 1999) pioneering outline of the existence of a deep hot biosphere has been convincingly confirmed by experimental drilling in several countries and in different substrates (Pedersen 1993; Frederickson and Onstott 1996).

Some studies merely confirmed the presence of unspecified subsurface bacteria, others identified such dominant species as *Thermoanaerobacter* and *Archaeoglobus* 1,670 m below the surface in a continental crude oil reservoir in the East Paris Basin (L'Haridon et al. 1995). Most impressively, bacteria related to *Thermoanaerobacter, Thermoanaerobium,* and *Clostriduim hydrosulfuricum* were obtained from water samples collected at 3,900–4,000 m in the Gravberg 1 borehole drilled to a depth of 6,779 m in the complete absence of any sediments in the granitic rock of the interior of the Siljan Ring, an ancient meteorite impact site in central Sweden whose drilling began at Gold's urging (Szewzyk et al. 1994).

Finally, a few paragraphs about the regional prospects for natural gas. Given less flexibility in gas transportation these concerns are more important than in the case of easily transportable oil. Huge reserves in the FSU and throughout the Middle East will easily provide not just for these two regions but will be increasingly exported, both through pipelines and as LNG, to Europe and to Asia, with China and India becoming major importers. Rising reserves in Africa present no supply problems for the continent. The European situation, in spite of limited amounts of the continent's gas reserves, is perhaps the most comfortable. As Odell (2001, p. 1) noted, "Western and Central Europe are really jam pots surrounded by wasps coming from all sides": rich resources of natural gas from Siberia, North Sea, and North Africa (and eventually from the Middle East) have already transformed the European energy supply (coal industries are collapsing, no nuclear stations are built), and they will allow the construction of a fully integrated continental gas transport network by the year 2020 when the fuel may supply 35% of Europe's energy.

Between January 2000 and January 2001, when the wellhead price of U.S. natural gas rose 380%, a spike surpassing the crude oil price rises of the 1970s, many analysts put out bleak appraisals of North America's gas prospects and began to question the common expectations of the fuel's dominance in the country's future energy supplies. Short-term aberrations caused by temporary supply shortages should not be mistaken for long-term trends. By August 2001 the wellhead price fell to the level just 50% above the 1999 mean, and the actual 2000–2001 increase in the average residential price was only about 30% (EIA 2001i). Moreover, the 25% decline in

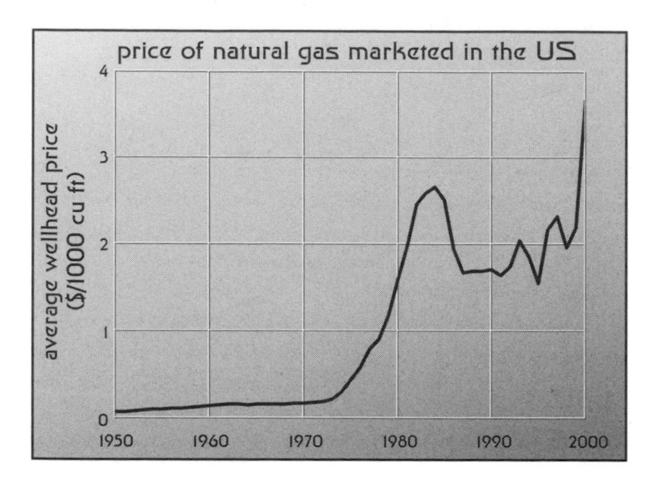

Figure 4.22
U.S. natural gas prices, 1950–2001. Plotted from data in USBC (1975) and EIA (2001a).

the continent's natural gas reserves since 1990 has been clearly related to low levels of exploratory drilling brought by very low wellhead prices that prevailed until the year 2000 (fig. 4.22). As expected, post-1999 price rises were translated into massively expanded drilling and supplies and prices are now widely expected to stabilize.

Still, given the high residential and industrial demand and the fact that more than 90% of planned electricity-generating units will be gas-fired, even vigorous exploration will not reduce the need for higher imports from Canada and eventually also from Mexico. Expansion of LNG imports, initially from Algeria but now primarily from Trinidad and Tobago and increasingly from spot sales of gas originating in every producing region (Australia, Middle East, and Africa), is another sign of mismatch between domestic demand and supply. This is perhaps the most apposite place to stress, as I did when discussing the future of oil, that the rate of mid- and long-term growth of demand can be greatly influenced by efficiency improvements, some being readily available, others requiring more research and commercialization.

Certainly the best example in the first category is the widespread availability of high-efficiency natural gas-fired furnaces. About 70% of all energy used for U.S. residential heating comes from natural gas and this demand accounts for nearly 25% of the U.S. gas consumption, with another 15% of the total use burned to heat commercial space (EIA 1999a). There are still millions of older furnaces around, burning gas with efficiencies of less than 60%. The federal minimum annual fuel utilization efficiency (AFUE) for household units is now set at 78% and new mid-efficiency furnaces are rated at 78–82% efficiency compared to 88–97% for high-performance units (Wilson and Morrill 1998; LBL 2001). These units are more expensive but their electronic spark ignition eliminates pilot lights and their exhaust gases are cool enough to eliminate chimneys and vent them through a wall with plastic tubing. A realistic assumption of replacing the existing low- and mid-efficiency furnaces within 15 years by high-performance units would then cut the North American residential demand for natural gas by at least 20–25%.

Increased use of natural gas for electricity generation in the United States (it consumed about 15% of total gas deliveries in the year 2000) and in the European Union has been accompanied by the installation of more efficient means of conversion. Combined-cycle plants use waste heat to produce more electricity and cogeneration uses waste energy to deliver heat or steam for other buildings or processes. Either process can raise the overall conversion efficiency well above 50% and their universal application in coming decades would add up to huge fuel savings compared to the average of about 35% conversion efficiency for standard boilers and generators.

An important concern regarding highly efficient, and highly popular, gas turbines, high levels of NO_x emissions (as much as 200 ppm) from uncontrolled combustion, has been solved by progressively more effective control measures. Emissions fell by an order of magnitude between 1975 and 1985, and the latest XONON technique, burning the fuel in the presence of a catalyst at temperatures below which the formation of NO_x takes place, cuts the nitrogen oxide emission to as low as 2 ppm (Catalytica Energy Systems 2001; fig. 4.23). I will add a few more details regarding efficient combustion processes in the next section when describing advances in coal-fired generation.

Finally, natural gas will be used increasingly to generate electricity indirectly via fuel cells and thanks to this kind of conversion it is now also poised to become an important automotive fuel (Srinivasan et al. 1999). Steam reforming of natural gas that yields hydrogen is fairly efficient (at least 70%) and modular phosphoric acid fuel cells (PAFC) equipped with reformers have already entered small-scale electricity generation (kW to MW range) in institutional and industrial settings (offices, schools, hospitals, hotels, and waste treatment plants) and larger vehicles. Their best cogeneration efficiency is nearly 85% and they can also use impure hydrogen as fuel but they need expensive Pt as a catalyst. In contrast, molten carbonate fuel cells (MCFC) work at about 650 °C and the heat and water vapor produced at the anode

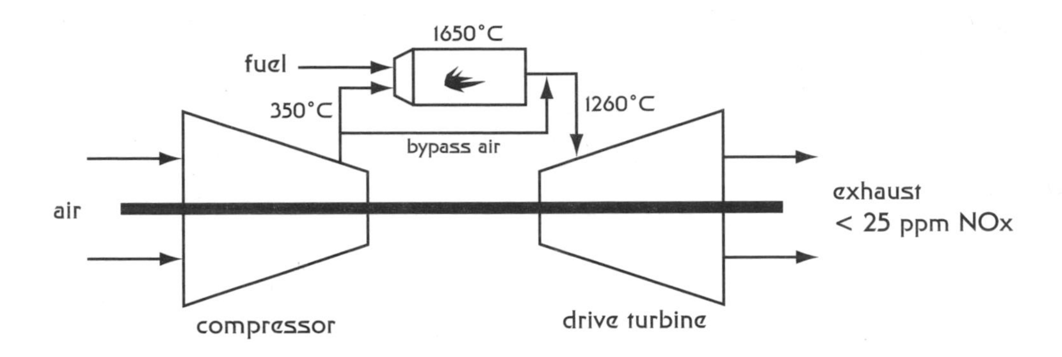

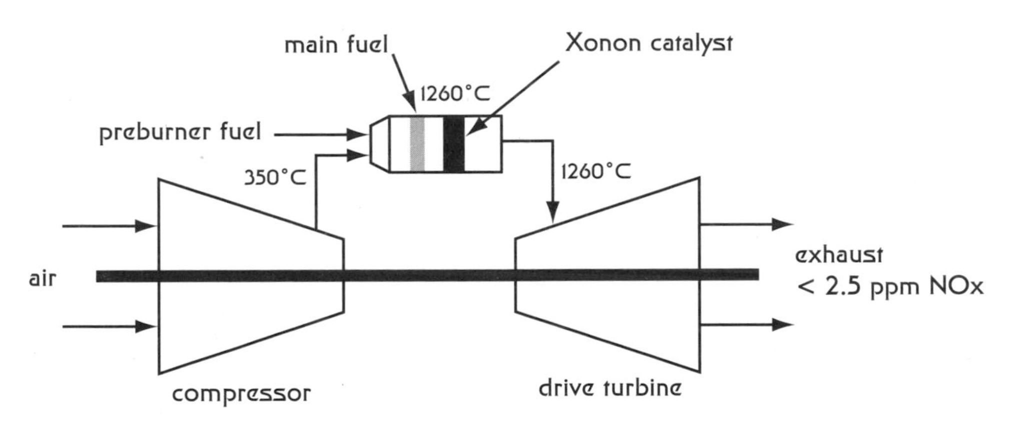

Figure 4.23
Comparison of traditional and XONON combustion. Based on figures in Catalytica Energy Systems (2001).

can be used for steam reforming of methane with inexpensive catalysts as well as for cogeneration. On the downside, high temperature accelerates corrosion and the breakdown of cell components. Stacks of up to 2 MW have been tested, working with efficiencies in excess of 50%, and future MCFCs may be used for a larger-scale electricity generation.

Internal combustion engines can, of course, run on natural gas, but methane's low energy density makes it an impractical transportation fuel. Methanol (CH_3OH), derived from CH_4 by conventional catalytic reforming of the feed gas and steam, is a much better choice. The compound is a liquid at ambient temperature and hence it has energy density superior to gas, it can be stored and handled much like gasoline but it is much less flammable (lower volume limit at 7.3% compared to 1.4% for gasoline) and, once ignited, it produces less severe fires (that is why it is the only fuel used in Indianapolis-type car races). Low emissions are the main advantage of burning methanol, and the high-octane fuel provides excellent acceleration. But methanol's energy density is only about half of gasoline's lower heating value and its cost is higher and hence most of today's methanol-fueled vehicles are fleet cars and buses.

The United States produces nearly one-quarter of the world's methanol and uses most of it to synthesize methyl tertiary butyl ether for blending in clean, reformulated gasoline (Methanol Institute 2002). Mass production of methanol (it could be made from any carbon feedstocks, including coal and wood) would diversify transport fuel supply but it would not improve the overall efficiency of the TPES. A much better choice is to use gas-derived methanol in fuel cells. Given the fuel's properties, methanol-fueled cells have a number of obvious advantages for automotive use but the need to reform the fuel (by passing it and water over a catalyst in a heated chamber and producing hydrogen) reduces the overall efficiency of methanol-fueled PAFC to less than 40%. Prototypes of direct methanol fuel cells (DMFC) were designed by the Jet Propulsion Laboratory in Pasadena, California. DMFCs eliminate the reformer and use lighter stacks, and their overall efficiency is comparable to that of proton exchange membrane fuel cells (PEMFC), which are now the leading contenders for automotive uses with Ballard Power Systems (2002) being their most important developer (for more on fuel cells and hydrogen systems see chapter 5).

High costs are the main obstacle to large-scale commercialization of even those designs that are technically ready for retail. Typical costs of PAFC and MCFC for stationary electricity generation in the range of 1–100 MW are projected to be comparable to the capital cost of conventional thermal conversions but while coal- and hydrocarbon-fired boilers and turbogenerators will be in service for more than 20 years, a fuel cell stack lifetime is only about five years (Srinivasan et al. 1999). Cost disparity is even greater in transportation: while gasoline and diesel engines cost about $10/kW, PEMFCs running on hydrogen from methanol are projected to cost $30/kW and methanol-fueled PAFCs and DMFCs $100/kW. There are, of course, partial compensations: even at no more than 30–40% efficiency, PAFC and DMFC would be 20–35% more efficient than typical internal combustion engines and low emissions make the arrangement more acceptable.

Consequently, PEMFC, developed by the Ballard Power Systems for Daimler-Chrysler, as well as by Toyota, Honda, and Ford were to be installed on tens of thousands of vehicles before 2004 (Srinivasan et al. 1999; Fuel Cells 2000; Methanex 2000). In reality, only limited marketing of fuel cell hybrid vehicles began in 2002 and the sales will remain modest for years to come. A note of caution: the rapidly advancing research and experimental applications of various kinds of fuel cells make it impossible to predict which system(s) will emerge as clear leaders in 10–20 years. I will make additional observations concerning the future of fuel cells in the next chapter when discussing the possible emergence of hydrogen-based energy system. In any case, there is no doubt that in the immediate future methanol-fueled cells are clearly major contenders, and that their succesful diffusion would make natural gas an important automotive fuel.

Fortunately, efficiently used supplies should be sufficient to meet the coming demand. Only a refusal to acknowledge the global riches of natural gas and indefensibly pessimistic assessments of coming advances in exploration and production of both conventional and nonconventional resources and gradual improvements in conversion efficiencies could lead to highly constrained appraisals of the fuel's role in global energy supply during the first half of the twenty-first century. EIA (2001f) sees the global output almost doubling by the year 2020 when the gas could supply close to 30% of all primary energy. More conservatively, Odell (1999) forecast the conventional global output going up by about two-thirds during the same period of time, and peaking at about 2.6 times the 2000 level by 2050; by that time the rising output of nonconventional gas could bring the total annual extraction to nearly 4 times the 2000 rate. There is a high probability that these forecasts will turn out to be excessively high but I am citing them only in order to stress that even if the actual totals were to be substantially lower (by, say, 20–35%), there is little doubt that natural gas will be the fastest-growing component of the world's fossil fuel supply in decades ahead.

What Will Be Coal's Role?

During the early 1970s, with cheap crude oil and the seemingly unstoppable rise of nuclear electricity generation, coal was widely regarded as the has-been fuel. But global output kept on increasing and by 1989 it was about 30% above the 1975 level. Even with the abrupt decline of production in the countries of the former Soviet Union and the Communist European countries that took place during the early 1990s, and with an equally unexpected drop in China's coal extraction before the decade's end, it was about 18% higher in the year 2000 than it was in 1975, and it provided nearly 25% of the world's TPES, the share virtually identical to that of natural gas.

Coal's output is thus far below the already noted (chapter 3) wildly optimistic expectation of more than double the production between 1975 and 2000 (Wilson 1980)—but far above the level expected by Marchetti's automatic substitution model that had the fuel slumping to 10% of the world's primary energy consumption by the year 2000 and providing a mere 2% of all commercial energy by the end of the new century's third decade (Marchetti 1987; fig. 3.23). Considering coal's disadvantages, the fuel's resistance to such a decline and its continuing global importance have been an even more remarkable phenomenon than has been the delayed dominance of natural gas. But the fuel's relatively robust post-1975 performance should not be seen as a harbinger of things to come. And not because of any resource-related concerns: unlike in the case of the two fuel hydrocarbons, coal resources are so immense that there is no possibility of shortages during the twenty-first or the twenty-second centuries.

Available estimates of global coal resources range between 6.2–11.4 Tt but it should be noted that the lower total changed little during the twentieth century as the first global estimate, offered at the Thirteenth International Geological Congress, was 6.402 Tt (McInnes et al. 1913) and IIASA assessment of 1996 ended up with 6.246 Tt (Rogner 1997). Secular fluctuations in global coal reserves reflect different classifications and up- or downgradings of national totals in major coal-producing nations (United States, China, and Russia) but, again, the 1913 total of 671 Gt of recoverable reserves is not very different from the recent aggregate of about 770 Gt of hard-coal equivalent (fig. 4.24).

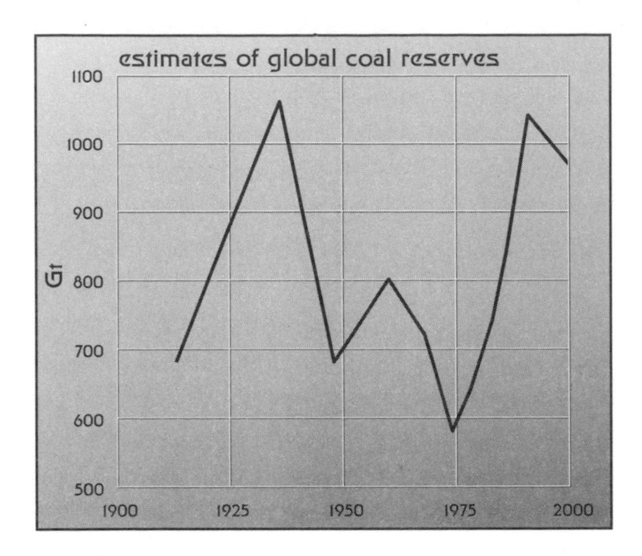

Figure 4.24
Changing estimates of global coal reserves, 1913–1996. Based on Smil (1991) and Rogner (2000).

This is in contrast with a substantial increase in the best estimates of global crude oil reserves that are now as much as three to four times higher than they were during the 1940s (see fig. 4.11). The fact that the total of proven coal reserves is based on a detailed exploration of less than 10% of the resource base reflects the simple reality that, given the size of coal reserves, there is no compelling need to engage in more probing examinations of total coal in place in the Earth's crust. Put differently, even perfect knowledge of global coal reserves would be irrelevant as most of the fuel in the ground will always remain undisturbed.

At a value near 230 (in the year 2000) the world coal's R/P ratio is more than 3 times as high as the analogical rate for natural gas and more than 4 times as large as the global R/P for crude oil. And while coal deposits are more widely distributed than are the hydrocarbon reservoirs, reserves of good quality coal are, contrary to a common perception, actually more concentrated: while the top five countries (Saudi Arabia, Iraq, UAE, Kuwait, and Iran) had about 64% of all oil reserves in the year 2000, the five nations with the largest coal reserves (the United States, Russia, China, Australia, and Germany) accounted for about 69% of the world's total (BP 2001). Coal's future use will be obviously determined by the state of the fuel's two largest remaining markets, electricity generation and iron and steel production.

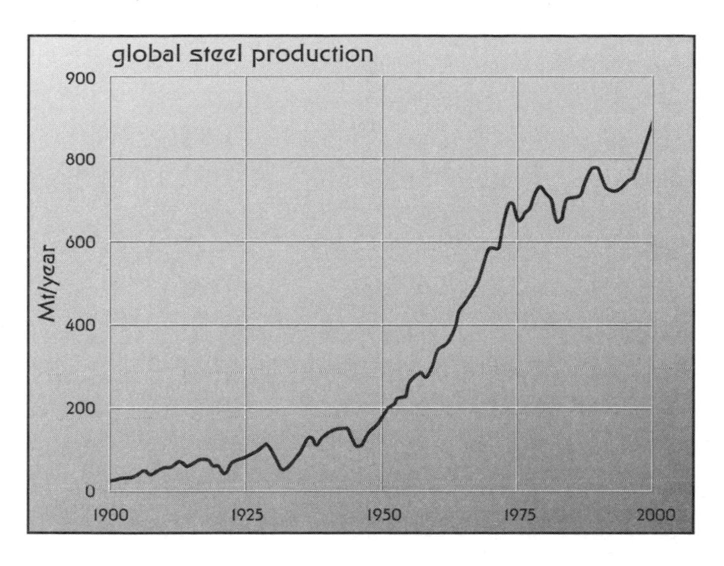

Figure 4.25
Global crude steel production, 1945–2000. Based on a graph in World Coal Institute (WCI 2000) and data in IISI (2001).

Coal will remain an indispendable input in iron and steel making for decades to come but the fuel's second largest global market is expected to grow only very slowly. These expectations are an inevitable outcome of much slower growth rates of steel industry: they fell to a mere 1.2% between 1975 and 2000, compared to about 6% between 1945 and 1975 (fig. 4.25). And at least three efficiency measures will reduce the future need for metalurgical coal. The first one is a more widespread use of direct coal injection into blast furnaces whereby 1 tonne of injected coal displaces 1.4 tonnes of coke. The second one is a gradual progress of direct reduced iron processes that dispense with blast furnaces and are energized mostly by natural gas (Chatterjee 1994). Finally, higher shares of steel are produced in electric arc furnaces as opposed to the still dominant (roughly two-thirds of the world output) basic oxygen furnaces whose operation needs about 630 kg of coal to make 1 tonne of steel (WCI 2000). Continuation of post-1975 growth combined with further efficiency gains would bring the global coal demand in the iron and steel industry to about 750 Mt by the year 2025, or only some 150 Mt above the 2000 level.

Given the abundance of coal, the fuel's future share of electricity generation, and its largest market in every country where the fuel is used will be decided above all by its competitive position vis-à-vis natural gas. This, in turn, will not be only a matter of the reasonable cost of coal conversions per se but will depend on our ability to limit a multitude of environmental impacts arising from coal extraction and combustion. Coal could become competitive in other markets, above all in household heating and cooking, only if the industry were to succeed in converting it into gases (or liquids) with not only an acceptable cost but, once again, with tolerable environmental consequences. These considerations have been important for most of the second half of the twentieth century, but they have become dominant in a world concerned about risks of global warming, as no fossil fuel generates more greenhouse gases per unit of converted energy than does coal.

Coal's future is thus not fundamentally a matter of resource availability or production cost but one of environmental acceptability. Some ingredients of the latter concern, above all the emissions of PM and sulfur and nitrogen oxides, have, or will shortly have, effective technical solutions. But, as yet, there are no effective, practical, large-scale means of large-scale elimination or sequestration of CO_2 after coal combustion (see chapter 5). The only viable partial solution to coal's greenhouse gas burden is to make its conversion more efficient and hence less CO_2-intensive. Best technical fixes may then greatly prolong coal's contributions to global energy supply—or even their relatively high success may not be enough in a world subject to exceptionally rapid global warming.

And there is also a critical national bias in appraising coal's future. Although China and India, the two countries with a high dependence on coal and a limited domestic availability of crude oil and natural gas, will try to reduce their reliance on coal, their expanding economies will almost certainly require higher coal extraction. Indeed, the U.S. EIA's forecast of global coal production credits the two countries with 92% of the total expected increase in worldwide use of coal by the year 2020 (EIA 2001d). Most of the remaining gains would come from coal-fired electricity generation elsewhere in Asia and in North America. Global coal fortunes will be thus overwhelmingly set by developments in China, India, and the United States.

Not suprisingly, those observers who see coal extraction and combustion as "a leading threat to human health, and one of the most environmentally disruptive human activities" (Dunn 1999, p. 10) believe that the fuel is in an inexorable retreat. Demonization of coal does the fuel an historic injustice. Modern civilization with its remarkable improvements in quality of life, was after all built on the conversion of this admittedly dirty fuel. Denigration of coal also precludes an impartial assessment of the fuel's prospects. Counterarguments stress the enormous advances made in more efficient conversion of coal and in the reduction of resulting pollutants—

changes that have already brought a considerable internalization of external costs (see chapter 3)—and extol further technical fixes. Clean coal technology (CCT) has been a mantra of the fuel's advocates for decades, and a considerable amount of government and private funding, particularly in the United States, has gone into developing better ways of preparing coal before combustion, burning it more efficiently, and minimizing or almost entirely eliminating both solid and gaseous emissions (Alpert 1991; DOE 2000b).

Acid deposition, widely considered as the most important environmental concern in most of Europe and in the eastern North America during the 1970s, could not have been reduced to its present level of being just one of many manageable problems without widespread installation of FGD. By the late 1990s about 230 GW of coal-fired capacity in North America and Western Europe (that is nearly 25% of all fossil-fueled capacity in those two regions) had either wet- or dry-removal FGD. At the same time, capture of SO_2 in the form of sulfates creates new challenges for the disposal of this bulky waste. Similarly, particulate by-products of coal combustion and air pollution control, including fly-ash, bottom ash, and boiler slag, add up to millions of tonnes every year.

Reuse rates for these products are high in the EU countries, ranging from about 40% for gypsum from FGD to more than 90% for fly-ash but the United States uses only about 33% of its coal ash, and just 10% of its FGD products (Giovando 2001). Proven and experimental applications range from the use of ash (most of it is a highly variable mixture of three oxides, SiO_2, Fe_2O_3, and Al_2O_3) as a structural fill, ingredient in concrete and paving materials, and stabilizer of coastal reefs to the use of FGD scrubber material as a pond liner and soil amendment (pH buffer) and in the construction industry. Given its high dependence on coal and its limited availability of landfills such reuse will become particularly critical in China.

But neither electrostatic precipitators nor FGD do anything to lower the emissions of nitrogen oxides, and both systems consume a small part of the electricity generated by the plant. Fly-ash removal needs at least 2–4% of the generated total, and, depending on the process, FGD can use up to 8% of the gross electricity generation and hence these effective controls actually lower a plant's overall fuel conversion efficiency. Just the opposite is badly needed because the average efficiency of fossil fuel-fired electricity generation has been stagnant in affluent countries since the early 1960s.

The U.S. record illustrates this inexcusable perpetuation of high inefficiency whereby two-thirds of all fuel are wasted (see fig. 1.17). This stagnation contrasts with remarkable improvements of energy conversion in industrial processes and households alike and it points to regulated monopolies uninterested in technical innovation as the main cause of this failure (Northeast–Midwest Institute 2000). As any other technical failure that can be addressed by readily available means, this one, too, presents enormous opportunities as aging plants, built during the two decades of booming demand for electricity during the 1950s and 1960s, will have to be replaced soon and as new combustion techniques and new plants have been either ready or will soon become commercial.

One commercial technique can virtually eliminate NO_x without external emission controls while raising overall conversion efficiency. Fluidized bed combustion (FBC) is an idea whose origins go back to the 1920s, and for decades its promise seemed to be greater than its actual performance (Valk 1995). During the FBC process, solids of any combustible material are suspended on upward-blowing jets of air with the resulting turbulent mixing allowing for more effective chemical reactions and heat transfer. No less important is the fact that the combustion takes place at just between 760–930 °C, that is well below 1,370 °C, the temperature at which NO_x begin forming due to the splitting of atmospheric N_2. In addition, finely ground limestone ($CaCO_3$) or dolomite ($CaMg(CO_3)_2$) could be mixed with combustion gases in order to remove as much as 95% (and commonly no less than 70–90%) of all sulfur present in the fuel by forming sulfates (Henzel et al. 1982; Lunt and Cunic 2000).

During the 1990s FBC was overshadowed by the widespread embrace of gas-turbine generation but the atmospheric version of the technique (AFBC) is now a fully commercial option, and most manufacturers of large boilers offer it as a standard package. AFBC units are now available in sizes up to 300 MW and are burning a variety of fuels ranging from coal to municipal waste, and the process is often competitive with natural gas-fired combined-cycle systems for new installations (Tavoulareas 1991, 1995; Schimmoller 2000). More than 600 AFBC boilers, with the total capacity of 30 GW, operate in North America; similar capability is installed in Europe and China has more than 2,000 small units. AFBC is 5–15% less expensive than the same size plant burning pulverized coal and using wet or dry FGD. Pressurized fluidized bed combustion (PFBC) produces gas that is capable of driving a gas turbine and operating it in a combined cycle. DOE's Clean Coal Technology Program led to the market entry of the first generation of PFBC, with about 1 GW of such capacity installed worldwide by the year 2000 (DOE 2001b).

Second generation PFBC will be integrated with a coal gasifier to produce a fuel gas that will be burned in a topping combustor, adding to flue gas energy entering the gas turbine. PFBC boilers will occupy a small fraction of space taken up by conventional units burning pulverized coal and the process—predicated on the introduction of new burners, corrosion-resistant materials, sorbents, and gas turbines, expected to be commercially available before 2010—will combine near-zero NO_x, SO_2, and PM emissions with overall 52% conversion efficiency (DOE 2001b). This performance would represent a 30–50% efficiency gain compared to conversion achieved in standard coal-fired power plants, an improvement that would go a long way toward reducing the CO_2 emission disadvantage of coal combustion.

Higher combustion efficiencies will be also achieved in supercritical power plants, operating with steam pressures of up to 35 MPa and temperatures of up to 720 °C (the adjective supercritical simply denotes operating conditions above 22.1 MPa, the pressure at which there is no distinction between the liquid and gaseous phases of water as they form a homogeneous fluid). Supercritical (up to 25 MPa) power plants were in favor in the United States during the 1970s but their construction declined rapidly in the early 1980s, largely due to early problems with availability. Europeans and the Japanese had a better experience and by the late 1990s there were 462 units with installed capacity of roughly 270 GW (Gorokhov et al. 1999). Current state-of-the-art allows pressures up to 30 MPa (by using steels with 12% Cr content) and efficiencies in excess of 45%; 31.5 MPa are possible by using expensive Austenite, and 35 MPa would be possible that Ni-based alloys, resulting in efficiencies just above 50% (Paul 2001). The U.S. DOE is now developing low-emission supercritical units that will combine conversion rates of up to 50% (by 2010) with advanced NO_x and SO_2 controls (Gorokhov et al. 1999).

Another old high-efficiency concept that has been embraced in Europe, Russia, and Japan and that has been neglected in North America is the use of a single primary heat source to produce both electricity and heat (Clark 1986; Elliott and Spurr 1999). During the early decades of the twentieth century this practice, generally known as cogeneration, was common even in the United States but after 1950 most industrial enterprises and large institutions (hospitals, universities, and business parks) began buying outside electricity while they continued to produce the needed thermal energy (steam, hot water) on-site. By the late 1990s cogeneration added up to only about 6% of the total U.S. electric generating capacity but that may rise to nearly 10% by the year 2010 (Smith 2001). In Europe, Denmark receives 40% and the Netherlands and Finland about 30% of electricity from combined heat and power, mostly from district energy systems. Efficiency gains are obvious: compared to standard 33–35% efficiencies of electricity generation, combined electricity generation and heat production may convert up to 70% of the fuel into useful energy.

Finally, higher efficiencies (above 45%) and lower NO_x emissions (halved in comparison to standard pulverized coal combustion) could be achieved by such advanced systems as integrated gasification/combined cycle (IGCC) that combines gas-turbine and steam-turbine cycles (Williams 2000; Falsetti and Preston 2001). As of 2001 fewer than 30 IGCC plants were operating in the United States, Europe, and Asia—but in the year 2000 their capital cost was about $1,300/kW compared to less than $500/kW with combined cycle gas turbines whose efficiencies are even higher (just above 50%) and that, a most important difference, emit only half as much CO_2 per generated kWh. Consequently, even an economically competitive IGCC, expected by the industry sometime before 2010 (Schimmoler 1999), would still have a critical greenhouse gas disadvantage.

As far as the national and continental perspectives are concerned, nobody sees any coal revival in the EU, only further decline. And a continuing decline of coal's importance is also expected in the countries of the former Soviet Union and Communist Europe. Some industry experts have concluded that the prospects for the U.S. coal-fired generation are equally poor. As the numbers of existing coal-fired plants will dwindle and as there are no signs of either weakening opposition against siting of new stations or of any enthusiasm to finance such new units, some forecasts see the country's coal-based electricity generating capacity slipping from 50% in the year 2000 to a mere 15% by 2020 (Makansi 2000). Such views are unduly pessimistic.

As already noted, there is an obvious need to replace a great deal of ageing electricity-generating capacity during the coming two decades. At the same time there is the security and abundance of coal's domestic supply and its low prices, which are not expected either to increase appreciably or to fluctuate wildly (NRC 1995). When these considerations are combined with new combustion techniques whose application will produce deep cuts or near-elimination of SO_x and NO_x emissions and improve conversion efficiencies that will reduce CO_2 generation, the fuel appears in a much more favorable light. That is why the EIA (2001e) projects a nearly 25% increase in the overall U.S. coal consumption during the first two decades of the twenty-first century, with coal-based generation still providing 44% of the total demand by 2020.

Prospects for China's coal should not be judged by the extraordinarily rapid decline in demand and extraction during the late 1990s. In 1995 coal provided just over 75% of the country's primary commercial energy, a share higher than at any time since 1972 (Fridley 2001). Yet four years later its share fell to just over 68%, a relative drop of 10% and the absolute decline of slightly more than 250 Mtce, or more than the total coal extraction of any other coal-producing nation with the exception of the United States. Even a brief continuation of this trend would necessitate massively increased oil imports as there is no way to bring in the requisite amounts of natural gas in such a short period of time. At the same time, it is clear that coal's share will be slowly declining.

The first reason for coal's gradual retreat is the ongoing restructuring of China's economy that helped to reduce the country's energy intensity in general and its need for urban and industrial coal combustion in particular (for details see chapter 2). The second one is the belated but now rather determined effort to reduce air pollution in China's large cities, a policy exemplified by the goal of coal-free Beijing by the time of the 2008 Olympics. In view of these trends many recent midterm Chinese and foreign forecasts of coal consumption, including EIA's prediction of a 2.4-fold increase by the year 2020, are unrealistically high (see also chapter 3 and fig. 3.12).

On the other hand, there is China's far-from-adequate supply of electricity and the very high reliance on coal-fired generation (World Bank 1995; IEA 1999; Fridley 2001). During the late 1990s China's per capita consumption of electricity was only about one-eighth of Taiwan's and one-tenth of Japan's mean and three-quarters of it came from coal combustion. Again, recent forecasts of some 20 GW of new generating capacity to be added every year during the next generation most likely exaggerate the actual additions. But expansion of coal-fired capacity running at just half that rate would still require at least 250 Mtce by the year 2015. A new, richer China, that as a matter of necessity will not be able to eliminate its dependence on coal in electricity generation, would do itself a great service by investing in any and all high-efficiency, low-pollution combustion techniques, be it AFBC, PFBC, IGCC, or cogeneration.

Perhaps only this much is certain. During the last quarter of the twentieth century coal use did not conform to the predictions of continuing rapid decline and the fuel still supplies nearly one-quarter of the world's primary commercial energy. Should it resume its pre-1974 rate of decline the fuel would provide only about 10% of the world's energy by 2025 and its share would become insignificant before the year 2050. But such a decline would have to be compensated by an increase of natural gas consumption whose rates would have to be substantially higher than the world has seen during the past generation when the fuel's relative gain was actually slower than during the preceding 20 years.

I view the possibility of a future combination of a *dramatically* re-accelerated decline and rise (of coal and gas, respectively) as highly unlikely and I see coal losing its current share of global energy use in only a gradual, and also temporarily bumpy, fashion. And the further ahead we look, the more important it is to consider not only the global role of natural gas but also the improvement and commercial penetration of nonfossil energies as well as the realistic possibilities for lowering overall energy use through a combination of technical advances, more realistic pricing and performance targets, and gradual changes in behavior.

5

Nonfossil Energies

There is at least one obvious certainty when examining the global energy supply on a civilizational timescale: our predominantly fossil-fueled society is bound to be a relatively ephemeral affair. Two qualifications are required to make that conclusion clear. First, there is no binding quantification of the key adjective. If predominantly means more than 75% of the global TPES, we are still well above that mark: in the year 2000 the share was about 82%, with some 320 EJ of the world's TPES coming from fossil fuels, 35 EJ from hydro and nuclear electricity and at least 35 EJ from biomass (see chapter 1). Second, I assume that even if some deposits of hydrocarbons were to be repleting (a consequence of their possible abiogenesis explained in chapter 4), the rates of their renewal would be too slow to match the pace of the ongoing extraction on a timescale of 10–100 years.

Whatever the actual longevity of our predominantly fossil-fueled societies will be, it will not be able to match the duration of human civilization. We have been around as settled farmers for roughly 10,000 years and the first city-centered societies with complex high cultures go back about 5,000 years. Only during the past three generations (i.e., 60–75 years), or for no more than 1.5% of civilization's time span, has humanity satisfied its varied energy needs by relying predominantly on fossil fuels. Even the combination of very large ultimately recoverable conventional fossil fuel resources, very high efficiencies of their conversion and very low energy demand of a stable global population would make it impossible to have a predominantly fossil-fueled global society lasting for more than a few hundred years. Large-scale recovery of nonconventional fuels could extend this span but only if the environmental impacts of this reliance would remain tolerable.

What comes after fossil fuels? This question should not be asked with either regret or unease. No doubt, both the much demonized coal and cleaner and more flexible hydrocarbons have served us admirably. Their conversions have furnished at least one-quarter of humanity (roughly two-thirds of this total in affluent countries, the rest being the high-income families in low-income countries) with a standard of living that was not available even to the richest few just a few hundred years ago. More important in terms of sheer numbers, for another 3 billion people these conversions have made the difference between miserable, precarious, and illiterate subsistence on one hand and a modicum of comfort and greatly improved life expectancy and basic educational opportunities on the other. Significantly, this still leaves out at least 1.5 billion people whose families have benefited from modern energies only indirectly and marginally and hence far too inadequately.

The inevitable and gradual transition from fossil fuels to nonfossil energies should not imperil the maintenance of what has been achieved and it should not make it more difficult to extend the benefits of higher energy use to those who will need it the most. The magnitude of renewable energies is not an issue, as there is an enormous flux of solar radiation reaching the Earth that can be tapped directly or after its natural conversion to flowing water, wind, waves, and biomass. As far as any earthly civilization is concerned, this energy source is renewable for the next 500 million years; that is, until the increasing radiation from the expanding Sun will begin heating the oceans. Huge volumes of the evaporated water will then migrate to the stratosphere, leaving only fierce, hot winds sweeping through the Earth's atmosphere (Smil 2002). Even the boldest sci-fi writers do not usually set their stories that far in the future.

What matters is the availability of renewable energy flows in time and space and their power density. Of all the renewable energy flows only solar radiation has a fairly high power density. Solar radiation reaching the top of the Earth's atmosphere amounts to 1,347 W/m^2 of extraterrestrial space: this flux is commonly called the solar constant but in reality it is subject to minuscule fluctuations. After prorating the flux over the planet's basically spherical surface and subtracting the energy absorbed by the atmosphere and reflected to space, the global mean of the total insolation—the solar radiation absorbed by surfaces—averages 168 W/m^2 (fig. 5.1). Maxima are in excess of 250 W/m^2 in the cloudless high-pressure belts of subtropical deserts; minima are obviously 0 during polar winters. Existing techniques allow us to convert solar radiation to electricity with power densities of 20–60 W/m^2. No other renewable flux has such a high average performance.

All indirectly harnessed solar flows are very diffuse to begin with. Their typical power densities are mostly below 10 W/m^2, and many average less than 1 W/m^2. And because most of the existing techniques of converting indirect solar energies are not very efficient (hydro generation is the only exception) they can produce the most desirable forms of useful energy (electricity and liquid fuels) with power densities

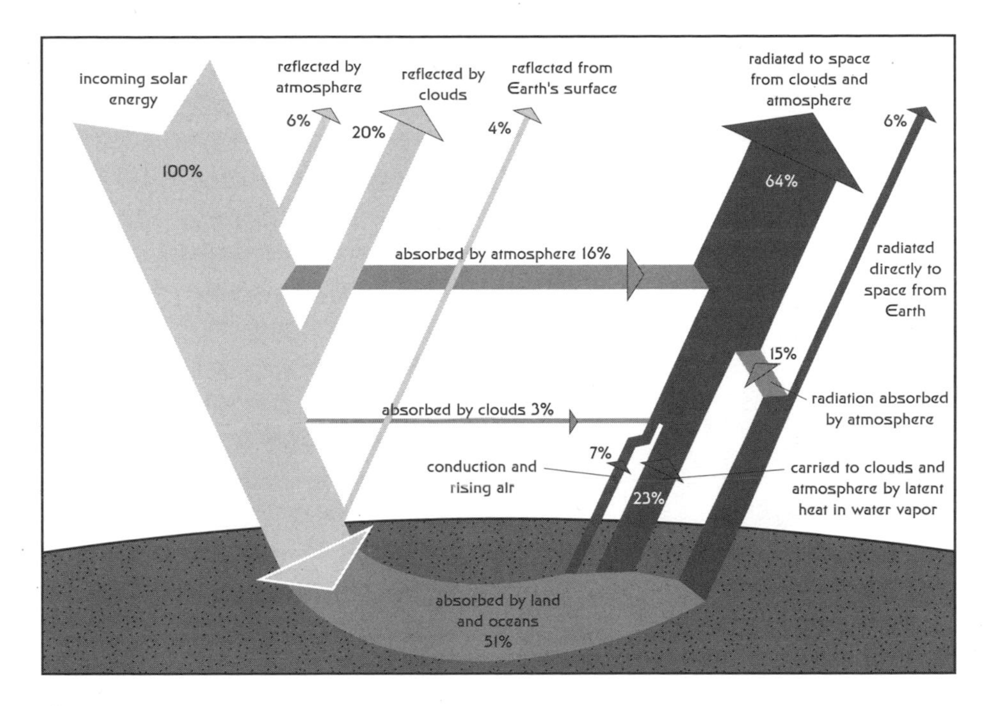

Figure 5.1
Partitioning of the incoming solar radiation. Based on Smil (1999a).

that are only a small fraction (mostly orders of magnitudes smaller) of power densities achieved by using fossil fuels. Typical rates illustrate these disparities. Power fluxes are 10–50 W/m^2 for capturing tides and the kinetic energy of rivers in their upper-course, between 5–20 W/m^2 for wind, just above 1 W/m^2 for most of the lower-course hydro generation requiring large reservoirs, and below 1 W/m^2 for biomass energies (fig. 5.2).

This is in great contrast to the extraction of fossil fuels and thermal generation of electricity. These activities that define the modern high-energy civilization produce commercial energies with power densities orders of magnitude higher, ranging mostly between 1–10 kW/m^2 (fig. 5.2). Power densities of final energy uses in modern societies range mostly from between 20–100 W/m^2 for houses, low energy intensity manufacturing and offices, institutional buildings, and urban areas. Supermarkets and office buildings use 200–400 W/m^2, energy-intensive industrial enterprises (such as steel mills and refineries) 300–900 W/m^2, and high-rise buildings up to 3 kW/m^2 (fig. 5.3). In order to supply these power densities, fossil-fueled societies are diffusing concentrated energy flows as they produce fuels and thermal electricity with power

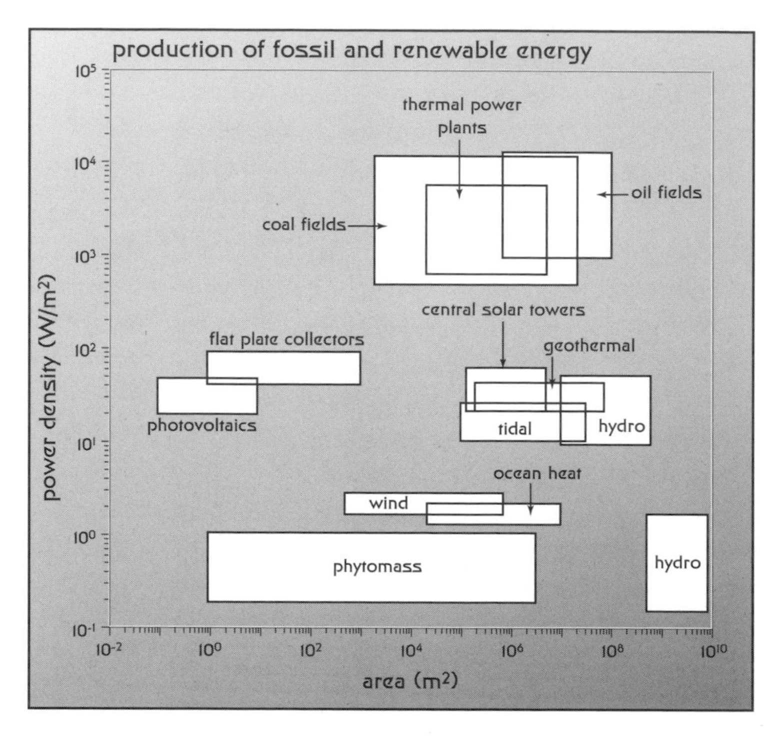

Figure 5.2
Power densities of fossil and renewable energy conversions show differences of up to four orders of magnitude between the two modes of primary energy supply. Based on data in Smil (1991).

densities one to three orders of magnitude higher than the common final use densities in buildings, factories and cities.

As a result, space taken up by extraction and conversion of fuels is relatively small in comparison with transportation and transmission rights-of-way required to deliver fuels and electricity to consumers. For example, in the United States the total area occupied by extraction of fossil fuels is less than 1,000 km^2 (a mere 0.01% of the country's area) while the land whose use is preempted by pipeline and transmission rights-of-way adds up to nearly 30,000 km^2 (Smil 1991). A solar society that would inherit the existing residential and industrial infrastructures would have to do the very opposite as it concentrates diffuse energy flows. Only for some final uses, most notably for heating and lighting energy-efficient houses, it could harness its renewable energies with the same power densities with which they would be used: distributed generation of electricity through photovoltaic conversion is the best ex-

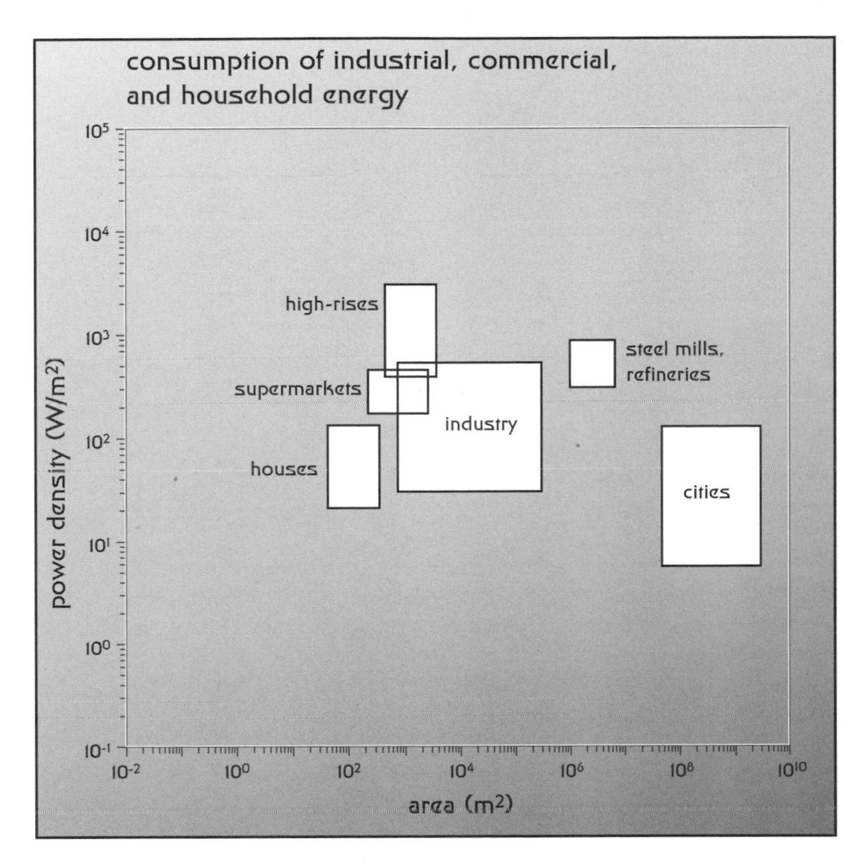

Figure 5.3
Power densities of industrial, commercial, and household energy consumption prorate mostly to 10^1–10^2 W/m^2. Based on data in Smil (1991).

ample of this possibility. But in order to energize its megacities and industrial areas any solar-based society would have to concentrate diffuse flows in order to bridge power density gaps of two to three orders of magnitude (figs. 5.2 and 5.3).

Mismatch between the low power densities of renewable energy flows and relatively high power densities of modern final energy uses means that any large-scale diffusion of solar energy conversions will require a profound spatial restructuring, engendering major environmental and socioeconomic impacts. Most notably, there would be vastly increased fixed land requirements for primary conversions and some of these new arrangements would also necessitate more extensive transmission rights-of-way. Loss of location flexibility for electricity generating plants converting direct or indirect solar flows and inevitable land-use conflicts with food production would be additional drawbacks. I will explain specific attributes of this power density mismatch when I will discuss both the advantages and the drawbacks of the commercial use of individual renewable energies.

Stochasticity of all solar energy flows is the second major challenge for any conversion system aiming at a steady, and highly reliable, supply of energy required by modern industrial, commercial, and residential infrastructures. Only solar energy transformed into chemical energy of phytomass through heavily subsidized photosynthesis (in order to eliminate any water and nutrient shortages and to protect the harvest from pest attack) can be harvested and used quite predictably. Without massive storages none of the renewable kinetic flows could provide large base loads required by modern electricity-intensive societies. Yet voluminous water reservoirs have been the only practical means of storing large quanta of almost instantly deployable energy: in spite of more than a century of diligent efforts to develop other effective storage all other options remain either inefficient or inadequate (fig. 5.4). I will return to this fundamental problem in chapter 6.

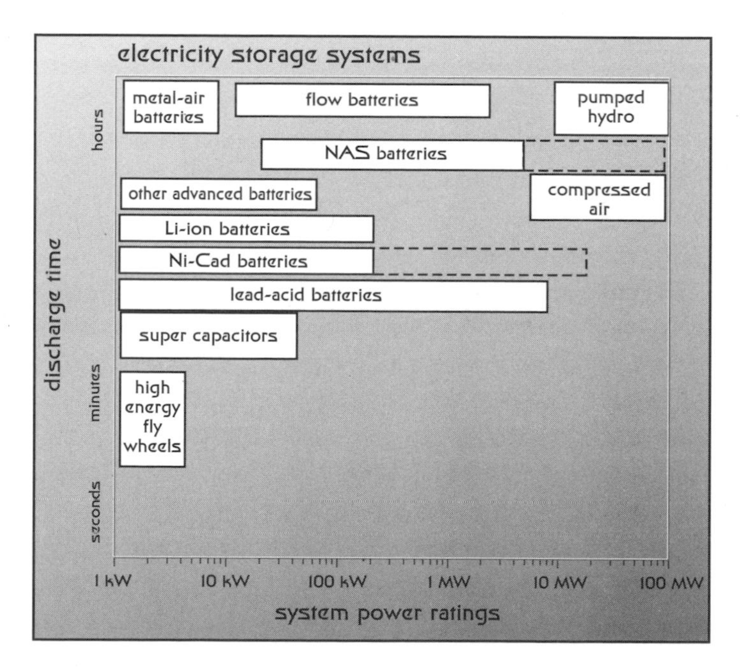

Figure 5.4
Power ratings and discharge times of principal large-scale electricity storages. Based on a graph in ESA (2001).

Finally, there are two renewable energy flows that do not arise from the Sun's thermonuclear reactions. First is the terrestrial heat flow resulting from the basal cooling and from the decay of radioactive elements (U^{235}, U^{238}, Th^{232}, and K^{40}) in the Earth's crust; the second is the tidal energy produced by gravitational pulls of the Moon and the Sun. Geothermal flow is minuscule not only in comparison to the direct solar flux but even in contrast to the least powerful indirect flows. Its power density averages merely 85 mW/m^2, but the flux is available everywhere all the time. As a result of plate tectonics and hot spot phenomena, many places around the Earth have a much more intensive heat flow through the crust, often delivered by hot water or steam and suitable for commercial extraction of geothermal energy. Tidal movements are predictably periodic but they reach high power densities only in a few locations around the world.

I will start the review of nonfossil energies by taking a closer look at the prospects for hydroelectricity, the most important commercial source of nonfossil energy. Then I will turn to the most important nonfossil source of a largely noncommercial energy, to biomass fuels ranging from wood and charcoal to crop residues and dried dung. These fuels still provide heat for hundreds of millions of households as well as for many small manufacturers in low-income countries and charcoal energizes even some larger industrial enterprises. Can these locally and regionally important but generally very inefficient conversions be modernized, or do we need fundamentally different ways of biomass energy uses in low-income countries? And should affluent countries try to elevate their use of biomass from being a marginal source of primary energy to becoming an important part of their diversified supply?

Appraisals of two rapidly rising modes of renewable electricity generation, harnessing of wind by modern turbines and conversions of direct solar radiation by thermal systems and by photovoltaics will come next. All remaining proposed and actually commercialized modes of renewable electricity generation will be swept into the penultimate section. I will close the coverage of nonfossil energies by looking what may be ahead for the option that was entirely unknown at the beginning of the twentieth century, that was practically demonstrated only during its fourth decade and commercialized less than half a century ago: at nuclear energy. Basic facts of its rapid rise and demise and failed forecasts of its dominance were already noted (in, respectively, chapters 1 and 3). No forecasts will be offered here, just reviews of factors that either boost or diminish the likelihood of a nuclear renaissance and its basic modalities, most likely a new generation of inherently safe, and also proliferation-proof, systems.

Hydroenergy's Potential and Limits

Running water turns turbines that supply almost 20% of the world's electricity. Relative importance of hydro generation is much greater in dozens of tropical countries where it is the dominant means of electricity production. When hydroelectricity is converted by using its energy content (as is commonly done when preparing the TPES statistics: see chapter 3) its global generation is now equivalent to nearly 10 EJ. But if the same amount of electricity were to be generated from fossil fuels, the world would have to find, extract, move, and burn nearly 30 EJ, or additional 1.3 Gt of steam coal. This would release at least 1 Gt of carbon and more than 25 Mt of sulfur into the atmosphere. These additional emissions would be equal to about 15% and 35% of today's global anthropogenic fluxes of these gases.

Hydro generation also has lower operating costs and longer expected plant life than any other mode of electricity production, a combination translating to often very low prices for the consumer. From an electric system's point of view, hydro generation in general, and spinning reserve (zero load synchronized to the system) in particular, is an excellent way to cover peak loads created by sudden increases in demand. And many reservoirs built primarily for hydro generation have multiple uses serving as sources of irrigation and drinking water, as protection against flooding and as resources for aquaculture and recreation.

Moreover, even conservative assessments of global hydro energy potential show that this renewable source of clean electricity remains largely untapped and that even when the worldwide capacity installed in large turbines would be doubled most of the energy in flowing water would still remain unused. The accounting goes as follows. Published totals of global surface runoff range from 33,500 to 47,000 km^3 a year, with the most likely average (excluding the ice flow of Antarctica but including the runoff to interior regions of continents) of 44,500 km^3 (Shiklomanov 1999). Assuming the mean continental elevation of 840 m, the potential energy of this flow is about 367 EJ, almost exactly the world's commercial TPES in the year 2000. If this natural flow were to be used with 100% efficiency, the gross theoretical capability of the world's rivers would be about 11.6 TW.

Many realities combine to lower this total. Competing uses of river flows, unsuitability of many environments to site any hydro stations, often drastic seasonal fluctuations of flow, and impossibility to convert water's kinetic energy with perfect efficiency at full capacity mean that although the exploitable capability (share of the theoretical potential that can be tapped with existing techniques) can be well in excess of 50%, for some streams it is commonly just around 30% for projects with installed capacities in excess of 1 MW. Worldwide totals of technically feasible capacities must be thus constructed by adding specific national assessments and the ICOLD (1998) puts it at about 52 EJ (over 14 PWh) of electricity, or roughly 14% of the theoretical total. As expected, continental distribution of this total is very uneven, with Asia claiming 47% and Latin America nearly 20%. China has the largest national total (about 15%), countries of the former Soviet Union come second with 12%. Not everything that is technically possible is economically feasible, and projects in the latter category add up to about 30 EJ worldwide (just over 8 PWh), or roughly three times the currently exploited total.

There are also large differences in shares of potential capacity that have been already exploited. Europe had the highest share during the late 1990s (more than 45%), followed by North America (almost 45%). Latin America tapped about 20% of its potential but Asia only 11% and Africa a mere 3.5% (ICOLD 1998). The three continents containing about 80% of humanity whose per capita consumption of electricity is only a small fraction of the average use prevailing in the affluent countries are thus in a welcome position of having plentiful hydro-energy resources whose development could help to modernize their economies and raise the average quality of life while preventing additional releases of carbon dioxide and sulfur and nitrogen oxides.

The encouraging and comforting message of the previous sentence would have been seen as an unremarkable truism during the 1960s and 1970s, when the world experienced its greatest boom in construction of large dams, with about 5,000 new structures built per decade (fig. 5.5). But it does not sound so at the very beginning of the twenty-first century. Now it would be seen by many as, at best, a misguided opinion or, at worst, as indefensible propaganda put out by industries and companies promoting hydro megaprojects now widely considered as environmentally harmful, socially disruptive, and economically dubious. Increasingly vocal opponents of dam building have argued for some time that in the United States, and in other countries with a large number of reservoirs (I hasten to add: not all of them built and used to generate electricity), the activity has gone far beyond any rational limit defensible not just from a broader ecosystemic viewpoint but also from a properly defined economic perspective.

Carrying this argument to its extreme, Devine (1995, p. 74) asked "Is there such a thing as a good dam?"—and he found it very revealing that even the executive director of the U.S. Committee on Large Dams answered it by naming just two

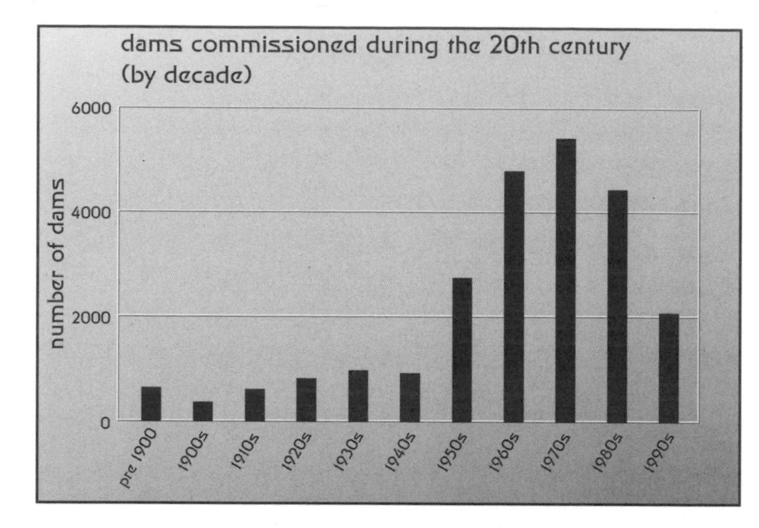

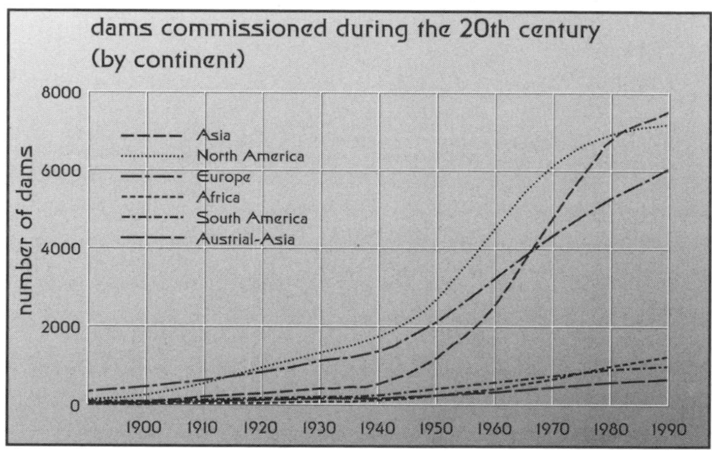

Figure 5.5
Worldwide commissioning of large dams: decadal totals and cumulation by continents, 1900–2000. Based on graphs in WCD (2000).

structures (Hoover on the Colorado and Grand Coulee on the Columbia), and then admitting that there may be others among 5,500 of the country's large dams that may deserve the label "but they just didn't come to mind." An even more remarkable fact is that a paper in the leading journal of hydro-energy engineering noted recently that "hydro power is still battling to be recognised as a renewable energy source in the United States" (Moxon 2000, p. 44). What seems logically obvious is not so in the world where dams are seen as incarnations of technical arrogance, economic mismanagement, and environmental destruction.

Recent studies showing that water reservoirs are significant sources of greenhouse gas emissions (mostly due to CO_2 and CH_4 from decaying vegetation) have partially negated even the last argument that seemed to weigh in favor of hydroelectricity as a substitute for fossil fuels. At the same time several widely publicized cases of mass population displacement caused by hydro projects focused attention on the undeniable plight of large numbers of people uprooted by dams. Huge cost overruns and questionable economic performances of many projects are two other points of contention. All of this questioning and negative publicity led to the establishment of the World Commission on Dams whose comprehensive report reviewed and confirmed many undesirable impacts of large dams and reservoirs and counselled that, in contrast to the practice that has prevailed for more than a century, any future dam projects should consider many social and environmental effects to be as important as economic benefits (WCD 2000).

These new realities make it less likely that the future exploitation of the huge energy potential that remains to be tapped in Africa, Asia, and Latin America will proceed at a pace matching, or even surpassing, the additions of new hydro-generating capacity during the second half of the twentieth century. During those decades the total number of large dams (including those built solely for water supply and flood control) rose more than ninefold, from less than 5,000 to more than 45,000, and when the installed capacity in hydro stations increased from about 80 GW to almost 700 GW (UNO 1956; WCD 2000; UNO 2001). A closer look at the reasons why the worldwide hydro development during the first half of the twenty-first century may not match the pace of the past 50 years should help to get a more realistic perspective on the industry's prospects.

Displacement of large numbers of usually poor people has been the most contentious matter. Until the 1950s most hydro-energy megaprojects did not require any mass relocation of people but as larger dams began to be built in densely populated countries, particularly in Asia, resettlement numbers began surpassing 100,000 individuals for a single reservoir and totals added up soon to millions of affected people. Construction of large dams displaced at least 40 million people during the twentieth century (some estimates go as high as 80 million), and during the early 1990s, when the construction began on 300 new large dams every year, the annual total reached 4 million people (WCD 2000). Decades of experience have demonstrated that the planners frequently underestimate the total number of people to be moved, as well as the complexity of the entire relocation process, and the variety of negative impacts on affected population; insufficient compensation to displaced families and their failure to restore income-earning capacity years, even decades, after the forced move, are other commonly encountered problems (Gutman 1994).

As expected, China and India, the two countries that have built nearly 60% of the world's large dams, had to relocate most people: more than 10 million in China and at least 16 million in India. By far the most publicized recent cases of mass population displacement have been those arising from the construction of a series of large dams on the Narmada River in central India (Friends of the Narmada River 2001) and from the building of China's Sanxia, the world's largest hydrostation on the Yangzi (Dai 1994). Sardar Sarovar dam on the Narmada would displace more than 300,000 people and the ensuing controversy was the main reason why the World Bank withdrew from the project's financing. Sanxia's 600-km long and up to 175-m deep reservoir will force relocation of at least 1.3 million people in an area where there is obvious lack of suitable land for rural resettlement, and in a country where most of the money appropriated for the resettlement may be defrauded by what is perhaps the world's most corrupt bureaucracy.

As is the case with the harnessing of all indirect solar flows, hydroelectricity generation operates with generally low power densities, and reservoirs have inundated large areas of natural ecosystems, including many unique, and highly biodiverse, forests and wetlands. Reservoirs behind the the world's large (higher than 30 m) dams now cover almost 600,000 km^2 (an area nearly twice as large as Italy) and those used solely, or largely for electricity generation cover about 175,000 km^2. This prorates to about 4 W/m^2 in terms of installed capacity (and to about 1.7 W/m^2 in terms of actual generation) but power densities of individual projects span three orders of magnitude. Dams on the lower courses of large rivers impound huge volumes of water in relatively shallow reservoirs. The combined area of the world's seven largest reservoirs is as large as the Netherlands and the top two—Ghana's Akosombo on the Volta (8,730 km^2) and Russia's Kuybyshev on the Volga (6,500

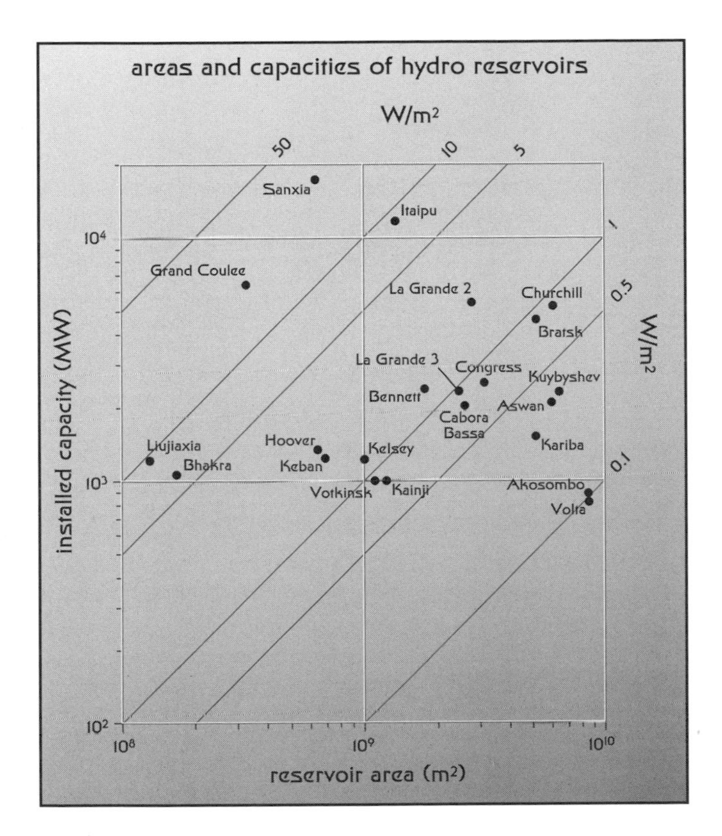

Figure 5.6
Power densities of the world's largest hydroelectricity-generating projects range over two orders of magnitude. Based on data in Smil (1991), ICOLD (1998), and WCD (2000).

km^2)—approach the size of such small countries as Lebanon or Cyprus. Power densities of these projects are below 1 W/m^2 (fig. 5.6).

In contrast, dams on the middle and upper courses of rivers with steeper gradients have much smaller, and deeper, reservoirs and their power densities are on the order of 10 W/m^2. Itaipu, so far the world's largest hydro project (12.6 GW), has a power density of 9.3 W/m^2, Grand Coulee (6.48 GW) rates nearly 20 W/m^2, China's Sanxia (the world's largest at 17.68 GW when finished in 2008) will have about 28 W/m^2, and some Alpine stations surpass 100 W/m^2 (fig. 5.6). The world's record holder is the planned Nepali Arun project whose 42-ha reservoir and 210-MW installed capacity translates to about 500 W/m^2. Because load factors of most hydro stations are typically below 50%, and because many projects are built to supply electricity during the hours of peak demand and hence have even shorter operating hours, effective power densities would be actually only 25–50% of the just cited theoretical rates.

Construction of dams and creation of water reservoirs has also had a profound cumulative impact on the world's rivers. Design specifications for large reservoirs add up to about 6,000 km^3 of water storage (WCD 2000) and other estimates are as high as 10,000 km^3 of water in all man-made impoundments, or more than five times the water volume in all streams (Rosenberg et al. 2000). This huge storage has dramatically increased the average age of river runoff and lowered the temperature of downstream flows. While the mean residence time for continental runoff in free-flowing river channels is 16–26 days, Vörösmarty and Sahagian (2000) estimated that the discharge-weighted global mean is nearly 60 days for 236 drainage basins with large impoundments. Moreover, mouths of several of the world's largest rivers show reservoir-induced ageing of runoff exceeding 6 months (Huanghe, Euphrates and Tigris, Zambezi) or even 1 year (Colorado, Rio Grande del Norte, Nile, Volta; fig. 5.7).

Fluctuating water levels of many tropical reservoirs create excellent breeding sites for malaria mosquitoes and once the schistosomiasis-carrying snails invade a new reservoir (or were not eradicated from the area before its inundation), it is very difficult to get rid of them. Most dams also present insurmountable obstacles to the movement of aquatic species, particularly of anadromous fish. New turbine designs and control systems aimed at raising dissolved oxygen levels in turbine discharges and improving the survival of fish during turbine passage that are now available (March and Fisher 1999) have come too late for many species in many rivers. Multiple dams have also caused river channel fragmentation that now affects more than 75% of the world's largest streams.

Finally, recent studies of a previously ignored environmental impact have helped to weaken the case for hydro generation as a substitute for fossil fuels in a world endangered by global warming: large reservoirs are also significant sources of greenhouse gases (Devine 1995; WCD 2000). Not surprisingly, reservoirs in boreal and temperate climates have relatively low specific emissions similar to those from natural lakes and equal to only a small fraction of releases from coal-fired plants. Tropical reservoirs are generally much larger sources but measurements show large spatial and annual variations: some reservoirs may release in some years more greenhouse gases per unit of electricity than does fossil-fueled generation (fig. 5.8). These fluctu-

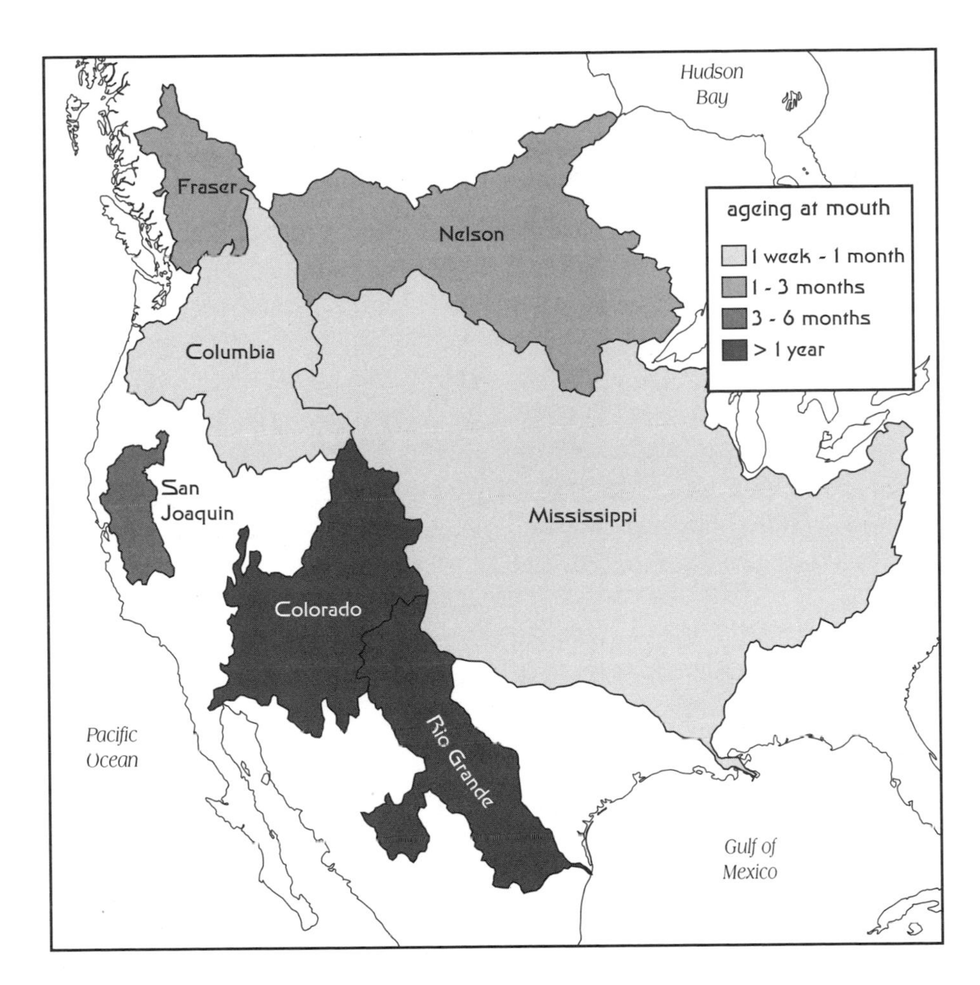

Figure 5.7
Aging of river runoff caused by dams in several major watersheds of western North America. Based on Vörösmarty and Sahagian (2000).

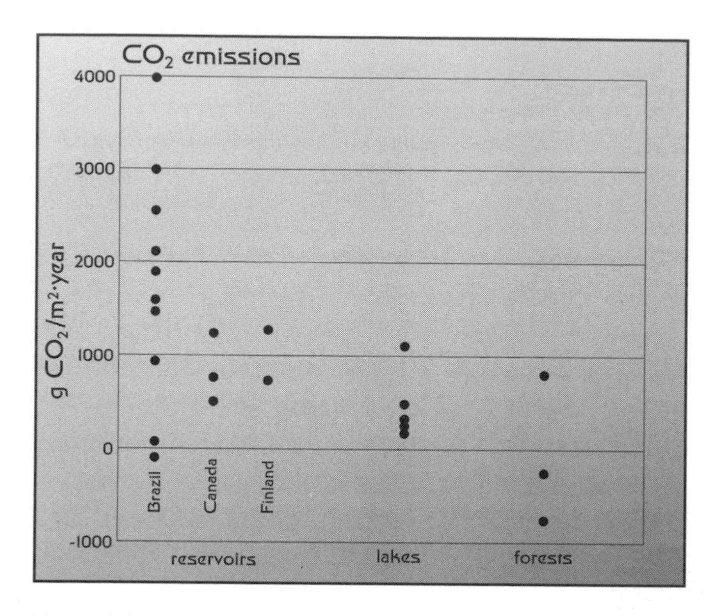

Figure 5.8
Generation of greenhouse gases by water reservoirs, lakes, and forests. Based on figures in WCD (2000).

ations make it difficult to assess actual contribution of reservoir GHG to the global warming: they may account for as little as 1% and as much as 28% of all anthropogenic emissions (WCD 2000).

And new studies keep uncovering the extent of many subtle and lasting environmental impacts. A veritable litany of other environmental impacts caused by large dams now includes massive reduction of aquatic biodiversity both upstream and downstream—for example, we now know that plant communities along the banks of dammed rivers harbor significantly fewer species than do communities bordering free-flowing streams (Nilsson, Jansson, and Zinko 1997)—increased evaporative losses from large reservoirs in arid climates, invasion of tropical reservoirs by aquatic weeds, reduced dissolved oxygen, and H_2S toxicity in reservoir waters.

But it is an old problem associated with dam construction that poses perhaps the greatest threat to the long-term viability of reservoirs: their excessive silting. Most reservoirs are built to last at least 100 years but natural and anthropogenic factors may accelerate their siltation and reduce their effective life span. The former factors include high erosion rates in the world's tallest mountain ranges and in the most erodible loess regions, and droughts followed by floods and landslides; the latter factors include deforestation and overgrazing and improper crop cultivation. High sediment loads also directly affect the operation of plants by throttling feeder tunnels, eroding guide vanes and runner blades, and increasing restoration costs and unscheduled downtime. Excessive sedimentation has been a growing problem in many U.S. reservoirs (Dixon 2000) but monsoonal Asia is the region with generally the highest erosion rates.

China's Loess Plateau, the word's most erodible area, and India's Himalayan rivers make the problem particularly acute in the two countries with the largest numbers of dams on the continent. Some large Indian reservoirs have lost 50% of their storage capacity in less than 30 years and by 2020 more than 20% of the country's reservoirs will have experienced that degree of 50% of loss due to sedimentation (Naidu 2000). In China, excessive silting of the Sanmenxia reservoir built on the Huanghe in the 1950s forced the opening of water outlets at the bottom of the dam and the downgrading of the project from 1.1 GW to 250 MW (Smil 1984). The best nationwide estimate is that the excessive accumulation of silt has eliminated an equivalent of about one-fourth of the reservoir capacity built between 1960–1989, causing a total direct cost of Rmb $20 billion (1989) (Smil 1996).

Deposition in reservoirs has effects far downstream as it cuts the global sediment flow in rivers by more than 25% and reduces the amount of silt, organic matter, and nutrients available for alluvial plains and coastal wetlands downstream. As a result, some coastlines are now eroding at rapid rates. Because of the Aswan High Dam (2.1 GW) the Nile Delta is receding by up to 5–8 m a year, and because of the Akosombo dam on the Volta river in Ghana (833 MW) coastlines of Togo and Benin lose 10–15 m a year (WCD 2000). Retention of silicates is also causing a shift from Si-shelled diatoms to non-Si phytoplankton in some estuaries and coastal regions (Ittekkot et al. 2000). Finally, ageing dams are showing more signs of inevitable deterioration and raise questions about the cost and methods of eventual decommissioning. Long-term degradation of concrete is caused by chemical reactions with surrounding water (waters with high acidity and sulfates dissolved in groundwater are particularly aggressive), loss of strength under loading, and reduced resistance to cycles of freezing and thawing (Freer 2000).

The ultimate life span of large dams remains unknown. Many have already served well past their designed economic life of 50 years but silting and structural degradation will shorten the useful life of many others. Problems with decommissioning of nuclear power plants have been widely studied and acknowledged (see the last section of this chapter) but the decommissioning of large dams remains a matter of fundamental uncertainties and vague speculations (Leyland 1990; Farber and Weeks 2001). Many dams do not have low level outlets and cannot be easily drained; downstream plunge pools excavated by open spillways could soon endanger the dam structure; controlled explosion to destroy dams might cause catastrophic floods. We are not designing large dams to last indefinitely—but neither are we building them in ways that could make it easier to decommission them in the future.

All of these now widely discussed and acknowledged realities have turned a significant share of the Western public sentiment against new hydro projects in general and against large dams in particular. Sweden has banned further hydro stations on most of its rivers and Norway (with excess hydro capacity) has postponed all construction plans. On November 25, 1997 the U.S. Federal Energy Regulatory Commission made its first-ever decision to order removal of the Edwards dam, which had blocked the Kennebec River in Maine since 1834 (Isaacson 1998). This decision spurred hopes for more such orders, particularly in the country's Pacific states where dams have drastically reduced or entirely wiped out formerly grand salmon runs. And, indeed, since 1998 the decommissioning rate for large U.S. dams has overtaken the construction rate (WCD 2000), and the modalities and consequences of dam removal have become a new field of ecological studies (Hart et al. 2002).

Higher sensitivity to resettlement problems and environmental consequences will make it harder to obtain the World Bank and the Asian Development Bank loans for major projects in countries with large untapped hydro-generating capacity. Consequently, such projects as Nepal's Pancheshwar on the Mahakali river (6 GW) and an even bigger plant on the Karnali (10.8 GW) may have to be financed privately. But even then these projects may be subjected to concerted opposition by national and international organizations such as the International Committee on Dams, Rivers, and People and the International Rivers Network (McCully 2002). But a wider perspective is necessary in order not to look ahead through the prism of recent attacks on dams.

This flood of negative comments on hydro generation has been an understandable reaction to years of largely uncritical appraisals of the industry: balanced assessments cannot be so negative. As noted in the first paragraph of this section, the amount of major air pollutants (above all SO_2 and NO_x) and CO_2 whose emissions were prevented, and whose often substantial economic and environmental impacts were avoided, by building dams rather than by burning fossil fuels has been very significant. Studies also show that no other method of electricity generation has a lower energy cost than does hydroelectricity. According to Hydro-Québec (2000) the energy payback—the ratio of energy produced during a project's normal life span divided by energy needed to build, maintain and fuel the generation equipment—is about 200 for hydro stations with reservoirs compared to about 40 for wind turbines, nearly 30 for electricity generated by burning forestry waste and, respectively, about 20 and just above 10 for heavy oil- and coal-fired thermal power plants.

No less important is the fact that reservoirs of many hydro-generating projects around the world have also provided other important services by supplying irrigation and drinking water, supporting aquaculture, and offering some protection against recurrent, and often very destructive, floods. The best available breakdown of principal uses of large reservoirs shows that multipurpose impoundments make up 25% of all reservoirs in Europe, 26% in Asia (excluding China), and 40% in North and Central America (WCD 2000).

And, yet again, I will note the fact that hydro generation has been the dominant means of electrification in scores of countries, with more than 20 of them deriving in excess of 90% of all electricity from water and with about one-third of the world's nations relying on dams for more than half of their supply (WCD 2000). Most of the nearly 2 billion people who still have no access to electricity live in these countries and further expansion of hydro generation would be a key step toward modernization. Not surprisingly, some forecasts see that hydro energy will account for slightly over 20% of nearly 700 GW of new electricity-generating capacity to be installed during the first decade of the twenty-first century (IHA 2000).

Most of this continuing expansion will take place in Asia, not only in China, where 371 new large dams were under construction during the year 2000, and India, but also in the countries of Indochina and also in Japan (where 125 new dams were being built in 2000). Sub-Saharan Africa has also a very large untapped potential but its development is predicated, as so much in that part of the world, on the end of civil conflicts and, in the case of some megaprojects, also on the construction of intercontinental transmission lines and on unprecedented international cooperation. Most notably, Paris (1992) proposed that 30-GW Inga station on the lower Congo be connected to the European grid via lines traversing the Congo, Gabon, Cameroon, Nigeria, Niger, Algeria, Tunisia, and the Mediterranean to Italy or, alternatively, reaching the continent via lines running through Rwanda, Uganda, Sudan, Egypt,

Jordan, Syria, and Turkey to the Balkans. Other proposals envisaged links spanning Euroasia, crossing the Bering Strait from Siberia to Alaska, and connecting Iceland and the United Kingdom (Partl 1977; Hammons 1992).

Asian, African, and Latin American countries have also large and dispersed water resources that could be tapped by small hydro stations. Most notably, small plants (defined as projects with less than 10, or sometimes 15, MW) can supply decentralized power for many of the 2 billion people whose villages and small towns are not connected to any grid, or they can be interconnected to become a foundation of a regional electricity supply. Small hydro stations produced about 115 TWh, or some 5% of the worldwide hydro generation, in 1995, nearly two-fifths of it in Asia (EC 2001b). Not surprisingly, the largest exploitable small-hydro potential is in China, Brazil, and India, but many new small stations can be also built (or the abandoned ones dating to the beginning of water power can be refurbished) in affluent countries.

With only small or no reservoirs (run-of-river plants) small hydro stations are vulnerable to seasonal fluctuations of water flow and their capacity factors may be less than 10%; their reservoirs may also silt rapidly and they may be abandoned as soon as a more reliable supply of electricity becomes available. That has been the precise experience in China. Maoist policies promoted many small rural projects and by 1979 the country had some 90,000 small hydro stations with average capacity of just 70 kW (Smil 1988). During the 1980s, as Deng Xiaoping's modernization became the undoing of most of the Maoist policies, several thousands of these stations were abandoned every year because of poor construction, silting, or desiccation of reservoirs and poor maintenance. Other countries with relatively large numbers of not-so-small (less than 1 MW) hydro stations include Germany, France, Italy, Sweden, the United States, and Japan.

With increasing shares of electricity coming from intermittent modes of generation (wind and PV) there should be also steady expansion of pumped-storage projects whose first use dates back to the 1890s in Italy and Switzerland. These stations, with two reservoirs separated by heads commonly in excess of 300 m and as much as 1260 m, are generally expensive to build. But they remain the most widespread means of energy storages for use during the periods of peak demand: they can start generating in as little as 10 seconds. Global capacity of pumped storages has surpassed 90 GW, or about 13% of all installed hydro power. Europe has one-third of the total, and Italy, Japan, and the United States have the largest numbers of high-capacity pumped storage. The two largest stations, 2.88 GW Lewiston and 2.7 GW Bath County, are in the United States (ESA 2001). But one of the world's largest projects, rated at 2.4 GW, was completed in China's Guangdong province in the year 2000: it is a part of the Dayawan nuclear station and its function is to cover peak demand for Hong Kong and Guangzhou.

Biomass Energies

As already noted in chapter 1, biomass fuels continue to be the most important source of heat in many countries of the poor world. Given the fact that imprecise statistics and uncertain conversion factors make it impossible to find the annual consumption of fossil fuels with an error smaller than about 5%, it is not surprising that plausible estimates of the global use of biomass energies can differ by more than 10%. FAO (1999) estimated that about 63% of 4.4 Gm^3 of harvested wood were burned as fuel during the late 1990s. With about 0.65 t/m^3 and 15 GJ/t of air-dry wood, this would be an equivalent of about 27 EJ. In some countries a major part, and even more than 50%, of all woody matter for household consumption is gathered outside forests and groves from bushes, tree plantations (rubber, coconut), and from roadside and backyard trees. For example, rural surveys showed that during the late 1990s nonforest fuelwood accounted for more than 80% of wood burned in Bangladesh, Pakistan, and Sri Lanka (RWEDP 1997). A conservative estimate of this nonforest woody biomass could raise the total to anywhere between 30–35 EJ.

Crop residues produced annually in poor countries added up to about 2.2 Gt of dry matter during the late 1990s (Smil 1999b). Burning in the field, recycling, and feeding to animals account for most of their disposal, and if about 25% of all crop wastes (mostly cereal straws) were burned by rural households this would add about 8 EJ. The most likely minimum estimate for the year 2000 would be then close to 40 EJ of biomass fuels, while more liberal assumptions, higher conversion factors, and addition of minor biomass fuels (dry grasses, dry dung) would raise the total closer to 45 EJ. This means that in the year 2000 the global TPES from all sources would have been roughly 410 EJ, with biomass energies providing about 10% of the total. For comparison, Hall (1997) estimated that biomass supplied as much as 55 EJ during the early 1990s, the WEC (1998) assessment of global energy resources used 33 EJ, and Turkenburg (2000) brackets the total at 45 ± 10 EJ.

China and India are the largest consumers of wood and crop residues in absolute terms: during the mid-1990s China's annual consumption was at least 6.4 EJ, India's about 6 EJ (Fridley 2001; RWEDP 2000). Brazil and Indonesia rank next, but in relative terms it is sub-Saharan Africa, where biomass continues to supply in excess

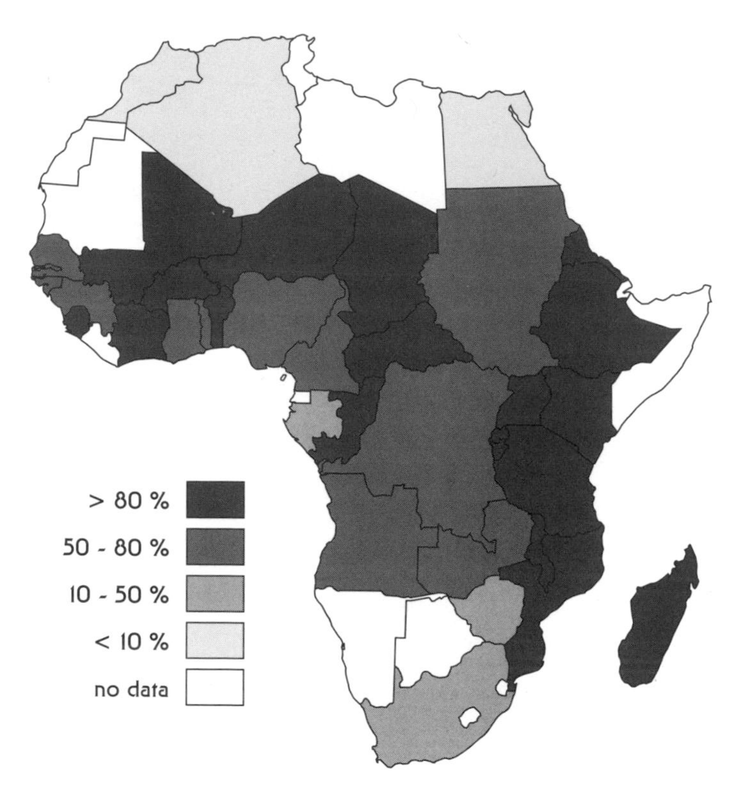

Figure 5.9
High shares of biomass fuels in national TPES make the sub-Saharan Africa the region that is still most dependent on traditional energies. Plotted from estimates by the UNDP (2001).

of 80% of fuel in most of the countries, that comes first, and the biomass shares are 30–50% in densely populated rural areas of South and East Asia (fig. 5.9). The most recent estimate of dependence on biomass energy shows its share above 80% in about 20 of the world's poorest countries and in excess of 50% in some 40 nations (UNDP 2001). Among the most populous modernizing nations the shares are still close to 15% in China, roughly 30% in India and Indonesia, and about 25% in Brazil (IEA 2001).

These shares tell us nothing about the abundance or scarcity of wood or straw in countries still heavily dependent on these fuels. High and widespread reliance on wood can be easily supported in the still heavily forested basin of the Congo. In contrast, even a modestly increased demand in the deforested interior of Asia translates into locally severe ecosystemic degradation as no additional biomass, be it shrubs or grass root mats, should be removed from arid and erosion-prone slopes. Consequently, in many places around the world, woody biomass should not be classed as a de facto renewable resource. Large-scale surveys of prevailing fuel availability show that, besides Asia's arid interior, the areas of worst biomass fuel deficits include large parts of the Indian subcontinent, and much of Central America.

Biomass (fuel wood, charcoal, and to a much lesser extent also some crop residues, above all cereal straw) provides only between 1–4% of the TPES in the world's richest countries, with Sweden (at more than 15%) being a major exception. Woody biomass consumed in the rich countries, mostly by lumber and pulp and paper industries and only secondarily for household heating (be it in once again very fashionable fireplaces or in high-efficiency wood stoves), amounted to no more than about 5 EJ in the year 2000. This clear geographical divide in the current use of biomass fuels leads to two obvious questions regarding the global future of biomass energies. Most importantly, what can be done to improve their supply for hundreds of millions of people in Africa, Asia, and Latin America who either have no access to alternative sources of energy or no means to buy them? And, much less acutely, how realistic, and how desirable, is it to use modern cultivation and conversion methods to produce more biomass fuels in both modernizing and affluent countries?

The most effective way to bring about the first condition is not just to take steps toward expanding supply but also to engage in widespread and intensive campaigns to design and adopt more efficient types of household stoves. This is an obvious requirement given the dismal efficiencies of traditional combustion that converts typically less than 10% of wood or straw into useful heat for cooking or space heating. Common aim of scores of programs aimed at designing and diffusing improved stoves has been to replace traditional open, or partially enclosed, fires with closed or shielded stoves built mostly with locally available materials and minimizing the use of more expensive parts such as iron plates and metal flues. Improved stoves were one of the iconic objects of two movements fashionable in development circles during the last quarter of the twenty-first century, of appropriate technology and the worship of small-is-beautiful (Schumacher 1973). In spite of a great deal of enthusiasm and interest, results of these efforts were largely disappointing (Manibog 1984; Kammen 1995).

As simple as they were, many designs were either still too expensive to be easily affordable or were not sufficiently durable or easy to repair. In the absence of noticeably large fuel savings many stoves were abandoned after a short period of use. More than a subjective perception was involved here: given the wide range of actual performances of both traditional and improved cooking arrangements and lack of uniform measurement standards, efficiency gains were hard to evaluate. Moreover, to make a real difference a good design is only a part of a much broader effort that must also include training of local craftsmen to build and repair such efficient units, active promotion of new stoves and, where needed, financial help in their purchase.

The first encouraging departure from this experience of recurrent failures came only with China's National Improved Stove Programme launched in 1982 and initially aimed at 25 million units to be build within five years. Gradually, this largely successful effort was transformed into a commercial venture and by the end of 1997 about 180 million stoves were disseminated to some 75% of China's rural households (Smil 1988; Smith et al. 1993; Wang and Ding 1998; fig. 5.10). These stoves, operating with thermal efficiencies of 25–30%, are credited with annual savings of close to 2 EJ of fuelwood and coal (Luo 1998).

Efficient use of any suitable waste biomass should be the second ingredient of the quest for better use of cellulosic matter in low-income countries. All crop residues

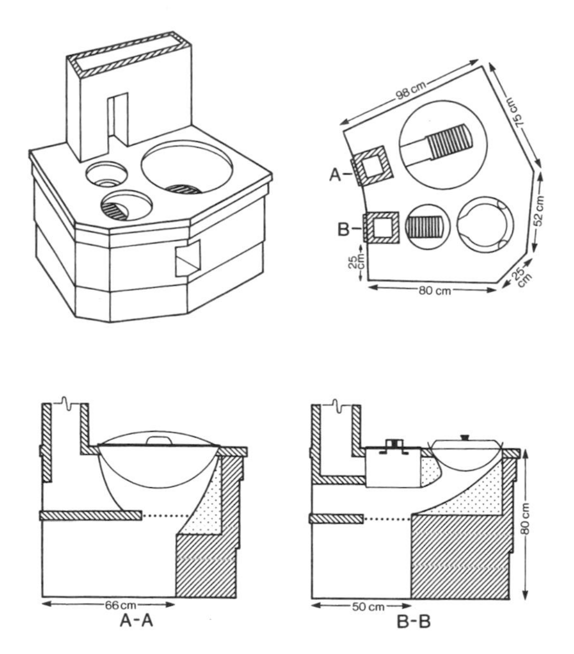

Figure 5.10
An improved cooking stove design from China: a triple-pot unit from Jiangsu (Smil 1988).

that do not have to be recycled (for protecting soil against erosion, replenishing nutrients, and retaining moisture) or fed to animals, and all accessible logging and lumber mill waste should be converted as efficiently as possible to heat, gas, or electricity. Finally, many poor countries have relatively large areas of degraded and marginal land, often on eroding slopes that are unsuitable either for highly productive cropping or for establishment of intensively cultivated tree plantations but that can be planted with appropriate native or foreign species of fast growing trees and shrubs. These extensive fuelwood plantations have only low-to-moderate yields but besides their considerable contributions to local energy supply they would also provide poles for rural construction, wood for crafts, leaves for animal feed, and highly valuable ecosystemic services of reducing slope erosion and retaining moisture.

China provides an excellent example of possible gains. Official estimates put the total of barren slopes suitable for afforestation at nearly 85 Mha, and after private woodlots were permitted once again in the post-Mao China, the area under fuelwood trees reached 5 Mha by 1985 and by 1990 these new plantings were yielding as much as 25 Mt a year (Zhang et al. 1998). These harvests have greatly alleviated previously severe fuelwood shortages, not only in the arid Northeast but also in remote, deforested, and poor regions of the rainy Southwest where annual yields of acacia or leucaena wood can be fairly high (Smil 1988). One global estimate put the total area of degraded land in the tropics at about 2 Gha, with nearly 40% of it suitable for reforestation (Grainger 1988). If all of that land were planted to low-yielding trees (producing just 2.5 t/ha), the annual harvest would be roughly 35 EJ, or about as much as is the world's current biomass energy consumption.

Expanded production of biomass that would go beyond the maximum possible use of phytomass waste and the afforestation of marginal land would have to rely either on intensified management of existing forests or on growing special energy crops, be they fast-maturing trees, perennial grasses, aquatic species, or carbohydrate-rich sugar cane or corn. Cogeneration of heat and electricity is the best way to convert these fuels to heat and electricity, and a variety of techniques are available to transform fuels to liquid or gaseous compounds (Smil 1983; Larson 1993; Klass 1998; Kheshgi, Prince, and Marland 2000). There are three fundamental objections to the pursuit of such a quest on anything but a modest scale: inherently low power density of photosynthesis; already very high human appropriation of the biosphere's net primary productivity (NPP); and the costs (be they in economic, energy, or environmental terms) of any large-scale intensive cultivation of biomass crops and their conversions to electricity or to liquid or gaseous fuels.

A number of recent global terrestrial NPP estimates averages roughly 55 GtC/year, that is about 120 Gt of dry biomass containing (using the mean of 15 GJ/t) some 1,800 EJ (Smil 2002). This prorates to less than 0.5 W/m^2 of ice-free land, a rate an order of magnitude lower than the common power densities for wind and water flows and three orders of magnitude lower than the radiation flux reaching PV modules in sunny locations (see fig. 5.1). In natural ecosystems a large part of NPP is consumed by heterotrophs ranging from bacteria to megaherbivores and hence the amount of phytomass actually available for energy conversion is only a fraction of the primary productivity. And the rates remain low even for the world's most productive natural growth or for the best agroecosystems.

Average NPP for mature tropical rain forest, the world's most productive tree ecosystem, is no higher than about 1.1 W/m^2 and only intensively cultivated crops can equal or slightly surpass that rate. Both an excellent harvest of Iowa corn, yielding 12 t/ha of grain and the same amount of stover, and rapidly maturing trees (poplars, willows, pines, eucalypti) intensively cultivated as short-rotation crops and producing exceptionally high harvests of 20 t/ha of dry-matter, above-ground phytomass a year will yield about 1.1 W/m^2. And even when converting these high yields with the highest possible efficiencies—above 50% for cogeneration of electricity and heat, and 55–60% for conversion to methanol or alcohol—the best achievable rates fall to just 0.5–0.6 W/m^2 for the final energy supply, one to two orders of magnitude below the power densities of electricity generation using water, wind, or solar radiation (fig. 5.11).

Any number of impressive examples could be given to illustrate the practical consequence of these extraordinarily low power densities; just two must suffice here. If we were to replace the world's coal consumption in the year 2000 (nearly 90 EJ) by harvesting woody biomass we would have to cultivate trees—even when those plantations had a high average yield of 15 t/ha—on about 330 million ha, an area larger than the combined total of remaining forested land in the European Union and the United States. And if the U.S. vehicles were to run solely on corn-derived ethanol (and assuming the recent average corn yield of 7.5 t/ha and roughly 0.3 kg ethanol/kg of corn) the country would have to plant corn on an area 20% larger than is its total of currently cultivated cropland! And even greater claims on arable land would result from such schemes as using plant oils to substitute for diesel fuel. In contrast, biodiesel produced from recycled cooking oils makes no such claims and its blending (typically 20% addition) with petroleum diesel lowers exhaust emis-

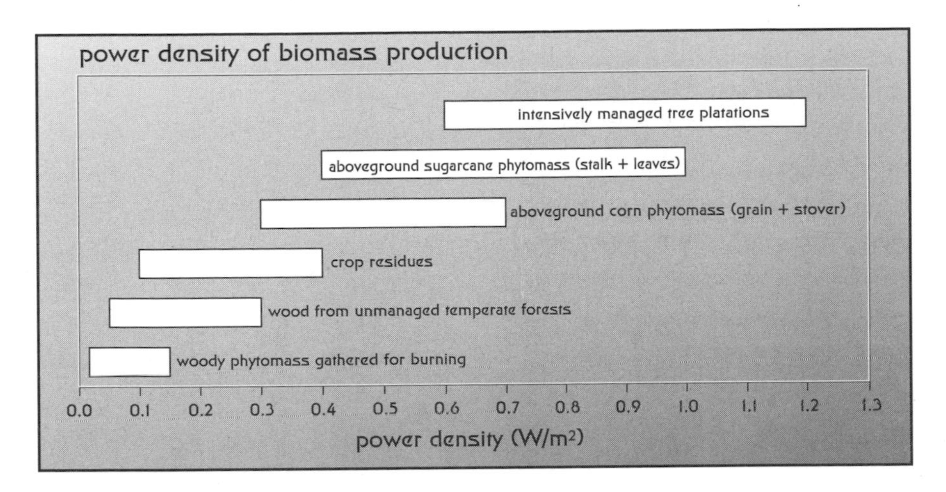

Figure 5.11
Power density of various kinds of biomass energy production. Plotted mainly from data in Smil (1983; 1991; 1999b) and Klass (1998).

sions—but only a minuscule share of the world's vehicles could be fueled by the waste from fried-food empires.

Obviously, any cultivation schemes on such grand scales are impossible as they would preclude any food production, eliminate wood harvests for timber and paper, and do away with many irreplaceable environmental services rendered by forests. In addition, considerable energy costs arising from the cultivation, harvesting, and collection of phytomass grown on such huge areas would greatly reduce energy gains of such biomass production. Consequently, I consider the recent maximum estimates of 150–280 EJ of *additional* biomass energy available by the year 2050 (Turkenburg 2000) as utterly unrealistic. But an even much smaller extent of intensive biomass production for energy is inadvisable as it would further increase the already surprisingly high human appropriation of annual terrestrial NPP, diminish the constantly shrinking area of ice-free land untouched by human actions, and weaken many ecosystemic services by compromising the biosphere's integrity.

Vitousek et al. (1986) calculated that during the early 1980s, humanity appropriated between 32–40% of the continental NPP through its field and forest harvests, animal grazing, land clearing, and forest and grassland fires. Mean values of a recent recalculation (Rojstaczer, Sterling, and Moore 2001) matched almost precisely the older estimate. Given the inherently low power densities of biomass production any truly large-scale reliance on phytomass energies would have to increase this already worrisome share by an uncomfortably large margin. Competition with food and fibre crops, destruction of remaining natural ecosystems, loss of biodiversity, and intensified environmental degradation far beyond the areas converted to biomass plantations are the main reasons why it is not desirable to use either agricultural land or natural or degraded forests, wetlands, and grasslands for large-scale cultivation of biomass crops. Using the existing arable land for biomass crops is a proposition that should be rejected outright as far as all but a few nations are concerned.

Most of the world's populous nations have a limited, and declining, availability of arable land and larger harvests needed to feed their growing populations will have to come from intensified cropping of the existing farmland (fig. 5.12). Very few countries, most notably the United States and Brazil, have relatively large areas of fairly productive agricultural land that could be taken out of food and fibre production and dedicated to the cultivation of biomass crops. Many studies have investi-

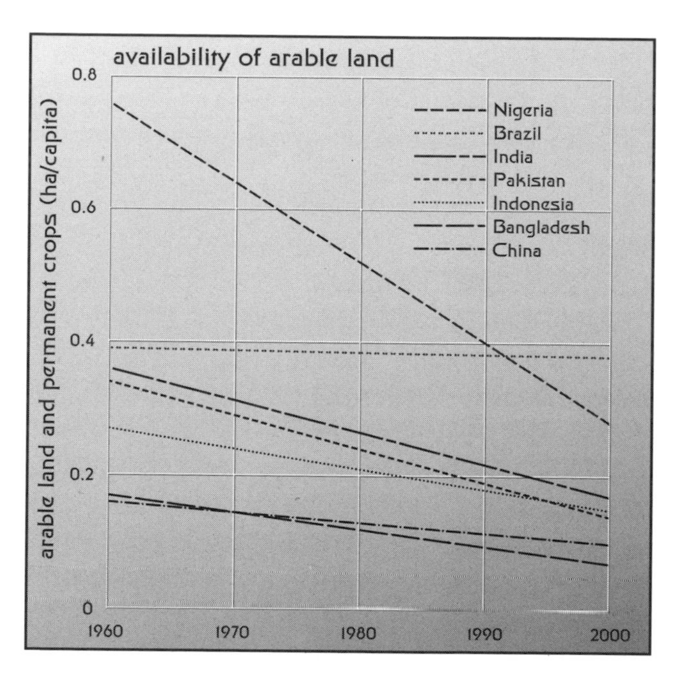

Figure 5.12
Decreasing availability of arable land in the world's most populous low-income countries, 1960–2000. Brazil, with its huge reserves of good *cerrado* farmland, is the only exception to the declining trend. Plotted from data in FAO (2001).

gated practical limits of this option for the United States with predictable outcomes (Smil 1983; Cook et al. 1991). As already noted, replacing all U.S. gasoline by corn-derived ethanol would require more land than is currently planted to all food, feed, and fibre crops. Consequently, the only practical option would be to use relatively small areas of very intensively managed energy crops (most likely less than 5% of the U.S. farmland)—but even if their average yield would be 25 t/ha (or 375 GJ/ha) output would be just over 3 EJ, basically equal to the currently used waste biomass (EIA 2001a).

While this could be practical it would be hardly desirable as the intensified use of cropland used for biomass energy would carry all the well-known requirements and impacts encountered in highly productive agriculture. Requirements include large-scale use of machinery, high amounts of inorganic fertilizers, particularly of nitrogen, supplementary water during dry spells or irrigation in drier areas, and pesticides and herbicides to combat heterotrophic infestations and weeds competing for the same water and nutrients. Environmental impacts range from excessive soil erosion, compaction of soils, and nitrogen and phosphorus leaking into waters and causing unintended eutrophication to the depletion of aquifers and agrochemical residues in aquifers, biota, and downwind air, further impoverishment of biodiversity, and concerns about large-scale cultivation of transgenic crops. None of these objections to massive biomass cultivation for energy has changed, and some have only intensified, since I analyzed them in great detail nearly two decades ago (Smil 1983).

And, of course, any intensive cultivation scheme predicated on substantial direct energy inputs to run field machinery and indirect (embodied) inputs needed for synthetic fertilizers and other agrochemicals and followed by additional energy-requiring conversion to liquid or gaseous fuel raises the obvious question of actual energy return. Heat and electricity from wood has the highest energy return but several studies have shown that the U.S. production of ethanol from corn is a net energy loser (Pimentel 1991; Keeney and DeLuca 1992), while the proponents of bioethanol stress the fuel's positive energy, and its environmental contributions as its combustion is less air-polluting than is the burning of gasoline and as it also helps to reduce CO_2 emissions (Wyman 1999).

Energy balance studies are not easy to compare as they use different assumptions regarding typical crop cultivation and grain conversion practices. The latest detailed account indicates that without energy credits for coproducts, ethanol produced from the grain grown in the nine Midwestern states has a minuscule energy gain (1.01)—but when the credits are taken for distillers' grain and corn gluten meal and feed, the ratio has a positive weighted mean of 1.24 (Shapouri, Duffield, and Graboski 1995). Improving energy balance of the U.S. corn ethanol has been due not only to advances in grain milling, enzymatic hydrolysis of the grain starch to glucose, and the sugar's subsequent yeast-based fermentation to alcohol but also to reduced applications of energy-intensive N (and P) fertilizers and to more efficient transportation.

The fundamental disadvantage of ethanol production based on yeast-mediated fermentation of monomeric sugars or polymeric glucose (starch) in field crops (sugar cane, sorghum, and corn) is that it can use only a very small fraction of available phytomass present in relatively expensive feedstocks. Most of the produced crop and tree phytomass is made up of two hydrolysis-resistant biopolymers, cellulose (40–60% of the total mass), and hemicellulose (20–40%), and their inexpensive commercial conversion would turn crop residues (above all corn stover and cereal straws), grasses, and forestry, and a large part of municipal solid waste into valuable feedstocks (fig. 5.13). Cellulosic feedstocks can be converted to sugars either by acid or enzymatic hydrolysis. The former process, known for more than a century, has not led to any commercial conversion. The latter technique, particularly the simultaneous saccharification and fermentation (SSF), is still very expensive due to the cost of producing the requisite cellulases, but it has a better prospect for eventual commercial diffusion (DiPardo 2000).

A key scientific breakthrough in expanding the resource base for the production of bioethanol and reducing the process costs was the introduction of genetically

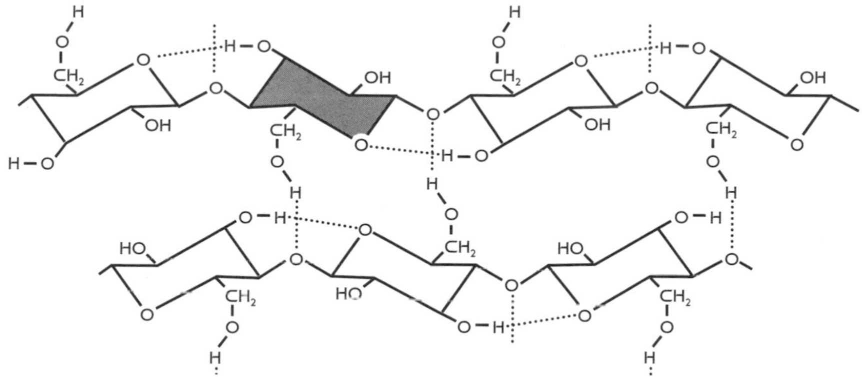

Figure 5.13
A short segment of cellulose, the dominant hydrolysis-resistant polymer of biomass typically consisting of 1,000–10,000 glucose units connected by H-bonds (Smil 2002).

engineered bacterium *Escherichia coli* K011 (Ingram et al. 1987; Wyman 1999). When endowed with genes from *Zymomonas mobilis* the bacterium is able to do what no yeast can: ferment pentoses (five-carbon sugars, xylose and arabinose) that are present in hemicellulose to ethanol. Availability of inexpensive processes for converting cellulose and hemicellulose to ethanol would add significantly to overall bioethanol production. The subsidized corn-based U.S. ethanol output rose more than ninefold between 1980 and 2000, from about 14 PJ in 1980 to about 130 PJ (I am converting the volume by using 21.1 MJ/L). DOE's forecast for ethanol from cellulosic feedstock is for about 68 PJ by the year 2020 in the reference case, and up to 224 PJ in the case of successful diffusion of SSF process (DiPardo 2000). For comparison, the U.S. gasoline consumption in the year 2000 amounted to nearly 16 EJ (BP 2001).

Until the cellulosic phytomass can be profitably converted to ethanol the average power density (the net energy gain) of the U.S. corn-based ethanol will remain a minuscule 0.05 W/m^2 (Kheshgi, Prince, and Marland 2000). Moreover, net energy gain on the order of 0.25 is small compared to the gain obtained from burning woody phytomass (easily above 10) but it can be argued that ethanol's portability outweighs, for example, wind electricity's superior energy gain. More importantly, even if a particular process has a positive energy balance, or if future farming and bioengineering advances will bring such gains, this is not a sufficient reason to proceed enthusiastically with crop-derived liquid fuels: a simple energy balance tells us nothing about the already noted long-term environmental costs of cultivating energy crops.

At least the ethanol production yields a high-quality fuel that also reduces need for any additives when used in internal combustion engines. In contrast, some biomass conversion techniques produce only low-quality fuels or are unreliable. Thermochemical gasification is the best example in the first category: the resulting producer gas has just 4–6 MJ/m^3, less than 20% of the energy content of natural gas. Anaerobic digestion of phytomass producing biogas (a mixture of CH_4, and CO_2) exemplifies the second category. Although the generated gas has about 22 MJ/m^3, its bacteria-mediated production is highly sensitive to the properties of feedstocks and to changes in many environmental variables, above all temperature, liquidity, alkalinity, pH, and C:N ratio of the feedstock (Smil 1983; Chynoweth and Isaacson 1987). As a result, repeated promises of biogas generation becoming a significant supplier of clean rural energy in populous Asian countries or a welcome extension of energy supply in Western nations have not materialized.

Most notably, China's massive program of constructing small household biogas digesters that began in the early 1970s, surpassed 5 million units by 1977 and had plans for the total of 70 million by 1985. But the biogas campaign began falling apart as soon as Deng Xiaoping's modernization began changing the conditions in China's countryside in 1980 (Smil 1988). Problems with often poorly built digesters that resulted in low rates of biogas generation on one hand, and renewed planting of private woodlots, greater access to coal from small mines, and more efficient household stoves on the other led to the abandonment of most of the units built before 1980.

In contrast, problems with manure disposal from large U.S. livestock enterprises have brought renewed interest in using biogasification as perhaps the best means to control offensive wastes. Large-scale, optimized processes produce fuel for running diesel engines or microturbines that generate electricity for dairy or beef operations, and nonodorous, nutrient-rich liquid for fertilizing nearby fields and processed solids for sale as an organic soil amendment (Goldstein 2002). Fewer than 100 digester-powered livestock farms operated in the United States in 2001, but given the fact that the country's animals produce about 200 Mt of waste solids a year and that the feeding operations are increasingly concentrated, the nationwide potential for biogas generation is large. But even after a much broader adoption the process will remain more of a pollution control measure than a major energy-producing enterprise.

As for the economics of biomass conversion to liquid fuels, most of the properly prepared accounts show that both wood-derived methanol and grain-derived ethanol are not competitive with liquids refined from crude oil, and that only sustained petroleum prices in excess of $40–50 per barrel would make their large-scale production profitable (Larson 1993; Kheshgi, Prince, and Marland 2000). Even the Brazilian production of alcohol from sugarcane, the most successful large-scale effort at substituting gasoline by biomass-derived ethanol, required tax incentives (Goldemberg 1996; Moreira and Goldemberg 1999). Although the production costs of cane-derived ethanol fell by nearly two-thirds as the production volume more than tripled between 1980 and 1995, they remained higher than the market price for gasoline, and taxes on the latter fuel have been used to subsidize the former. Subsidies were withdrawn only in 1999.

But, in contrast to the disputed energy benefits of the U.S. corn alcohol, Brazilian cane ethanol has always had a favorable energy balance even when there is no credit taken for bagasse, the fibrous residue that remains after pressing the stalk and that can be best used for cogeneration (Geller 1985). This is because of high natural yields (averaging about 65 t/ha during the late 1990s) of this tropical grass in what are some of the world's best regions for its cultivation, and because the presence of N-fixing endosymbiotic bacteria in the plant's stems reduces or entirely obviate the need for energy-intensive N fertilizers. Even so, the net energy gain of the country's ethanol prorates to just 0.31 W/m^2 (Kheshgi, Prince, and Marland 2000). Ethanol from a lower-yielding and heavily fertilized U.S. sugarcane grown in Florida or Hawai'i would have a much lower gain, and a marginal, or negative, energy balance.

Improved forest management in order to harvest more biomass would then seem a better option than growing energy crops on arable land. For example, full stocking of U.S. commercial forests could raise the annual productivity from 2–4 t/ha to 5–10 t/ha, but only a small share of 200 Mha of that growth could be managed in such intensive ways. Upgrading one-quarter of U.S. commercial forestlands by 3 t/ha would produce annually an additional 3 EJ, or only about as much as the recent use of forest phytomass wastes (EIA 2001a). And even then there may be many conflicts with other, and less replaceable, uses of forests. And grasslands are even less suitable places for biomass energy production. Because of their high natural productivity wetlands have been considered as prime areas for future biomass farms but two realities make this a most undesirable choice. When intact or only slightly affected by human actions wetlands are the most complex freshwater biomes, containing much higher biodiversity than lakes and streams (Majumdar, Miller, and Brenner 1998).

Consequently, investment in their preservation offers some of the best rewards per unit of protected area. Moreover, wetlands on every continent have been already extensively converted to croplands or were destroyed or flooded by construction of reservoirs and regulation of streams. And although the global area of grasslands is orders of magnitude larger and these ecosystems are far from being as biodiverse as wetlands, similar arguments apply in trying to keep these lands from conversions to biomass cultivation. Remaining natural grasslands are unique ecosystems optimized to thrive in often semiarid or arid climates and any major change in the existing plant cover can cause rapid and irreversible changes. Indeed, a large share of the world's grasslands used for pasture has been already degraded by overgrazing and excessive soil erosion, with such effects as silting of streams and downwind transport of dust extending far beyond the immediately affected areas.

Electricity from Wind

There is, of course, nothing new about capturing wind for useful work. Windmills have a long history as providers of locally important mechanical energy. After 1600 progressively better designs and higher conversion efficiencies made wind power a major source of primary energy in parts of Northwestern Europe and during the late nineteenth century also on the windy U.S. plains (Smil 1994a; Sørensen 1995). Initially a significant share of rural North American electricity on isolated farms was generated by wind but cheaper, and much more reliable, coal- and water-based electricity delivered by new transmission lines gradually eliminated wind as a source of useful energy. Until the early 1970s there was no real interest in designing better wind machines in order to turn them into efficient and competitive producers of electricity. Post-1973 interest in renewable energies changed that but the most impressive advances came only during the late 1990s.

The first modern boom in wind energy was launched by the U.S. tax credits during the early 1980s as the total installed capacity rose 100-fold (Braun and Smith 1992; AWEA 2001a). By 1985 the country's wind turbines had installed capacity of just over 1 GW and the world's largest wind facility was at the Altamont Pass in California. Its 637 MW of installed capacity was generating about 550 GWh a year, enough to provide electricity for about 250,000 people (Smith 1987). The above figures mean that electricity was generated only during 10% of the time, while the rates for fossil-fueled plants are commonly 65–70% and the best nuclear facilities have load factors in excess of 90%. These low load factors were partially caused by frequent problems with turbines and before any major improvements were made the expiration of tax credits in 1985 ended this first wind wave.

By the early 1990s two kinds of technical advances began changing the prospect: better turbine designs, with blades optimized for low speeds, and larger turbine sizes (Ashley 1992; Øhlenschlaeger 1997). The average size of new machines rose from a mere 40–50 kW in the early 1980s to over 200 kW a decade later and during the late 1990s the commercial market was dominated by turbines between 500–750 kW, and the first machines with power of more than 1 MW began to enter the service (Øhlenschlaeger 1997; BTM Consult 1999). New designs under development are for machines as powerful as 4–5 MW, with rotor diameters of 110–112 m (McGowan and Connors 2000). But, in contrast to the 1980s, new U.S. installations have been growing slowly compared to rapid advances in several European countries, above all in Germany, Denmark, and Spain where new laws guaranteed

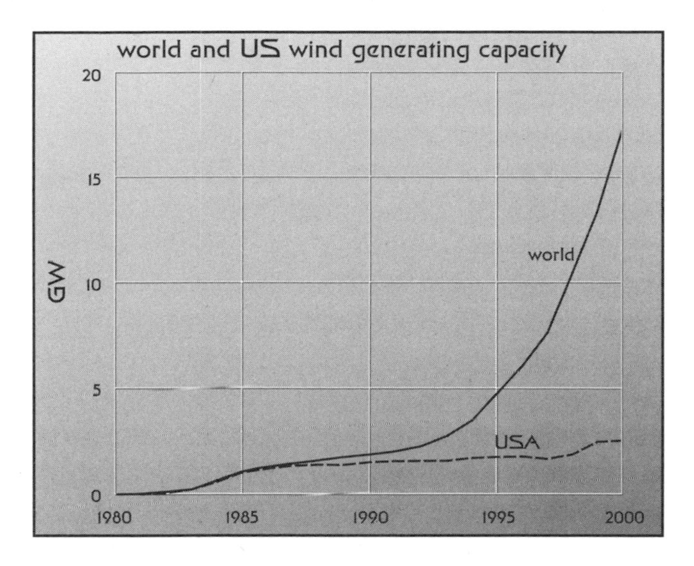

Figure 5.14
Electricity-generating capacities installed in wind turbines: U.S. and global totals, 1980–2000. Plotted from data in AWEA (2001a) and BTM Consult (1999).

fixed price for wind-generated electricity. The Danish government has been particularly active in promoting wind power and the country now has the highest per capita installed capacity and it dominates the world export market in efficient wind turbines.

At the end of 2000 the country had 6,270 wind turbines with the capacity of 2.417 GW and during that year wind produced 4.44 TWh, or 13.5% of Denmark's total electricity consumption (DWIA 2001; fig. 5.14). Moreover, Danish wind turbine manufacturers captured about 50% of the global market (65% when joint ventures are included), with total exports adding up to nearly 2.9 GW by the end of 2000, and with Germany being the largest importer. Germany continues to be the world leader in absolute terms, with more than 9,000 machines and about 6.1 GW of installed capacity at the end of 2000 (AWEA 2001a). But in relative terms wind electricity provided only 2.5% of Germany's domestic demand.

U.S. wind-generating capacity rose from 1.039 GW in 1985 to 2.554 GW by the end of 2000 and it was expected to more than double during 2001 (fig. 5.14; AWEA 2001b). California has half of all installed capacity, but the largest turbines (four 1.65 MW Vestas V66) are now operating, with another 46 smaller (600 kW) units, at Big Spring in Texas (Giovando 1999). Some of the world's largest wind projects got underway during the year 2000, led by the 300-MW Stateline wind farm on the border of Washington and Oregon (PacifiCorp Power 2001). Spain, with 2.235 GW, was close behind the United States, and India, with 1.1167 GW, was a distant fifth, followed by the Netherlands, Italy, and the United Kingdom (each less than 500 MW in total). Global capacity of wind turbines reached 1 GW in 1985, 10 GW in 1998 (fission plants did that in 1968), and with 17.3 GW at the end of the year 2000 it was more than double the 1997 total and nearly nine times the 1990 aggregate (fig. 5.14; BTM Consult 1999; AWEA 2001a).

Thanks to these rapid gains wind generation has been lauded as one of the fastest-growing energy sources and the production and installation of increasingly larger wind turbines is being praised as one of the world's most rapidly expanding industries (BTM Consult 1999; Flavin 1999; Betts 2000; Chambers 2000; AWEA 2001a). Wind-driven electricity generation is seen as the most promising of all new renewable conversions, being far ahead of other solar-based techniques both in terms of operational reliability and unit cost. Some experts argue that, at the best windy sites, even unsubsidized wind electricity is already competitive with fossil-fueled generation, or even cheaper than coal- or gas-fired production, and hence we should go for a more aggressive maximization of wind's potential (BTM Consult 1999; AWEA 2001a; Jacobson and Masters 2001a).

Not surprisingly, some plans foresee 10% of the world's electricity demand generated by wind by the year 2020 (Betts 2000). Denmark wants to get 50% of its electricity from renewable energy (mostly from wind) by 2030. The European Union is aiming at 10% of electricity, or about 40 GW of installed power, from wind by the year 2010 and 100 GW by 2020 (Chambers 2000). In order to meet the October 2001 legal requirement to generate 10% of the country's electricity coming from renewables by the year 2010, British plans call for 18 offshore sites, with total installed power of up to 1.5 GW. The regional target for Western Europe is 66 GW by 2010 and 220 GW by the year 2020, with about one-third of that capacity placed offshore (Olesen 2000).

Are we witnessing the real takeoff of a globally significant, if not quite revolutionary, mode of electricity generation or will this new wave of enthusiasm bring only some nationally or regionally important gains without any substantial modification of the world's electricity supply? As with all solar energies, the available resource is no obstacle to even the boldest dreams. Lorenz (1976) estimated that some 3.5 PW, or about 2% of all insolation, are needed to drive the atmospheric motion. We do not know what share of this flux could be harnessed without any significant changes of planetary circulation but practical considerations limit the height of our machines to the lowermost layer of the atmosphere. If only 1% of the total flux could be converted to electricity, the global capacity would be some 35 TW, or more than 10 times the year 2000 total installed in all fossil, nuclear, and hydro stations. A much more restrictive estimate taking into account only wind speeds above 5 m/s up to 10 m above ground, together with some arbitrary siting constraints (mainly visual intrusion), put the global wind power potential at about 6 TW (Grubb and Meyer 1993).

Whatever is the actual, and undoubtedly huge, global total, it is available at very low power densities and it is distributed very unevenly both in space and time. Many windy sites, including large parts of America's Great Plains and Pacific coast, have mean annual wind speeds of 7–7.5 m/s that produce power densities of 400–500 W/m^2 of vertical area swept by rotating blades 50 m above ground (fig. 5.15).

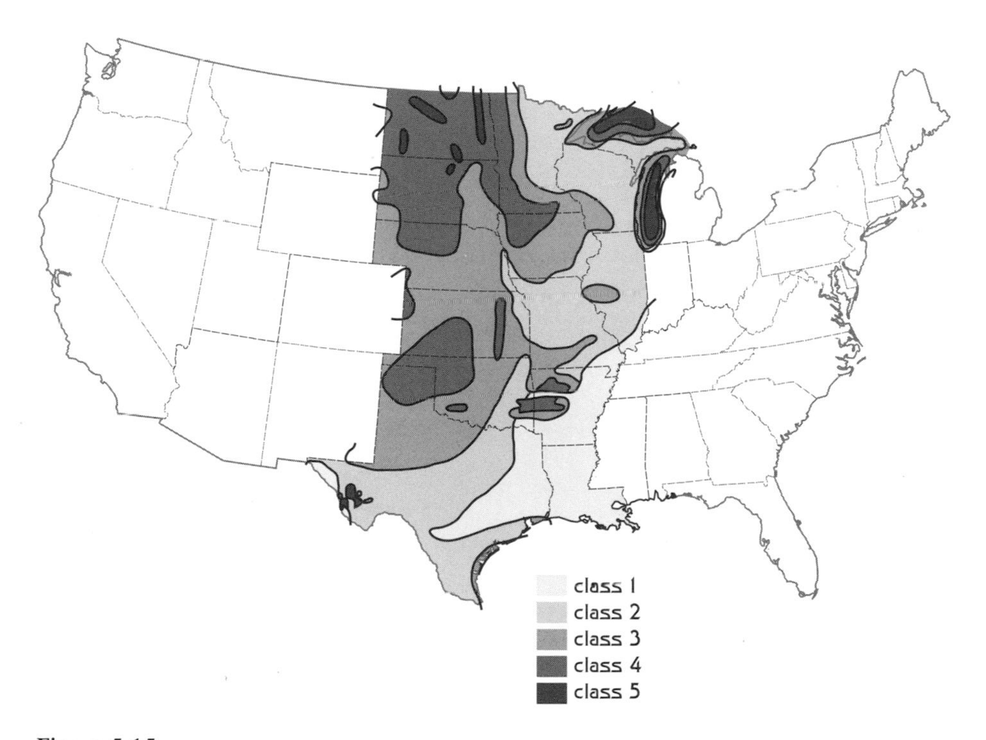

Figure 5.15
Annual average wind power on the Great Plains of the United States. Simplified from a map in Elliott et al. (1987), available at <http://rredc.nrel.gov/wind/pubs/atlas/maps/chap2/2-01m.html>.

Translating these rates into horizontal power densities cannot be done without a number of nested assumptions. Sufficient spacing must be left between the units in order to eliminate upstream wakes and to replenish kinetic energy. Spacing equal to five rotor diameters is enough to avoid excessive wake interference but at least twice that distance is needed for wind energy replenishment in large installations. Moreover, no more than 16/27 (59.3%) of wind's kinetic energy can be extracted by a rotating horizontal-axis generator (the rate known as the Betz limit) and the actual capture will be about 80% of that limit.

This means that, for example, machines with 50-m hub height with 10 × 5 diameters spacing would have to be 500 m apart. In locations with wind power density averaging 450 W/m^2 (mean value of power class 4 common in the Dakotas, northern Texas, Western Oklahoma, and coastal Oregon) these machines would intercept about 7 W/m^2, but an average 25% conversion efficiency and a 25% power loss caused by wakes and blade soiling reduce the actual power output to about 1.3 W/m^2 (Elliott and Schwartz 1993). More favorable assumptions (power density of 700 W/m^2, turbine efficiency of 35%, and power loss of a mere 10%) would more than double that rate to about 3.5 W/m^2. Actual rates are highly site-specific. Rated performance of California's early wind projects was almost exactly 2 W/m^2 (Wilshire and Prose 1987), the Altamont Pass grouping averages about 8.4 W/m^2 (Smith 1987), and the most densely packed wind farms rate up to 15 W/m^2; in contrast, more spread-out European wind farms have power densities mostly between 5–7 W/m^2 (McGowan and Connors 2000).

Low power density of wind is not in itself an insurmountable obstacle to tapping the large resource. This is due to two specific facts. First, as just noted, wind power densities of good sites are easily an order of magnitude higher than those of biomass energies and, unlike in the case of hydrogeneration or biomass crops, all but a small share of areas occupied by wind facilities (commonly less than 5% of the total) can still be used for farming or ranching. For example, a study using a variety of land exclusion scenarios (due to many environmental and land use reasons) concluded that windy land available in power class 4 and above is about 460,000 km^2, or about 6% of all land in the contiguous United States and that it could support generating capacity of about 500 GW (Schwartz et al. 1992). Even under the most restrictive specifications the Dakotas alone could supply 60% of the total U.S. 1990 electricity consumption (Elliott and Schwartz 1993).

More realistically, generating about 20% of the country's electricity by wind would require less than 1% of land in the contiguous United States, of which less than 5% would be actually taken up by wind turbines, associated equipment, and access roads, with the remainder used as before for cropping or ranching. Moreover, Denmark, the Netherlands, and Sweden have already installed large wind turbines offshore. Denmark's first small (4.95 MW) offshore installation was completed in 1991, 40-MW Middelgrunden, visible from Copenhagen and on a webcam (DWIA 2001), was connected to the grid in 2000, and the country has bold plans for 4 GW of offshore wind power by the year 2030 (Madsen 1997). The Netherlands has two offshore wind projects in the Ijsselmeer, and Dutch researchers are proposing to get all of the country's electricity from 4,000 individual 5-MW machines sited in a 900-km^2 offshore formation (Betts 2000). In the United Kingdom the country's first offshore wind project in the Blyth Harbour (North Sea, off the Northumberland coast) includes two of the world's most powerful wind turbines, 2-MW Vestas V66 (Blyth Offshore Wind Farm 2001). The best offshore sites to be developed during the coming years have power densities between 10–22 W/m^2.

What is much more important in contemplating the eventual impact of wind-generated electricity is that the resource is highly variable both in space and time, in the latter case on a daily as well as seasonal and annual basis (Sørensen 1995; McGowan and Connors 2000). Moreover, these fluctuations and hence the expected availability of wind-powered generation, are only imperfectly predictable, and peak wind flows only rarely coincide with the time of the highest demand. Inevitably, these realities complicate efficient commercial utilization. Many densely populated areas with high electricity demand experience long seasonal periods of calm or low wind speeds and hence are utterly unsuitable, or only marginally suited, for harnessing wind's energy. They include, just to cite a few notable examples, virtually the entire southeastern United States, Northern Italy, and Sichuan, China's most populous province. In contrast, most of the best sites are far from major load centers in thinly populated or uninhabited regions.

For example, North Dakota may be the best state to locate wind turbines in the United States, but because of its distance from major load centers it had only 500 kW of wind power, or less than 0.01% of the country's installed total, at the end of the year 2001 (AWEA 2001b). Construction of long-distance, high-voltage lines (preferably DC) would be needed to transmit large quantities of power from such locations to the regions of major demand. Relatively large-scale wind projects could be immediately helpful only in places where there are already major high-capacity interconnections capable transmitting both longitudinally (taking advantage of differences in peak demand) and latitudinally (taking advantage of seasonal differences in demand).

While it may be true that large shares of potential U.S. wind power are within 8–30 km of existing 230-kV lines (Shaheen 1997) a much more relevant fact is that the country, unlike Europe, does not have any strong East–West high-voltage interconnections that would make it possible to transmit tens of GW from the center of the continent to the Atlantic or the Pacific coast. Consequently, a study of seven U.S. utilities indicated that only in some cases does it appear possible to accommodate power transfers of 50–100 MW that could supply local load without significant upgrades of transmission lines. And while the construction of such dedicated lines may be equal to only a small part of the overall cost of new turbines needed to produce large amounts of power (Hirst 2001), acquiring the rights-of-way for these infrastructures may not be easy. Europe, with its much more developed interconnections and shorter distances between its windy sites and large load centers is thus in a much better position to take advantage of new wind power capacities.

Wind's seasonal and daily fluctuations are mainly the function of differential heating of the Earth's surfaces and as such they are broadly predictable but cannot be forecast accurately not just a year but even a week ahead, and sometimes even an hour, ahead. Average annual wind speeds in successive years at the same site can differ by up to 30%, and displacements on the order of 10 m can result in 20% higher wind speeds and, as the power goes up with the cube of the speed, in 70% higher outputs. Even in generally windy locations there could be a twofold difference in total wind energy available for extraction during the least windy and the windiest month (fig. 5.16), and monthly departures from expected long-term means can surpass 25%. For example, Big Spring, the largest wind facility in Texas, generated 120–128% of budgeted output during the first two months of 1999 but then the faster-than-expected winds were followed by lower-than-average velocities and capacity factors dropped by nearly 30% in March (Giovando 1999).

Moreover, safety considerations make it impossible to convert the fastest winds in rapidly moving cyclonic systems (hurricanes, tornadoes) to electricity and hence all modern wind turbines are designed to cut out at speeds around 25 m/s. Needless to say, these realities preclude using wind for baseload electricity generation in any isolated or weakly connected system. Of course, with a larger number of turbines at a given facility and with a greater number of more widely dispersed wind facilities within a system the intermittency of individual turbines becomes less important but even then the poor predictability of medium- to long-term availability of wind power limits the use of wind energy for baseload supply where sudden drops of 20–30% are intolerable. Mismatch between the times of maximum wind-driven generation

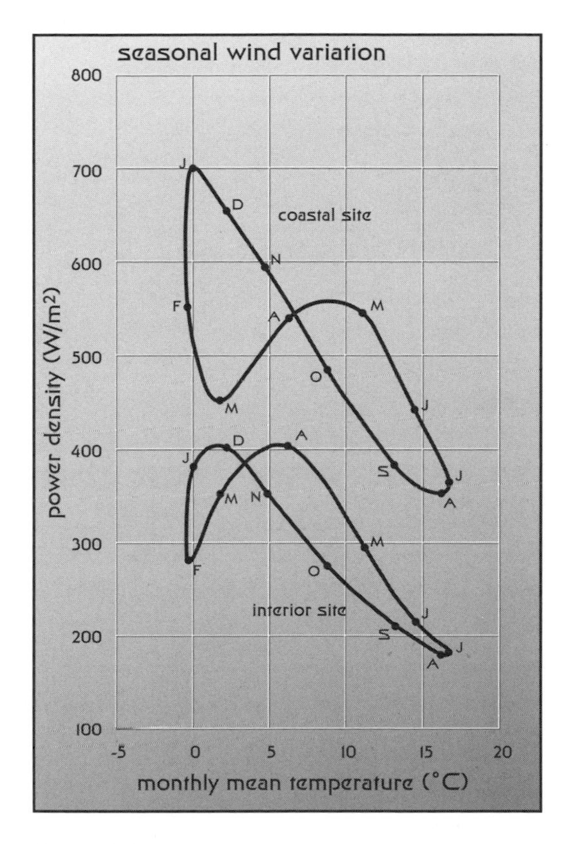

Figure 5.16
Seasonal variation of wind flows at a coastal and inland Danish site. Based on a figure in Sørensen (1995).

and peak electricity demand is another important complication. The former takes place often during the midday when demand is low, the latter peaks in late afternoon and evening when winds die down.

Steam turbogenerators and gas turbines are among the most reliable machines operated by modern society and unscheduled outages are rare. In contrast, wind turbines commonly experience sudden power losses on the order of 10–25%. Careful investigations have shown that insects caught on the leading edges of blades during conditions of low wind and high humidity are responsible for these unpredictable drops in power of wind turbines operating in high winds. Sudden power losses caused by blade soiling may be as large as 50%, and the only way to restore

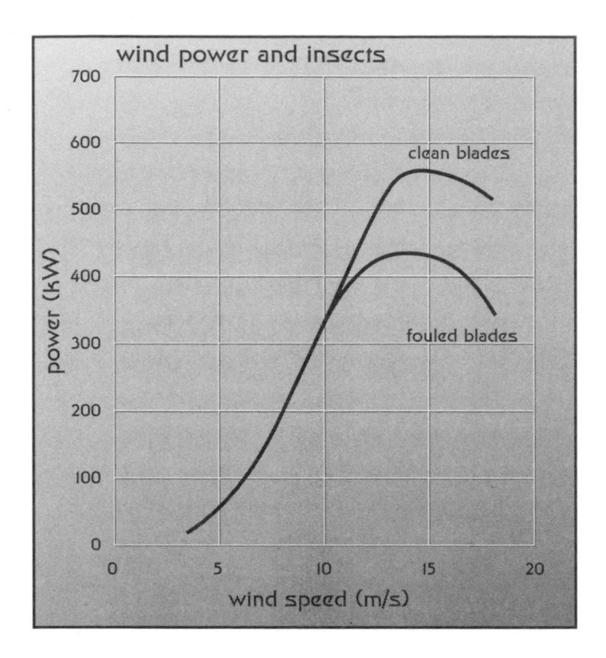

Figure 5.17
Insects foul the leading edges of turbine blades and cause a drop in available power: actual power levels at the same wind speed on different dates are well replicated by power curves for the turbines with rough and clean blades. Based on graphs in Corten and Veldkamp (2001).

the normal flow is to wash the blades (Corten and Veldkamp 2001; fig. 5.17). Accumulations of ice and dirt on blades could bring on the same sudden high-wind power losses, as does insect contamination.

Finally, there are the negatives most often listed by environmentalists and by people living in the proximity of wind machines: bird strikes, noise, interference with electromagnetic waves, and the esthetic aspect of large wind turbines. High blade tip velocities necessary for high conversion efficiencies in the earlier models of wind machines made them noisy and increased local bird mortality. Better designs can almost entirely eliminate the gearbox noise by insulating the nacelle and reduce substantially the aerodynamic noise caused by wind coming off blades. Appropriate exclusion zones and offshore siting are the only effective ways to deal with the noise annoyance and interference with TV reception. Bird kills can be minimized by careful site selection and slower-moving blades of turbines optimized for low speeds of wind help birds to avoid strikes.

Esthetic objections are more difficult to deal with, and some would argue that wind power's acceptance could hinge on how the public views the multitudes of wind turbines on the landscape (Pasqualetti, Gipe, and Righter 2002). There is no doubt that large wind machines—tips of 2.5-MW Nordex blades are 160 m above ground, compared to 93 m (from the pedestal to the torch tip) for the Statue of Liberty (fig. 5.18)—will be seen by many people as unsightly. They are certainly visible from further away than the much maligned high-voltage lines. And while large numbers of smaller, but much more densely sited, machines snaking along

Figure 5.18
Prototype of the world's largest wind turbine, 2.5-MW Nordex machine, commissioned at Gravenbroich (Germany) in the spring of 2000. Both the tower height and the rotor diameter are 80 m. The photograph is available at <http://www.windpower.dk/pictures/multimeg.htm>.

ridges or dotting mountain slopes or coastlines may be seen by some people as actually enhancing a particular landscape, others, including many who have sought out such locations for their unspoiled vistas, will object to such additions and will feel they are living or walking near an industrial park.

Such objections may be dismissed by pointing out that wind turbines produce no CO_2, reject no warm water, and emit no particulate matter or acidifying gases to society. When the concern is about the long-term integrity of the biosphere, then some visual intrusion is decidedly unimportant compared to the environmental impacts arising from fossil-fueled generation. But, as demonstrated by analogical comparisons of air- and water-pollution impacts of coal-fueled and nuclear electricity-generating plants (see the last section of this chapter), such objective appraisals may not sway many subjective perceptions.

Final, and perhaps the most important, consideration that has to be kept in mind when trying to assess the prospects of wind-generated electricity is the long-term performance of individual turbines and large wind facilities. As with any other engineered system this knowledge can be gained only by installing many GW of wind turbines and operating them for 10–20 years. Only then will we be able to quote accurately long-term availability and reliability rates, and quantify the costs of preventive maintenance (the experience so far points to about 1 c/kWh), unscheduled outages and catastrophic failures brought by man-made mistakes and natural phenomena. There is no shortage of events in the latter category including strong cyclonic systems, unusually long periods of extremely heavy icing (akin to the Québec–New England ice storm of January 5–9, 1998), protracted spells of high atmospheric pressure with arrested winds, or years with unprecedented abundance of insects.

By the end of the year 2000 wind turbines accounted for about 0.5% of the world's installed electricity-generating capacity, and most of the wind turbines had been operating for less than five years and machines with capacities in excess of 1 MW only since 1997. Both the share of the total capacity and accumulated generating experience do not yet allow any firm conclusions about the reliability of supply and the long-term operating cost. As far as the quoted and claimed costs per average kW of installed capacity are concerned, I see merits in arguments of both the promoters and critics of wind power (DeCarolis and Keith 2001; Jacobson and Masters 2001b). Guaranteed prices, tax preferences, and a variety of other governmental subsidies have undoubtedly helped to make wind power competitive but it is unfair to hold this against the wind industry as all of these advantages have been enjoyed in some forms and for much longer time by fossil-fueled and nuclear generations as well (for details see chapters 2 and 6, as well as the last section of this chapter).

On the other hand, if the real cost of electricity from many wind projects is now as low as 4 c/kWh, and if some recent U.S. contracts were at less than 3 c/kWh (AWEA 2001b), then its producers should be making large profits (costs are 5–6.5 c/kWh for coal- and natural gas-fired generation) and wind turbines should be dominating newly installed capacities both in the United States and in Europe. That is not the case: in the year 2000 wind turbines generated merely 0.15% of the U.S. electricity (EIA 2001a). More realistically, McGowan and Connors (2000) put the typical cost of wind-powered generation at 6 c/kWh, still 1.5–3 times the average spot price of electricity. But prices of 4–5 c/kWh are expected soon, close to the cost of new natural gas-fired units (2–4 c/kWh).

The claimed cost advantage should be even higher in many low-income countries of Latin America, Africa, and Asia where a large share of the insufficient supply of electricity is generated from imported crude oil at a cost surpassing 10 c/kWh. And just a few machines could make a great deal of local difference. Wind turbines can boost capacity totals in affluent countries only if the machines are installed in large multiples in wind farms; in contrast, just two or three 500-kW turbines would be sufficient to supply basic electricity needs to almost 100,000 people in Africa or Asia. Combination of these two advantages should have resulted in a global boom of wind farm construction but in reality wind remains a marginal source of electricity in modernizing countries and existing plans foresee only a modest market penetration during the coming 10–20 years.

Most notably, China, with its high dependence on coal, has only 265 MW installed in wind machines (less than 0.01% of its total capacity) and its Program for Development of New Energy and Renewable Energy Industry calls for just 2% of the TPES coming from all forms of renewables by the year 2015 (AWEA 2001a). India's wind-generating capacity was nearly five times as large, 1.267 GW in 2001, but that total was less than 3% of estimated potential and no firm commitment to major expansion has been made (CSE 2001). In Africa, Morocco has a 50-MW plant, Egypt has a 30-MW facility on the Red Sea, and in Latin America there are operating or planned wind farms in Costa Rica and Argentina. Clearly, wind energy will not become a significant global contributor until the Asian countries, which should account for the largest part of all newly installed electricity-generating capacity up to 2050, will embrace it at least as enthusiastically as the Germans and the Danes have done since the late 1980s.

Direct Solar Conversions

Direct conversions of solar radiation harness by far the largest renewable energy resource but their efficiency and capital and operating costs have, so far, kept them from making a commercial breakthrough comparable to the one experienced by wind power since the early 1990s. To be more accurate, I should add terrestrial commercial breakthrough, because space applications, where high cost is much less of a concern, have been spectacularly successful as photovoltaic arrays power scores of communication, meteorological, Earth-observation, and spy satellites (Perlin 1999). But on the Earth it was not photovoltaics, the technique with perhaps the most promising long-term potential among all of the renewables, but rather concentrating solar power projects that became the first grid-connected generators of commercial electricity.

These central solar power (CSP) systems use focused sunlight to generate superheated steam for conventional turbogenerators. Consequently, these techniques can be easily operated by using fossil fuels as well as solar radiation. Their hybrid nature clearly increases their availability (as they can work in cloudy weather and after sunset) and reliability while lowering their operating cost by allowing more effective use of the installed equipment. The CSP systems differ in the way they collect solar radiation: the three distinct ways use troughs, towers, and dishes (Dracker and De Laquill 1996; NREL 2001a). Trough systems use linear parabolic concentrators that focus sunlight onto a tubular receiver running the length of a trough. Better heat transfer rates are achieved by using synthetic oils but such arrangements obviously require heat exchangers in order to generate steam for a turbogenerator. Focusing sunlight along a line rather than to a point results in relatively low concentration ratios (10–100) and hence only in modest temperatures (100–400 °C).

In contrast, power towers collect sunlight with fields of sun-tracking mirrors (heliostats) and concentrate it on a tower-mounted receiver where it heats a circulating liquid. Their two most obvious advantages are high concentration of sunlight (ratios between 300–1,500) that produces high working temperatures (500–1,500 °C) and minimized transport of heated fluids. Parabolic dishes also track sunlight but concentrate it on a receiver at their focal point. This produces high concentration ratios (600–2,000) and temperatures above 1,500 °C but if the method were to be used for larger-scale electricity generation heat collected at individual focal points would have to be transported to a centrally placed engine or turbine.

Peak solar-to-electric conversion efficiencies range from just over 20% for troughs, 23% for power towers, and 29% for dishes but annual rates are considerably lower, ranging 10–18% for troughs, 8–19% for towers, and 16–28% for dishes (Dracker and De Laquill 1996; SolarPACES 1999). With best efficiencies and the sunniest sites troughs and towers could thus have peak power densities around 60 W/m^2 of collecting surface, in less sunny locations the densities would go down to less than 40 W/m^2. Spaces between collectors, service roads, and structures will naturally reduce those rates. Mojave Desert trough systems have overall peak power generation densities of less than 15 W/m^2 (Smil 1991).

Solar capacity factors (fractions of the year the technique can generate electricity at rated power) are mostly between 20–25% (and up to 30%) for troughs and dishes but those central receivers equipped with hot salt thermal storage can potentially have rates in excess of 60%. In spite of their relatively poor ratings parabolic trough systems are the commercially most mature technique, and until the late 1990s the nine plants operating at three sites in the Mojave Desert near Barstow in California accounted for more than 90% of the world's thermal solar capacity.

These plants were built between 1984 and 1990 and their total installed capacity is 354 Mwe with unit sizes between 14 and 80 MW. Luz International, the company that built the plants, went out of business in 1991 but all plants keep operating under new arrangements and selling electricity to Southern California Edison under long-term agreements. Other trough CSP projects include small plants in Egypt, Iran, Morocco, Spain, Greece, and India and much large trough facilities are planned for Iran (289 Mwe), Northern Morocco (178 Mwe), Egypt, Mexico, and Rajasthan (Aringhoff 2001).

If troughs are still rare, solar towers are real oddities. The biggest one, a 10-MWe pilot plant near Barstow was built in 1982 and worked until 1988 by directly generating steam; after retrofitting for the use of a more efficient heat-transfer medium (molten nitrate salt used as both the receiver and thermal storage fluid) the plant reopened (as Solar Two) in 1996 and worked until 1999 (fig. 5.19). Smaller experimental towers were built in Spain and Israel and larger projects, using air-based systems, are planned by a German Phoebus consortium. Nor is there any abundance of dish/engine solar plants. The largest set of six 9- and 10-kW German-built units has been operating since 1992 at Plataforma Solar de Almería in Spain, and both smaller and larger dishes (up to 50 kW) connected to Stirling engines have been tested in the United States, Europe, and Australia where there are plans for a 2-MW plant.

Figure 5.19
Aerial view of Solar Two, the world's largest solar tower project of the 1990s. Photograph by Joe Florez available at <http://www.energylan.sandia.gov/photo/photos/2352/235227d.jpg>.

Unlike CSP systems, PV generation of electricity has no moving parts (and hence low maintenance requirements), operates silently at atmospheric pressure and ambient temperature with a minimal environmental impact and it is inherently modular. All of these advantages could be translated into commercial diffusion of terrestrial photovoltaics only after the cells reached conversion efficiencies close to, and in excess of, 10% and as the average price of PV modules (typically 40 PV cells) and arrays (made up of about 10 modules) was reduced by nearly an order of magnitude between 1975 and 1995 (Zweibel 1993; Service 1996; Maycock 1999; fig. 5.20). Maximum theoretical efficiency of PV cells is limited by the range of photon energies, while the peak practical rates are reduced above all through reflection from cell surfaces and leakage currents.

Large efficiency gaps remain between theoretical, laboratory, and field performances as well as between high-purity single crystals, polycrystalline cells, and thin

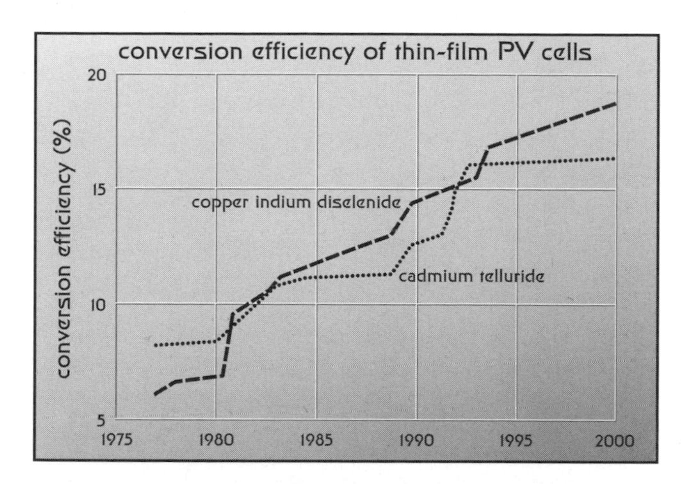

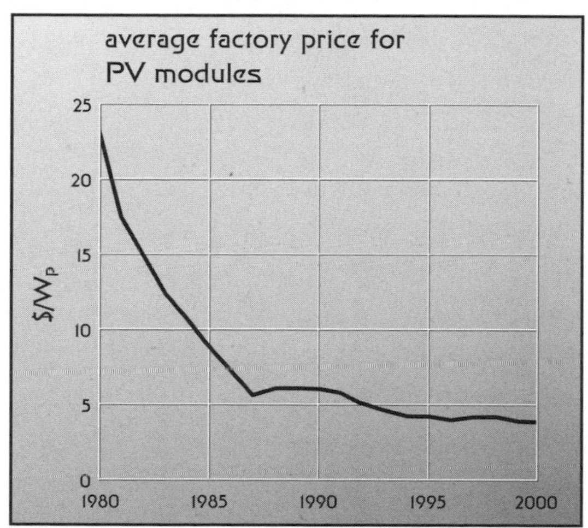

Figure 5.20
Increase in conversion efficiency of two kinds of thin-film cells and decrease in average factory price for PV modules, 1975–2000. Based on graphs and data in Zweibel (1993), Service (1996), NREL (2001b), and PES (2001).

films (Dracker and DeLaquill 1996; Goetzberger, Knobloch, and Voss 1998; Perlin 1999; Markvart 2000). Thin PV films are made of amorphous Si or of such compounds as gallium arsenide (GaAs), cadmium telluride (CdTe), and copper indium diselenide ($CuInSe_2$). Theoretical single-crystal efficiencies are 25–30%, with more than 23% achieved in laboratories. Lenses and reflectors can be used to focus direct sunlight onto a small area of cells and boost conversion efficiencies to more than 30%. Stacking cells sensitive to different parts of the spectrum could push the theoretical peak to 50%, and to 30–35% in laboratories.

In contrast, actual efficiencies of commercial single-crystal moduls are now between 12–14%, and after a time in the field they may drop to below 10%. Thin-film cells can convert 11–17% in laboratories but modules in the field convert as little as 3–7% after several months in operation. Multijunction amorphous Si cells do better, converting at least 8% and as much as 11% in large-area modules. Thin-film microcrystalline silicon is perhaps the best candidate for future mass applications (Shah et al. 1999). With an eventual 17% field efficiency PV cells would be averaging about 30 W/m^2.

Notable takeoff in annual shipments of cells and modules in general, and of grid-connected residential and commercial capacities in particular, came only during the late 1990s (Maycock 1999). Worldwide annual shipments of PV cells and modules reached 0.5 MW of peak capacity (MWp) in 1977, they were nearly 25 MWp a decade later and during the 1990s they rose from about 43 to 288 MWp as the typical price per peak W fell 30% for modules and 25% for cells (fig. 5.21). With nearly 14 MWp shipped in 1990 and just over 88 MWp sold in 2000 the share produced by U.S. manufacturers remained around 30% of the world total (EIA 2001a). At the century's end the largest PV cell and modules producers were BP Solarex in the United States, Japan's Kyocera and Sharp, and Germany's Siemens: these four companies shipped nearly 60% of all new capacities (Maycock 1999). Cumulative installed capacity in the year 2000 amounted to about 500 MWp in the United States and almost 1 GWp worldwide: recent large additions and the average PV cell life span of some 20 years mean that more than 98% of all capacity installed between 1975–2000 is still in operation.

During the early 1990s the four most important markets for PV converters were communication industries (about one-fifth of the total), recreation uses (mainly camping and boating), home solar systems (remote residential generation preferable to expensive hookups to distant grid, and village generation in low-income countries) and water pumping (EC 2001c). By the year 2000, grid connections, with about

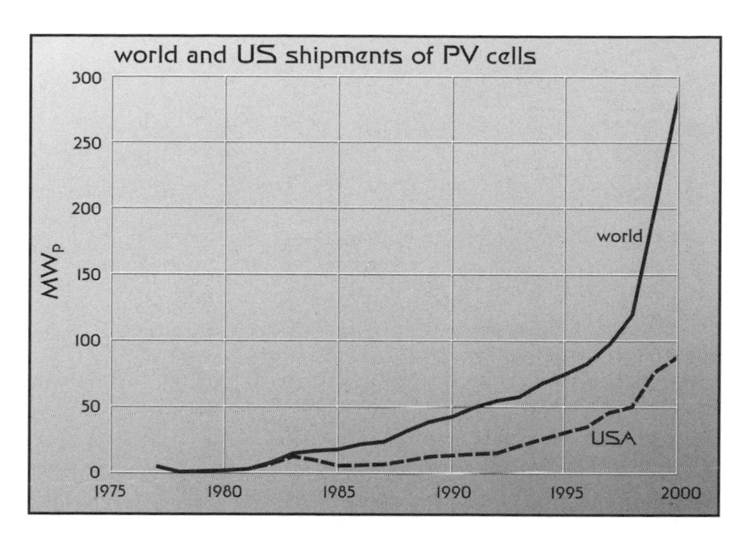

Figure 5.21
World and U.S. PV shipments, 1975–2000. Plotted from data in EIA (2001a) and PES (2001).

20% of all new capacity, had the single largest share of the total market, and grid-connected modules in small and medium-to-large projects are forecast to claim nearly one-third of all capacity added by 2010, followed by solar home systems and communication uses (EC 2001c). Recent growth of PV capacity at 15–25% a year has led to many optimistic forecasts. The U.S. DOE aims at 1 million rooftop PV systems with about 3 GWp by 2010 (compared to just 51,000 installations and 80 MW in 2000) and Japan plans to install annually ten times as many rooftop units by 2010 as it did in 1999.

Worldwide, Maycock (1999) foresees 200 MWp/year connected in 2005 and 700 MWp/year added in 2010. Looking even further ahead, the National Center for Photovoltaics believes that up to 15% (that is some 3.2 GW) of new U.S. peak electricity generating capacity required in 2020 will be supplied by PV systems, and that at that time cumulative PV capacity will be about 15 GWp in the United States and 70 GWp worldwide (NCP 2001). But it is salutary to recall that in 1989 one of the leading PV developers at that time, Chronar Corporation, forecast that 40 GWp will be in place by 2000 (Hubbard 1989)—and less than 1 GW was actually installed.

During the late 1990s it became clear that silicon-based PV may not be the only way of converting sunlight to electricity. New PV cells, photoelectrochemical devices based on nanocrystalline materials and conducting polymer films, offer a welcome combination of cheap fabrication, and a bifacial configuration that allows them to capture light from all angles and greater flexibility of practical use with, for example, transparent versions of different color made into eletricity-producing windows (Grätzel 2001). Electrochemical cells might be the best way to accomplish water cleavage by visible light (theoretical efficiency of converting the absorbed light into hydrogen is 42% in tandem junction systems, and rates around 20% may be eventually practical), while experimental dye-sensitized nanostructured materials produce electricity with efficiencies of 10–11%, and bipolar cells are about twice as efficient.

Finally, I should mention solar energy's fusion analogue, space-based solar power satellites (SPS) first proposed by Glaser (1968). Satellites located in a geosynchronous orbit would convert sunlight around the clock and transmit microwaves to the Earth where they would be absorbed by a rectenna, converted to DC, and fed to an electrical grid. There appears to be no insurmountable physical barrier to the realization of this idea; most notably, microwaves below 10 GHz are subject to minimal damping from the atmospheric water vapor. But scaling up the techniques able to beam the generated energy to the Earth and then to convert it back to electricity on scale required by the projects would need a number of innovative breakthroughs, as would the task of ferrying the enormous mass of material needed in space.

There has been no shortage of elaborations and modifications of this idea, and during the 1980s NASA spent more than $20 million analyzing the engineering and costs of a fleet of 30 SPS (NASA 1989; Landis 1997). Interest in the idea declined during the 1990s and only a low-intensity exploration of the SPS system is now going on in Japan (Normile 2001). Advantages of the Lunar Solar Power (LSP) system, capturing the solar radiation on the moon, are predicated on the successful production of requisite components primarily from lunar materials. Proponents of the idea argue that the efficiencies of the LSP would be very large indeed (Criswell and Thompson 1996)—but I believe that this scheme should be rightly put in the category of energy projects whose probability to contribute to the world's energy supply will remain nil for at least the next two generations. A few more items belonging to this category of futuristic energy supply will be noted as I review the prospects of other renewable energies.

How Promising Are Other Renewables?

This will be an eclectic but brief review looking at conversion techniques whose commercial applications have been around for more than a century but whose development and diffusion has received increasing attention only during the past generation as well as bold new proposals that are unlikely to make any notable commercial difference even after another two or three decades of improved designs and experiments. Geothermal energy is clearly in the first category, while the second group subsumes several intriguing proposals for converting the ocean's kinetic and thermal energies. And I must add that there are a number of proposed conversion curiosities and oddities that will receive no attention in this section beyond a simple mention.

These include proposals for harnessing melting waters of Greenland's glaciers and transmitting the generated electricity to Europe via Iceland through DC cables (Partl 1977), wind power through kites (Goela 1979) and solar chimneys, tall, plastic-covered structures erected amidst large solar radiation-collecting areas, and creating the air flow in order to drive wind turbines (Richards 1981). The first two options remain entirely on paper, while the last technique was actually tried in the early 1980s when the German Federal Ministry for Research and Technology paid for building the first (and last) large solar chimney in Manzanares in Spain that worked for about seven years. I cannot resist mentioning one more unorthodox proposal: creating giant genetically engineered galls and implanting them on tree trunks where they would perform a reverse of photosynthesis and generate hydrogen (or CH_4) that would be tapped by a collector pipe and led to a house! The author of this scheme concluded that it would create a proper interface between vast solar collection systems, the forests, and an efficient energy transportation and distribution system, the natural gas pipeline net (Marchetti 1978, p. 9).

But back to reality. As already noted, average geothermal flux is only a tiny fraction of the solar flow. Sclater et al. (1980) calculated the total heat loss through the ocean floor at 33.2 TW and through the continental crust at 8.8 TW; these rates prorate, respectively, to about 95 and 65 mW/m^2 and add up to about 1,325 EJ per year. Geothermal energy in place is, of course, much larger (by at least five orders of magnitude) and its theoretically extractable total depends on its final uses and on temperature and depth cutoffs. Only vapor- and liquid-dominated hydrothermal resources can be used directly for electricity generation but these flows reach the Earth's surface, or are accessible through drilling very near it, only a limited number of locations. In contrast, plentiful hot magma that intrudes into the top 10 km of the crust in many regions around the world contains an order of magnitude more of heat than do hydrothermal flows, and hot dry rocks within the same depth have an order of magnitude more heat than does the near-surface magma (Mock, Tester, and Wright 1997).

But while drilling to depths of more than 7 km to reach rock temperatures in excess of 200 °C (average thermal gradient is 25 °C/km) is now possible it would be economically prohibitive to do that in order to inject water for steam generation. Hot magma at accessible depth is limited to regions of ongoing or geologically recent volcanic activity, hydrothermal resources with flows below 100 °C are suitable only for industrial process or household heating, and the most economical way to use the hottest hydrothermal flows is to tap those that reach the surface or require only shallow wells. Consequently, today's conversion techniques could tap only tiny fractions of the enormous global geothermal flows: at best they could harness about 72 GW of electricity-generating capacity and enhanced recovery and drilling improvements currently under development could enlarge this total to about 138 GW (Gawell, Reed, and Wright 1999). The largest shares of this potential are, as expected, along the tectonically active Pacific margins of the North, Central, and South America, and Asia.

Geothermal electricity generation began at Italy's Larderello field in 1902 (fig. 5.22); New Zealand's Wairakei was added in 1958, California's Geysers came online

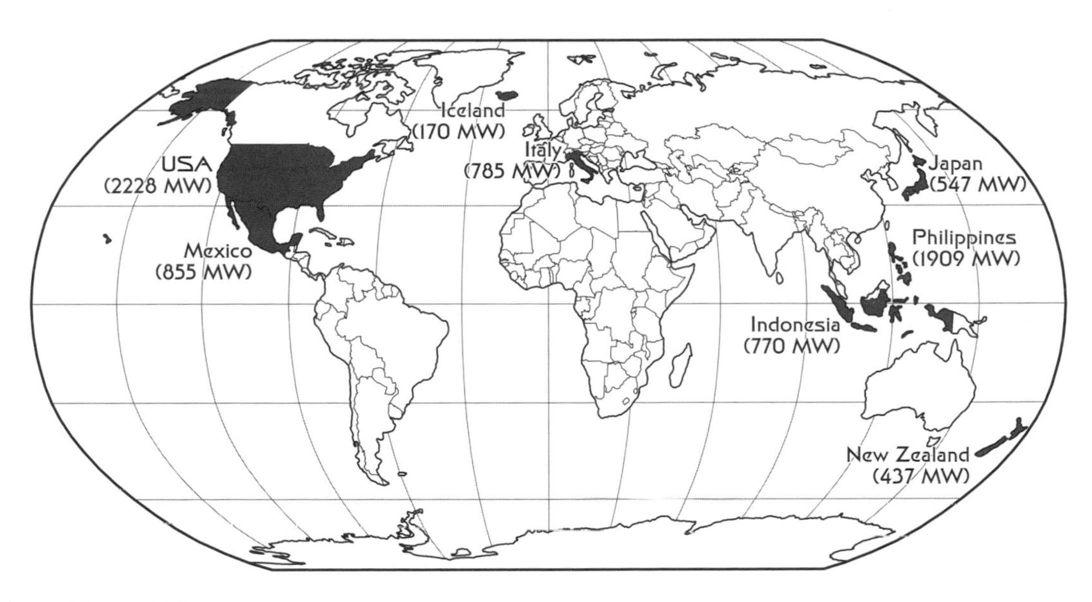

Figure 5.22
Locations of major geothermal electricity-generating plants show a predictably high degree of association with the margins of tectonic plates and with active volcanic areas. Based on data in IGA (1998).

in 1960 and Mexico's Cerro Prieto in 1970. Only since the 1970s has geothermal generation expanded beyond these four pioneering high-temperature vapor fields. By the end of the 1990s the United States had the highest installed capacity (nearly 2.9 GW), followed by the Philippines (1.8 GW), Italy (768 MW), Mexico, and Indonesia (IGA 1998). The global capacity of about 8.2 GW is only about 11% of the total that could be harnessed with existing techniques, and even if we were to develop the prospective potential of 138 GW it would represent less than 5% of the world's electricity-generating capacity in the year 2000. Clearly, geothermal energy will remain a globally marginal, although nationally and locally important, source of electricity. But it can contribute more in the distributed supply of industrial and household heat: for example, U.S. geothermal heating capacity is already more than twice as large as is the installed capacity in geothermal electricity generation.

Geothermally assisted household heat pumps, preferably closed-loop systems storing summer heat and releasing it in winter, are a particularly efficient option (Mock, Tester, and Wright 1997). If they were used in U.S. households with no access to natural gas they could save nearly 100 GW of peak winter electric capacity during the late 1990s. Geothermal energy should also be used more widely by industries (from secondary oil recovery to timber drying), greenhouses and in expanding aquaculture. The combination of electricity-generating and heating applications could make the greatest difference in about 40 low-income countries (many of them islands) with large geothermal potential and shortages, or outright absence, of other energy resources (Gawell, Reed, and Wright 1999).

Before 1975, proposals for converting oceans' energies were dominated by various tidal schemes; afterwards came designs for harnessing kinetic energy of waves as well as the relatively small difference in the temperature between surface and deep water. Global tidal friction dissipates about 2.7 TW and the maximum locally extractable rate depends on the tidal range (Smil 1991). The world's highest tides are in Nova Scotia's Bay of Fundy (7.5–13.3 m for spring tides), and other places with high tides include Alaska, British Columbia, southern Argentina, the coast of Normandy, and the White Sea (Wayne 1977). Theoretical power capacity of the world's best 28 sites adds up to 360 GW (Merriam 1977) but high capital cost of structures partly submerged in salt water has stopped all but one large project, the 240-MW French Rance tidal plant near St. Malo. None of the Bay of Fundy long-studied large-scale projects (with capacities of 1–3.8 GW) ever progressed beyond a planning stage (Douma and Stewart 1981).

Blue Energy Canada (2001) is now promoting the use of new submerged or surface-piercing Davis turbines that would be suspended from tidal bridges. Slow-moving blades of these turbines would pose little danger to marine life and would cope better with silt flows. The company predicts low capital cost ($1,200/kW for large facilities) and high capacity factors (40–60%) and it believes that the global potential to be exploited by this technique is more than 450 GW. Its first large design is for the Dalupiri plant built across the San Bernardino Strait in the Philippines whose 274 Davis turbines would have a peak capacity of 2.2 GW. Any meaningful assessments of this technique will have to wait for its commercial applications.

During the 1970s ocean thermal energy conversion (OTEC) was touted enthusiastically as "the only renewable source of power with the potential of satisfying the growing needs of the world's population" or even more memorably as "the OTEC answer to OPEC" (Zener 1977, p. 26). Some forecasts envisaged that the electricity produced by the OTEC would be also used on board of large factory ships to synthesize ammonia (Francis and Seelinger 1977). Theoretically available power is impressive: global potential is in excess of 200 TW (Rogner 2000) and energy extractable by an ideal heat engine working with top temperature of 25 °C (surface of subtropical and tropical ocean) and 5 °C (water at a depth of 900 m) is 2.87 J/g, an equivalent of a hydroelectric plant operating under a nearly 300-m head (fig. 5.23). In reality, all OTEC designs are trying to extract useful power from temperature differences seen as unusable in conventional thermal generation and hence costs are bound to be high and efficiencies low (OTA 1978).

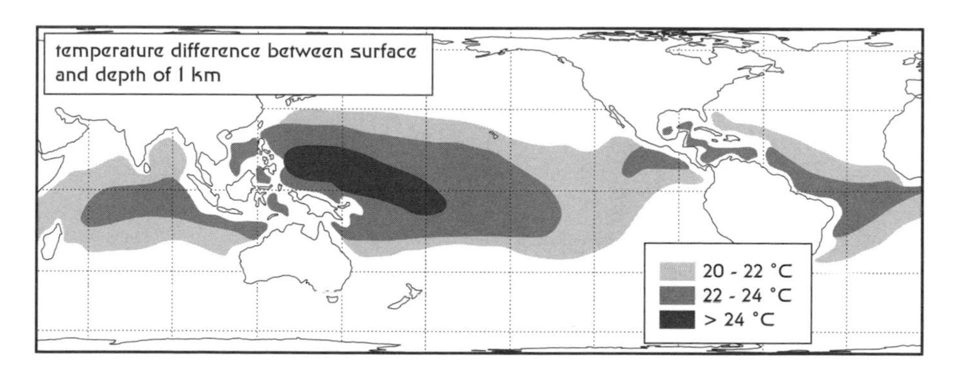

Figure 5.23
Ocean areas where the temperature difference between surface and depth of 1 km is greater than 20 °C. The eastern tropical Pacific has the best potential. Based on a map in NREL (2001c).

Two basic approaches to extracting the low temperature difference (at least 20 °C are needed) are a closed cycle using warm water to evaporate a working fluid with a low boiling point and an open cycle where seawater itself is boiled in a vacuum chamber. The resulting vapor would then drive a turbogenerator and deep, cold seawater would be used to condense the working fluid. The necessity of pumping huge amounts of water through the plant inevitably reduces the already low conversion efficiency. Tellingly, George Claude's first small OTEC prototype located off of Cuba generated about 22 kW while consuming 80 kW to run its equipment (Claude 1930).

Nevertheless, a great deal of research and development effort were invested in OTEC schemes during the late 1970s and the early 1980s (Hagen 1975; OTA 1978; NREL 2001c). A small (50 kW gross, 15 kW net) barge-mounted unit was operated intermittently for a few months off Hawai'i (for the last time in May 1993); larger heat exchangers were tested on a converted tanker and Tokyo Electric and Toshiba-designed closed-cycle plant worked for nearly a year off Nauru (Penney and Bharathan 1987). Eventually, large OTEC plants were to produce not only electricity but also fertilizer NH_3, fresh water, and food. One of OTEC's leading promoters claimed in 1987 that his company, Sea Solar Power, was able to supply power to many tropical countries at an acceptable price (Anderson 1987)—but no nation took up that wishful offer. Some enthusiasts may still believe that OTEC will be useful when oil gets expensive, particularly for tropical islands without any land energy resources but with high ocean temperature differences close to the shore. Cuba, Haiti, and the Philippines are excellent examples in this category (NREL 2001c), but all of these places may be much better off by turning to other fluxes of renewable energies.

I cannot see OTEC as anything but an intriguing armchair source of energy and an engineering curiosity. This leaves waves, whose global kinetic energy is nearly as large as the tidal flows (Rogner 2000), as perhaps the most promising source of ocean energy. Decades of research, particularly in the countries of Atlantic Europe and in Japan, have turned out a variety of conversion techniques. The first notable design was Salter's Duck and later came various oscillating water column devices. The former is made up of a series of segmented cam lobes that are rocked on a large spine by waves, with pumps connected to the lobes, sending pressurized water through small pipes to a generator (Salter 1974; Isaacs and Schmitt 1980). The oscillating devices contain a chamber with an underwater opening and the water oscillation inside the chamber forces air through an electricity-generating turbine (Kingston

1994). The Japanese began testing a prototype of an oscillating generator anchored in 40 m of waters in 1998 (JAMSTEC 1998).

The first commercial electricity-generating station based on this principle, was a land-based 500-kW LIMPET (Land Installed Marine Powered Energy Transformer) that was connected to a grid in November 2000 on the Island of Islay off Scotland's west coast (Wavegen 2001). A pair of 375-kW prototype semisubmerged, articulated offshore wave energy converters built by Ocean Power Delivery Ltd. was scheduled for trials in 2002 (Yemm 2000). The company sees as many as 900 devices and 700 MW installed by 2010 and LIMPET designers claim that there is enough recoverable wave power around the United Kingdom to exceed all domestic demand and that harvesting less than 0.1% of the world's potential could supply more than five times the global energy demand—if such conversions could be done economically. Needless to say, using short-term operations of installations of less than 500 kW as bases for scaling the costs of a TW-scale global industry is ridiculous, and it would be surprising if wave devices were to have cumulative worldwide installed capacity exceeding a few GW by 2020.

Infrastructural Changes and the Future Hydrogen Economy

Higher shares of renewable conversions will obviously need more than just investing in new turbines, collectors, or cells. Electricity produced by these conversions will expand the supply of the most desirable form of energy but it will create four major challenges. The first arises from the relatively low power density of renewable energy flows: higher reliance on these flows will bring a much higher share of distributed (decentralized) generation. The second is the consequence of inflexible locations of large concentrations of renewable conversions. The third results from the combination of the inherently intermittent nature of renewable conversions and the lack of any practical large-scale means of electricity storage. This reality creates complications for meeting both base-load and peak electricity demand in systems dominated by renewable conversions. The fourth challenge is due to the need of converting a part of renewably generated electricity to high-density portable fuel required for many transportation uses, a reality that makes it highly desirable to think about a gradual shift to hydrogen-based conversions.

Decentralization dreams tap into a deep stratum of discontent and unease concerning some overwhelming features of modern urban–industrial society. Their long historical tradition was perpetuated during the twentieth century by such disparate thinkers as F. D. Roosevelt, who believed that decentralized electric power will empty sprawling slums and put people back on the land (Novick 1976), and the pioneers of a nuclear age. In 1946 these enthusiasts concluded that “nuclear power plants would make feasible a greater decentralization of industry” as they could be located where needed in order to greatly reduce transmission costs (Thomas et al. 1946, p. 7). None of those predictions was turned into reality.

Distributed generation of electricity had eventually become appealing, both technically and economically, thanks to the changes in the optimum size of plants and advances in managing of transmission and distribution networks. An interesting example in the latter category is the net metering whereby small producers (even households) can send electricity they generate to the grid at the time they do not need it and withdraw it later when they require it. Increased sizes of boiler-turbogenerator units were one of the hallmarks of the evolution of electric systems since their conception until the early 1970s (see fig. 1.9). Optimum size of generating units grew eventually to about 1 GW but then the falling demand for electricity and technical advances (above all the rise of gas turbines) sharply reversed the trend as units of 50–250 MW filled most of the new orders.

The most important reason for this reversal has been the fact that the installed cost of simple aeroderivative natural gas-fueled turbines in the 150-MW capacity range fell to as little as U.S.$300(1995)/kW and the cost of combined-cycle systems with capacities below 300 MW had dropped to U.S.$350(1999)/kW while their efficiencies rose to close to 60% (Linden 1996). And the downsizing has gone much further, as hydrocarbon-fueled microturbines with capacities in the 25–250-kW range can provide economically generated electricity for many applications, including peak shaving for large industrial and commercial users and reliable supply for isolated communities (Scott 1997). Capstone MicroTurbines (2002) became the first company to get certification for meeting the grid interconnection standard with its 30- and 60-kW machines that can be used in hybrid vehicles as well as for stationary microgeneration. Mostly oil-fueled reciprocating engines for stationary cogeneration, averaging about 2 MW, are another increasingly popular means of distributed production of electricity (Williams 2000).

Actual contributions by fuel cells are still marginal but their potential is promising. Natural gas–fueled PAFC with cogeneration can deliver efficiencies close to 80%, which makes it already competitive, in spite of its still high installed cost, for such high-load users as hospitals where the waste heat is used for supplying large volumes of needed hot water. PEMFC running on reformed natural gas may become the principal means of high-efficiency distributed generation in the 100 kW–2 MW range, while in low-income countries a great deal of rudimentary electrification could be accomplished by PV-battery systems in a kW range.

Besides reaching a wide range of markets with high efficiencies that distributed generation allows, thanks to its high modularity, for faster completion of new capacities, it can also reduce, or even eliminate, the cost of grid connections, cut transmission and distribution losses, and produce waste heat that can be used locally for many low-temperature applications. Distributed generation can also reduce the overall reserve capacity because less capacity is likely to fail at once with a large number of smaller units and improve reliability thanks to lowered exposure to major grid failures. Using natural gas turbines and fuel cells, wind and PV systems can also produce electricity with significantly less pollution than in large central fossil-fueled plants. And, obviously, distributed generation will further accelerate the demise of the vertically integrated utilities that have dominated the electricity market throughout the twentieth century (Overbye 2000).

But all of this does not mean that the entire system can shift rapidly from the traditional centralized pattern to a completely distributed, self-powered generation—or that such a shift would be actually desirable. Reliable base-load supply will always be needed and while fuel cells could supply part of that demand, intermittent sources cannot. Moreover, about 90% of the population increase that will take place in today's low-income countries during the first half of the twenty-first century will be in cities where hundreds of millions of poor urban families will not have the luxury of putting up a few PV modules on their roofs. And most of the families in the world's megacities (with more than 10 million people) will have to be eventually housed in high-rises and supplying those buildings with locally generated renewable energies is either impractical or impossible: power density mismatch is simply too large. What is desirable is a gradual emergence of much more complex arrangements where large plants will retain a major role but where their generation will be complemented by significant contributions from medium and small capacities as well as from institutional and household-based microgenerators.

The location imperative is obvious. Unlike the fossil-fueled electricity-generating plants that can be flexibly located either near large load centers (to minimize transmission distances) or far away from densely populated areas (in order to minimize air pollution levels), GW-sized concentrations of wind turbines, PV cells, or wave-converting machines must coincide with the highest power densities of requisite renewable flows. The necessity of locating renewable conversions at these sites will, in almost every contemplated instance of such a development, call for new long-distance transmission links as well as for higher transmission capability (in W) and capacity (the product of capacity and distance) of the existing lines. For example, even a cursory survey of the existing U.S. HV links shows either complete absence of HV lines or only a rudimentary capacity to move electricity from the Southwest (the region of the most advantageous central solar and PV locations) to the Northeast, or from the windy Northern Plains to California or Texas.

And although the U.S. transmission capacity nearly doubled during the last 25 years of the twentieth century when it is divided by peak load, the resulting ratio, which represents the distance to which a given plant can expect to sell energy during the conditions of peak demand, has been falling since 1984 and the trade area of a given generator is now only about 70% of the mean in the mid-1980s (Seppa 2000). Even more notable is the fact that California's ratio, which is far above the national mean (about 480 vs. 285 km), is much lower than that for the U.K.'s national grid (about 960 km). Clearly the United States, as well as Canada, lag behind Europe in having high-capacity interconnections able to move electricity from the best locations for wind or solar generation to large load centers along the Atlantic coast, in the Midwest and in California.

Similar absence or inadequacies of high-capacity HV links between regions endowed with high power densities of direct and indirect solar flows and densely populated load centers can be found in every large country that is contemplating higher reliance of renewable conversions. For example, China's windiest and sunniest areas are, respectively, the northern grasslands in Inner Mongolia and the arid, unpopulated western interior of Xinjiang and Tibet, while its largest load centers are along the eastern and southern coast (the capital, Beijing, being an exception as it is relatively close to Mongolia). And, to add just one more set of examples, high wave power densities along the coasts are found in such remote, and often also extremely inhospitable, areas as Northern Scotland, Western Ireland, Alaskan islands, southern Argentina, and Chile. And proposals of transcontinental transmission megaprojects delivering electricity from the Congo Basin to Europe were already noted in the hydro energy section earlier in this chapter.

These realities do not make large-scale concentrations of renewable power impossible but they make them certainly considerably more expensive. In contrast, the inherent intermittency of both direct and indirect solar flows is much less amenable to technical solutions. Of course, the problem would be most acute in a system dominated by highly distributed direct conversions of solar radiation and relatively least worrisome when the supply would be composed of a variety of highly interconnected renewable conversions. Similarly, consequences of unresolved intermittency would be much less of a concern in construction and agricultural tasks whose operation has been long accustomed to weather-caused interruptions while they would be intolerable in a multitude of electricity uses in modern manufacturing and service sectors.

Overcoming solar intermittency requires either a grid large enough to move electricity across time zones or building a storage (or relying on a fuel-based backup) sufficient to cover the demand both during the predictable nightly interruptions and much less predictable spells of heavily overcast sky. Long-distance latitudinal connections would be able to take advantage of different peak-generation times while imperfectly predictable spells of weak diffuse sunlight, caused by cyclonic systems ranging from brief thunderstorms to powerful hurricanes and blizzards or by massive frontal systems, would require considerable storage capacities. Either case would obviously require major investment to insulate the society from undesirable effects of intermittency and only a careful consideration of these costs will make it clear how expensive it would be to operate a system with reliability comparable to the performance that is now routinely achieved by a fossil-fueled generation.

Unfortunately, storing electricity is not an easy task. Surveys of progress in large-scale techniques of stationary electricity storage come up always with the same theoretically appealing options and almost always with some promising new developments in battery design (Kalhammer 1979; Sørensen 1984; McLarnon and Cairns 1989; ESA 2001). But, so far, there has been no breakthrough in designing either high-density, high-efficiency, low-cost batteries for smaller applications or inexpensive high-power devices that would fit the multi-MW to GW-sized systems of modern electricity generation. Only the latter achievement would make it possible to decouple the timing of generation and consumption of electricity and it would allow entire systems to rely for periods of time solely on solar- or wind-generated electricity and individual consumers to be grid-independent for many hours. In contrast, storage needed for bridging power (when switching from one source of generation to another) and for power quality control is needed only for seconds, at worst for a few minutes.

Although it is now a significant share of the total hydroelectric capacity, pumped storage, expensive to build and requiring special sites to locate, cannot be seen as the principal means of energy storage in a predominantly renewable system. In some mountainous countries, pumped storages may take over a significant share of the needed load but the option is obviously inapplicable in large areas of the densely populated lowlands of Asia. Its only commercial analog, compressed air storage, uses low-cost, off-peak electricity to force air into underground caverns or mines and uses it later to supply peak power on a short (minute's) notice. There are only three compressed air storage plants. German Huntorf (290 MW) built in 1978, McIntosh in Alabama (1991; 110 MW), and Norton in Ohio, by far the largest (2.7-GW) unit scheduled for operation in 2003 and taking advantage of a limestone mine more than 600 m underground (ESA 2002).

Commercial use of one of more than half a dozen battery types for large-scale electricity storage is limited by one or more disadvantages (ESA 2002; fig. 5.24). Flow batteries (Regenesys, vanadium redox, zinc bromine) have low energy density while high-energy density metal–air batteries are difficult to charge and have very short life cycles. Sodium sulfur, lithium-ion, and various advanced battery designs are very expensive and common lead-acid batteries have limited cycle lives when deeply discharged. Among the latest commercial projects is a 15-MW Regenesys Technologies storage plant at a British station and TVA's similar 12-MW unit in Mississippi. Prototypes of new $Mg/Mg_xMO_3S_4$ batteries have been tested recently; the advantage of a Mg anode is obvious as the light element is plentiful, safe to handle, and environmentally benign. Aurbach et al. (2000) believe that it should be possible to get eventually large, rechargeable versions whose practical energy densities would be greater than 60 Wh/kg, 50% or more above the standard Ni–Cd and lead–acid units, but no commercialization is imminent.

All of these realities favor hydrogen as the most desirable energy carrier in the predominantly nonfossil, high-tech civilization and fuel cells as the most versatile convertors of that fuel. Fuel cells, fundamentally simple exothermic electrochemical devices that combine hydrogen and oxygen to produce electricity and water, are an old (William Grove in 1839) invention that languished for generations as a curiosity without any practical use. A fundamental change took place only during the early 1960s when NASA began installing fuel cells on its spacecraft. Alkaline fuel cells (AFC) have been the primary source of electricity on every program from Gemini to the Space Shuttles. During the last two decades of the twentieth century numerous advantages of fuel cells have been extolled in many technical as well as popular publications (Appleby and Foulkes 1989; Hirschenhoffer et al. 1998; Srinivasan et al. 1999).

Absence of combustion means elimination of any air pollutants save for water vapor. Absence of moving parts makes for very quiet operation. Electrochemical oxidation of hydrogen makes it possible to surpass the thermodynamic limits that restrict efficiencies of heat engines and convert commonly 40–60% and even more

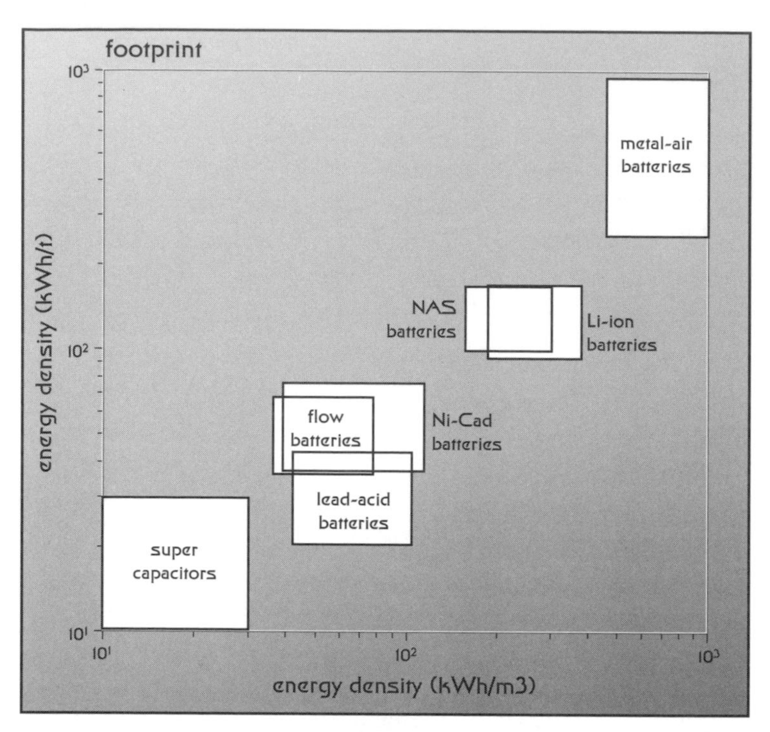
footprint
energy density (kWh/t)
energy density (kWh/m3)
10³
10²
10¹
10¹
10²
10³
metal-air batteries
NAS batteries
Li-ion batteries
flow batteries
Ni-Cad batteries
lead-acid batteries
super capacitors

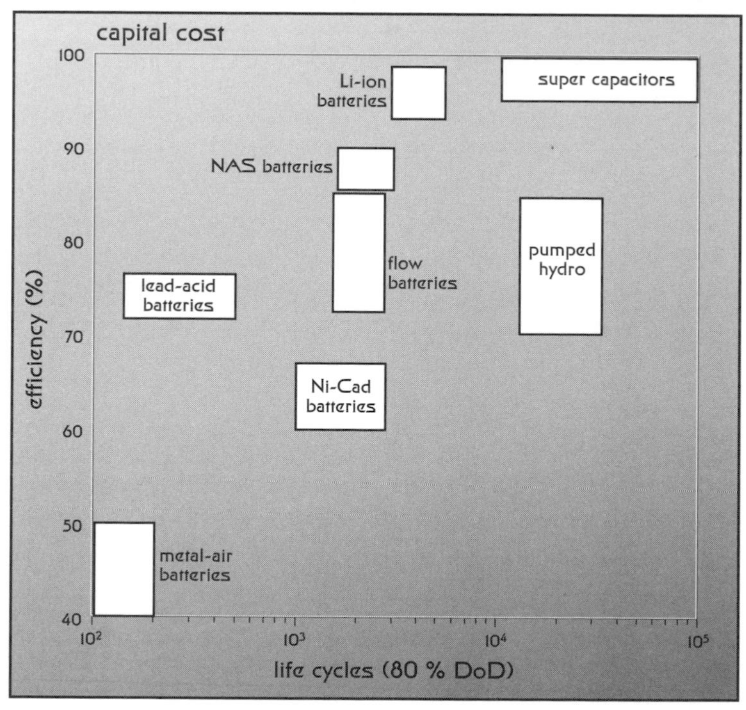
capital cost
efficiency (%)
life cycles (80 % DoD)
100
90
80
70
60
50
40
10²
10³
10⁴
10⁵
Li-ion batteries
super capacitors
NAS batteries
pumped hydro
flow batteries
lead-acid batteries
Ni-Cad batteries
metal-air batteries

than 70% of hydrogen into electricity (theoretically in excess of 90% for the lower heating value). High energy density of liquid hydrogen (120 MJ/kg compared to 44.5 MJ/kg for gasoline) favors mobile applications, and modularity means that the cells could be eventually used to power devices ranging from laptops and lawn mowers to heavy earth-moving equipment and multi-MW electricity-generating plants.

Half a dozen major fuel cell designs are now under intensive, and accelerating, research and development in North America, Europe, and Japan. All basic cell designs are compared in figure 5.25. NASA's widely deployed AFCs were just noted, and the PEMFC (proton exchange membrane), PAFC (phosphoric acid), MCFC (molten carbonate), and DMFC (direct methanol) were already mentioned in the last chapter when I was appraising the sources of the future demand for natural gas. Solid oxide fuel cells (SOFC) operate at even higher temperatures than do MCFC (about 1,000 °C) and, unlike all other designs, use solid ZrO_2–Y_2O_3 electrolytes. Cell stacks up to 100 kW have been tested and because of the high temperature conversion efficiencies with cogeneration may be in excess of 80% (Srinivasan et al. 1999). New electrolytes, including the recently developed solids based on proton-conducting acidic salts (Haile et al. 2001), and many other advances in materials used in fuel cells and techniques employed in their manufacturing—from membranes that are less permeable to methanol to cheaper fabrication processes—are needed before less expensive and durable cells begin claiming significant shares of various niche markets (Steele and Heinzel 2001).

By the end of the twentieth century the only fuel cell designs that were successfully commercialized were AFCs in space and small capacity PAFCs for some types of stationary generation, both representing important but relatively narrow-niche markets. As already noted in chapter 4, some forecasters are confident that the initial phase of a truly massive diffusion of fuel cells installed in passenger cars and in other road vehicles will begin taking place even before 2005 and that it will be in full swing before 2010. Even if these claims turned out to be correct and were followed by further rapid growth of mobile applications it might be premature to conclude that the first decade of the twenty-first century is the last decade of automotive internal combustion engines. These machines, with more than a century of improving

◀ **Figure 5.24**
Comparisons of energy densities, efficiencies, and life cycles for batteries ranging from the classical, inexpensive lead-acid units to modern flow batteries. Based on graphs in ESA (2001).

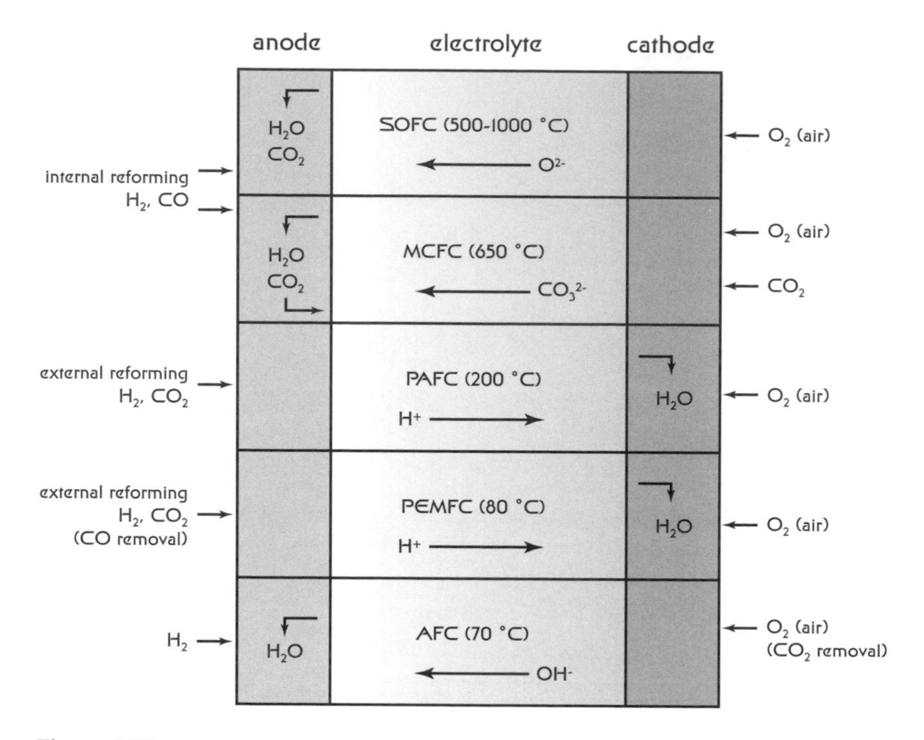

Figure 5.25
Anodes, electrolytes, and cathodes of five types of fuel cells. Based on Steele and Heinzel (2001).

performance and enviable reliability, can be made still more efficient and more reliable.

In addition, the first generation of PEMFC-powered cars will use reformers so they could be energized by various hydrogen-rich liquids rather than by the pure gas itself. Not surprisingly, there is a large degree of uncertainty regarding the coming size of national and global fuel cell markets: recent forecasts have differed by up to an order of magnitude and the divergence in projections has been actually increasing. Even if some more optimistic forecasts were correct, one million U.S. cars powered by fuel cells in 2010 would be less than 0.5% of all vehicles at that time. Consequently, we are not at the threshold of a new era dominated by fuel cells and hydrogen but this lack of any immediate need to contemplate such an arrangement has not discouraged a veritable growth industry devoted to speculations about the parameters of what has become generally known as the hydrogen economy.

As with any such fundamental transition there would be advantages and complications, and, expectedly, advocates and critics of hydrogen economy have been accentuating one or the other. Their writings range from sci-fi-like musing to serious engineering appraisals. Earlier detailed explorations are those of Bockris (1980) and Winter and Nitsch (1988). Among the recent volumes Hoffmann (2001) provides a comprehensive overview of the history and progress of the grand idea, while Dunn (2001) is a good example of an advocate account, Norbeck et al. (1996) of a detailed technical examination of automotive options, and Ogden (1999) of a general review of challenges inherent in building hydrogen infrastructure. All too commonly popular accounts of a hydrogen-dominated world betray deep scientific ignorance as they describe the gas as a near-miraculous energy source rather than identifying it as a still very costly and far from perfect energy carrier.

A recent account (Anderson 2001, p. 21) went even further and its astonishing claim is worth quoting at length:

> Hydrogen is the most common element in the universe, while oxygen is equally easy to obtain. So the fuel supply is infinitely renewable . . . There is a further piquancy. The easiest way to produce hydrogen is by electrolysis . . . It could be that in the longer-run, the fuel cell itself would generate the electricity to produce the hydrogen to run the fuel cell, even recycling the water which the fuel cell creates to be re-used for further hydrogen: the dream of perpetuum mobile realised at last.

By far the most piquant element of this embarrassingly erroneous account is that the author wrote it after consulting "one of the few people I know who is scientifically literate" who also assured him that "within 20 years at the outside" the West will be no longer dependent on oil and that energy prices will plummet everywhere as fuel cells will take over (Anderson 2001, p. 20). Historical evidence demonstrates convincingly that it takes much longer than less than two decades to make complex transitions to new energy sources and carriers (see chapters 1 and 3). In the specific reference to hydrogen Ausubel (1996), assuming average atomic H/C ratios of 0.1 for wood, 1 for coal, 2 for crude oil, and, obviously, 4 for methane found that the logistic growth process points to a global methane economy after 2030 and to a hydrogen-dominated economy, requiring production of large volumes of the gas without fossil energy, taking shape during the closing decades of the twenty-first century (fig. 5.26).

Today's mass-produced hydrogen is obtained mainly by steam reforming of methane and secondarily by partial oxidation of hydrocarbons. Steam reforming now requires as little as 103 MJ/kg H_2, with most of the gas used in the Haber–Bosch

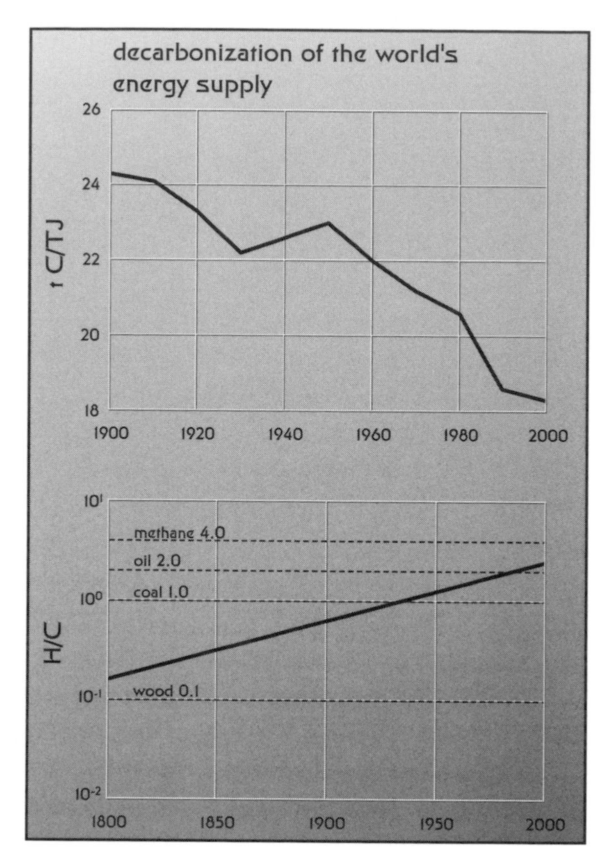

Figure 5.26
Two perspectives on decarbonization: declining carbon intensity of the global TPES and the rising H:C ratio, 1800–2100. Based on data in Smil (2000a) and a graph in Ausubel (1996).

synthesis of ammonia and with the second most important use in oil refining (Ogden 1999; Smil 2002). Energy cost of electrolysis is equivalent to between 70–85% of the higher heating value of the gas, or between 100–120 MJ/kg H_2 (Ogden 1999). Even if we were to assume that inexpensive electricity from renewable sources will be available in large amounts within 20 years, allowing us to electrolyze water by using indirect solar flows rather than by burning fossil fuels, that welcome advance would not translate immediately into a hydrogen-dominated world.

The main reason is the absence of the requisite infrastructure, the lack of a widespread, reliable, and safe system for storing, distributing, and delivering the gas (Ogden 1999; Jensen and Ross 2000; Schrope 2000). This absence creates the most obvious problem for hydrogen-powered vehicles: manufacturers will not make and promote cars that people could not fuel easily—but energy companies are not eager to invest into a new expensive infrastructure in the absence of any demand. The two most obvious incremental steps are to begin the transition with hydrogen-fueled cars equipped with on-board reformers and to engage in incremental diffusion of pure hydrogen vehicles by converting first many specialized vehicle fleets that use centralized fueling facilities.

As already noted, the first-generation PEMFC-powered vehicles have reformers and, depending on design, they will burn methanol, gasoline, or even naphtha. But putting in place a methanol infrastructure is hardly the best way to move into a hydrogen economy and, a broader problem, reformers cannot operate instantly and require a few minutes of warm-up before the catalytic conversions can take place. That is why proceeding via the niche route directly to hydrogen appears more rational. In modern urban societies there is no shortage of fleet vehicles that could be converted to hydrogen and fueled in centralized facilities: buses, taxis, numerous delivery trucks, city, and government vehicles. These fleets would be a significant foundation for gradual introduction (tax-preferred, or environmentally regulated) of passenger cars.

With accessible hydrogen, fuel-cell cars would not be the only choice as internal combustion engines need only a minor modification to run on hydrogen. This arrangement would be much less efficient than fuel-cell drive but making it a part of hybrid drive would bring the performance close to the cell level while retaining the device that has served so well for more than a century. Moreover, combination of internal combustion engines with a moderate storage capacity in hybrid vehicles, commercially introduced by Honda's and Toyota's pioneering designs (Insight and Prius, respectively) will provide highly competitive alternatives to fuel cell–powered vehicles even when the emphases were to be solely on energy conversion efficiency and on specific emissions. Consequently, it may be highly premature to conclude, as do many promoters of this new technique, that hydrogen-powered fuel-cell vehicles are the ultimate automotive design. Farrell and Keith (2001, p. 44) argue that the more polluting heavy-duty freight vehicles, as well as cargo ships, are better candidates for gradual introduction of transportation hydrogen than the passenger cars, and that "thoughts of hydrogen as a transportation fuel for the general public should be deferred for at least a decade."

But even if the enthusiasts were to prevail, the conversion to hydrogen-fueled cars will be slow. In order to progress beyond limited urban driving ranges, hydrogen-powered vehicles compatible to today's cars would have to carry enough fuel to cover 400–600 km without refueling. This is one of the two most intractable problems of using hydrogen for mobile application (Schlapbach and Züttel 2001). Uncompressed hydrogen occupies 11,250 L/kg while standard high-pressure steel tanks require about 56 L/kg H, and new carbon-composite tanks under development can halve that rate. But such high-pressure vessels (up to 45 MPa) are obviously dangerous as standard equipment in passenger cars. Liquefied hydrogen takes up 14.1 L/kg but energy costs and logistics of keeping the gas below −241 °C are not trivial.

That is why hydrogen adsorption in solids of large surface areas (such as nanostructured graphitic carbon), or the absorption of the gas by metal hydrides (various Pd, La, Zr, Fe, Mg, and Ti compounds) are more desirable storage options. For practical use the proportion of hydrogen reversibly absorbed by hydrides should be at least 4–5%, and preferably above 6%, of the total mass, compared to 0.6–3.6% achieved so far; higher, but unreplicated, claims were made for the storage in carbon nanotubes. Figure 5.27 contrasts the volumes of compressed, liquid, and hydride-absorbed hydrogen storage needed to drive 400 km with a fuel cell–powered car optimized for mobility.

Safety is the other key problem of a hydrogen economy. For decades the gas has been handled safely by chemical enterprises but this experience does not translate automatically to a mass public distribution system made up of hundreds of thousands of transfer points and millions of users. Hydrogen is safer than gasoline in

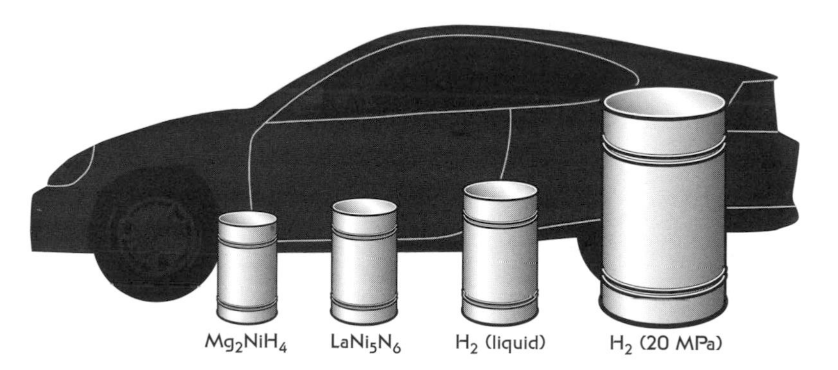

Figure 5.27
Volumes occupied by 4 kg of hydrogen compacted in four different ways; a silhouette of a compact car is used for a comparison. Based on Schlapbach and Züttel (2001).

two important ways: as it is very buoyant it disperses quickly from the leak (while gasoline puddles and its fumes persist) and it is nontoxic. On the other hand, it has a lower flammability limit (4% by volume vs. 5.3% for gasoline) and a much wider range of flammability (4–75% vs. 1–7.6% for gasoline). Even more importantly, hydrogen's ignition energy is an order of magnitude lower than for gasoline or methane (0.02 mJ vs. 0.24 mJ). But what will matter in the final instance is the public acceptance, and that can come only when the fuel's handling and distribution will be demonstrably no more risky than are today's ubiquitous gasoline or natural gas systems.

These musings about infrastructural demands of emerging or contemplated energy systems conclude my review of renewables but no assessment of future nonfossil options would be complete without a critical look at nuclear energy. Some observers of energy affairs might argue that given the parlous state of that industry this should not take more than a brief paragraph. If I were to agree with them I would do it as follows. Although nuclear power plants and their fuel are not such an easy target for terrorists as the media reporting would lead us to believe (Chapin et al. 2002), public acceptance of fission reactors and hence the prospect of their renewed construction—dim before September 11, 2001, because of a combination of economic, environmental, and safety reasons—look even less appealing after the terrorist attack on the United States.

The fast breeder option, as I made clear in chapter 3, has been shelved, if not completely discarded. And fusion research continues to pursue, doggedly and forlornly, its ever-receding 50-year horizon of commercialization. But these realities do not justify the abandonment of nuclear energy—they only make a strong case for critical and deliberate examination of new reactor designs, better waste disposal methods, and for keeping the nuclear option alive.

Is There a Future for Nuclear Energy?

When the question is asked so loosely the answer is obviously yes. Most of more than 400 stations (438 were in operation by the end of 2001) of the first generation of nuclear power plants whose combined capacity now produces over 16% of the world's electricity will be with us for several decades before being gradually decommissioned. Moreover, new stations are being completed every year (six were connected to the grid in 2000) and other projects are under construction (35 in December 2001) or plans are being made for further expansion of fission generation in several countries, most notably in Asia (IAEA 2001b; WNA 2001a). The question should be posed more specifically: is there a long-term future for nuclear fission as a major supplier of the world's primary energy? Or even more pointedly: will nuclear fission contribute a higher share of the global TPES by the year 2025, or 2050, than it does at the very beginning of the twenty-first century?

The best answer at the time of this writing (2002) is that it could—but most likely it will not. There are three obvious reasons for this conclusion. Most importantly, no utility in any Western country, with the exception of France, has placed any new orders of fission reactors during the last two decades of the twentieth century. All of the 39 orders for nuclear plants placed in the United States after 1973 had been cancelled, and even Electricité de France, which began building its last two reactors at Civaux in 1991 (the second one completed in 1999), has no new units on order. As an early nuclear renaissance appears unlikely, this reality increases the chances of eventual demise of the industry in North America and Western Europe through ageing and neglect. And as these two regions account for about two-thirds of the world's installed nuclear capacity, this trend, unless outweighed by massive new installations elsewhere, may be the harbinger of a global decline.

Moreover, several European countries have gone a step further and legislated binding deadlines for the abandonment of all nuclear generation. Sweden, one of the pioneers of nuclear generation, was the first country to forsake formally a nuclear future. Its 12 reactors, put into operation between 1972 and 1985, added up to 9.8 GW of generating capacity and made the country the world's leader in per capita installed nuclear power. But in 1980 Swedish Parliament, based on the result of a referendum, decided to phase out all of the country's nuclear power plants by the year 2010 and concentrate on developing a new energy system based, as far as possible, on renewable and indigenous resources with minimized environmental impact (Kaijser 1992). Barsebäck 1 was the first reactor to be shut down (in November 1999) and the closure of its twin should come by the end 2003 (Löfstedt 2001). Germany is another major European producer set on the path of closing down its nuclear stations before their expected life span.

And, finally, reactors under construction and in various planning stages will not bring any large net additions of generating capacity. Reactors to be commissioned between 2002 and 2006 will add about 26 GW, and as of the end of 2001 only Russia, India, and three East Asian countries (Japan, South Korea, and China) had plans for further expansion of their nuclear capacities (WNA 2001b). These projects, slated for completion between 2006 and 2015, add up to about 18 GW. If all of them will go ahead there may be an additional 44 GW on line by the year 2015, or only about 12% more than the 2001 total. The net addition will depend on the extension of nominal design lifetime of existing reactors. Original licences for 30–40 years of operations have been already extended in some cases to 50–60 years, and lifetimes up to 70 years are envisaged for some reactors, but others will be undoubtedly decommissioned before 2015.

Whatever the actual net expansion before 2015, it will still be a part of the first generation of commercial nuclear reactors whose diffusion began in the 1960s (or the second generation of all reactors, the first one being the prototypical design for submarines and first stations of the 1950s). Should there be a major commitment to a new generation of nuclear power plants based on substantially or radically new designs? Arguments against new reactor designs and in support of letting the nuclear era expire after the existing reactors will come to the end of their useful life have been repeated too many times to warrant anything but the briefest summaries of the three key concerns.

The most fundamental is the one seeing nuclear power as a categorical evil. Hence its acceptance as a leading mode of energy conversion would require, at best, that the civilization concludes a Faustian bargain with this dangerous technique, exchanging an inexhaustible source of energy for a vigilance and longevity of social institutions unlike any encountered during the millennia of organized human societies (Weinberg 1972). At worst, a worldwide adoption of nuclear generation as the dominant mode of electricity supply would lead to environmental and military catastrophes. The first ones would be brought about by severe accidents whose chances increase with the rising number of operating stations, the latter ones would be an inevitable consequence of unpreventable proliferation of nuclear weapons aided by the spread of commercial nuclear generation.

Not far behind this categorical rejection is the persistent public concern about the safety of everyday nuclear plant operations and about the security of both temporary and long-term repositories of radioactive wastes. Facts are clear: the public is exposed to more radiation from natural sources than from a normal operation of nuclear reactors (Hall 1984); the Three Mile Island accident in 1979 was the only U.S. release of radiation beyond the confines of a commercial plant; and no properly shielded Western reactor could have ever experienced the Chernobyl-type disaster. But none of these facts has been able to make the Western public opinion accept new nuclear power plants as benign neighbors. More informed segments of the population point beyond the probabilities of operation accidents to the dismal record of managing long-lived wastes.

Management of the U.S. radioactive waste has been a provisional affair that does not appear to be moving toward any early effective resolution (Holdren 1992; GAO 2001). Spent fuel generated at more than 100 commercial reactors is stored temporarily at the reactor sites in special pools that, at the older plants, were not designed to hold the lifetime output of the reactor waste. As for the safety of these storages at individual plants, Wilson's (1998, p. 10) suggestion that "modest human attention—not much more than a night watchman" might suffice was shown to be naïve in the wake of the terrorist attack of September 11, 2001, and the subsequent realization that the production of dirty bombs using relatively easily accessible radioactive waste was definitely on the list of al-Qaeda designs (Blair 2001). Reprocessing of spent fuel, the route followed by the United Kingdom, France, and Japan, is not a solution as it produces bomb-grade plutonium, and only new recycling techniques may produce fuel that would be useless for weapons.

The protracted, frustrating, and still inconclusive story of the U.S. Department of Energy's Yucca Mountain Project provides little confidence that the challenge of long-term disposal of radioactive wastes will be resolved soon (Flynn and Slovic 1995; Whipple 1996; Craig 1999; Macilwain 2001; GAO 2001). The project was designed to accommodate spent fuel enclosed in canisters in tunnels about 300 m below the desert floor yet also some 300 m above the water table in reconsolidated volcanic ash strata that have been undisturbed for millions of years—but even after two decades of studies worth some $7 billion there is still no assurance that the first shipments of temporarily stored spent fuel will take place by 2010 or 2015. Last, there is the matter of real costs of nuclear electricity. During the 1990s the performance of nuclear plants, in the United States and worldwide, has definitely improved (Kazimi and Todreas 1999). Operating costs were lowered (to as low as 1.83 c/kWh in 2000) as average capacity factors rose (in the United States from about 75% in 1990 to nearly 88% a decade later) and unplanned capability losses declined.

Operating costs among plants vary by as much as a factor of two but the best stations are quite competitive with any fossil-fueled or renewable alternatives. Of course, this good news applies only after ignoring excessive capital costs and massive past, and substantial continuing, public subsidies (the Concorde's profitable operation rested on the same foundations). Consequently, no recent sale of stations by vertically integrated U.S. utilities divesting themselves of the electricity-generation business has come close to recovering the construction costs (Lochbaum 2001). And, to cite a British example, in 2001 the state-owned British Nuclear Fuels Limited had £356 million of shareholders' funds but £35 billion of total liabilities, most of them tied to the decommissioning of its reprocessing plants, and the U.K. Atomic Energy Authority had a further £8.7 billion of liabilities (UKAEA 2001). Another important consideration is the declining interest in nuclear studies: continuation of the recent trend in enrollments would bring serious shortages of qualified personnel to operate the remaining plants 20 or 30 years from now.

What will it take to mitigate, if not to reverse, these negatives? Nothing can erase the enormous past subsidies sunk into the development of nuclear energy and it is hard to imagine that the industry could operate without such fundamental exceptions as the protection given to it by the Price–Anderson Act. Beyond this, the same conclusions keep reappearing decade after decade: a nuclear option requires increased public acceptance, which, in turn, rests on greater scientific literacy, on effective alleviation of safety concerns, and on demonstrating that new nuclear plants could be competitive without any additional government subsidies (Rose 1979; GAO 1989; Rossin 1990; Beck 1999). But one development that should have worked strongly in favor of nuclear generation has not made the expected difference.

Concerns about the widespread and costly impacts of global climate change caused by anthropogenic emissions of greenhouse gases emerged as one of the leading public policy matters of the last two decades of the twentieth century and nuclear power offers a perfect solution as it produces no greenhouse gas emissions beyond those associated with the production of materials needed to build the plants and mine the uranium ore. But even when the full energy chain is evaluated nuclear generation produces only about 9 g CO_2/kWh, an order of magnitude less than do coal-fired power plants, and also less than the PV generation (IAEA 2001c; fig. 5.28). If all of the electricity generated by nuclear plants was to be produced by burning coal, the world's CO_2 emissions would rise by about 2.3 Gt C, an equivalent of more than one-third of the total produced by fossil-fuel combustion in the year 2000. Curiously, this impressive total of avoided emissions is almost never mentioned in current debates about the management of greenhouse gases.

In addition, nuclear generation does not obviously produce any SO_2, globally the most important cause of acid deposition, or NO_x, gases contributing not only to acid precipitation but also in the formation of photochemical smog. And high energy density of nuclear fuel—1 kg of uranium can generate 50 MWh, compared to 3 kWh/kg of coal—and high power density of generating stations means that the land claimed by nuclear sites is one to three orders of magnitude smaller than the totals needed by renewable conversions. But none of these incontestable environmental benefits seem to count in favor of nuclear power as others concerns, notably those about safety, security, and radioactive waste disposal, override them.

Concerns about operation safety began to be addressed with added urgency by many studies and by improved reactor designs right after the Three Mile Island

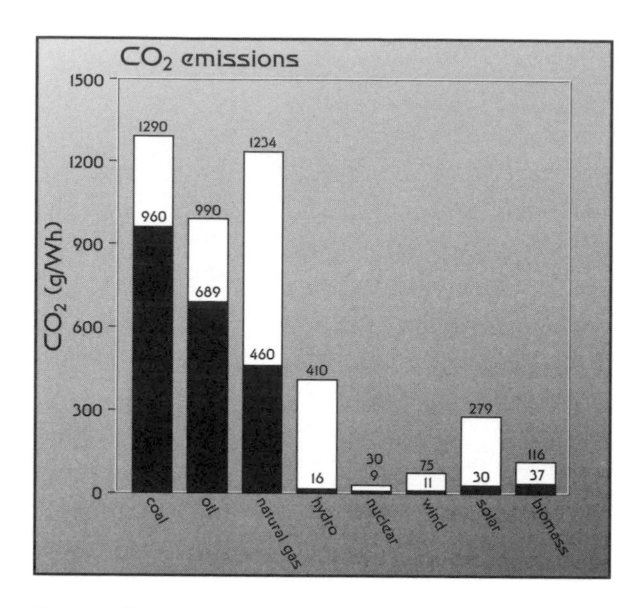

Figure 5.28
Calculations of CO_2 emissions for the full cycle of energy production show nuclear generation having a lower rate than wind and biomass. Based on a figure in IAEA (2001c).

accident in 1979. The mishap did not cause either any casualties or any delayed radiation-induced increases in mortality but it was a key factor in speeding up the end of further expansion of nuclear capacities in the United States. In 1980 David Lilienthal, the first Chairman of the U.S. Atomic Energy Commission, called on nuclear engineers to devise effective technical fixes (Lilienthal 1980). Within the next five years several groups of physicists and engineers, including those who were among the key creators of the first nuclear era, reexamined the design of light water reactors. Their conclusion was that inherently, or passively, safe nuclear reactors are feasible and that they could provide the basis for what they termed is to be the second nuclear era (Firebaugh 1980; Lester et al. 1983; Weinberg et al. 1984).

New designs for inherently safe nuclear generation were to eliminate any possibility of core meltdown. In PIUS (Process Inherent Ultimately Safe) reactors, the entire primary system (including the steam generator) was to be submerged in a large pressurized pool of water containing boric acid (boron being a strong neutron absorber). Modular HTGR (High Temperature Gas Reactor) designs had such a low specific power output that even after the failure of all heat-removing systems the reactor would still cool itself by conduction, convection, and radiation. Modular design was to be also a key feature of new, inherently safe, fast breeders. Nearly two decades after these designs were proposed we are still waiting for any one of them to become commercialized as attempts to lay intellectual foundations of a new nuclear era continue. The U.S. Department of Energy initiated its Generation IV program in 1999; OECD framed the future of nuclear energy within a fashionable sustainable development perspective (OECD 2000), and engineering companies in several countries keep refining better reactor designs (Lake, Bennett, and Kotek 2002).

Helium-cooled reactors whose core contains hundreds of thousands of billiard-ball-sized graphite spheres filled with poppy-seed-sized particles of uranium oxide have emerged as perhaps the most promising design. Its relative simplicity would allow for small (120-MW) modules and hence for more flexible deployment with lower costs due to standardized factory production. And there is an even more far-reaching proposal. Teller et al. (1996) outlined the concept of a new fission reactor producing electricity at the 1-GW level that is emplaced at least 100 m underground and could operate for 30 years without human access after the fuel loading and the start of fission. The breeding reactor would consist of a small ignition region containing enriched U^{235} embedded in a much larger breeding-burning section of U^{238} or, preferably, Th^{232}. Helium would be the coolant removing the generated heat to the surface, and the reactor vessel would serve as the spent-fuel burial cask.

Research on inherently safe and economic reactors is going on in at least nine countries. So far, none of these innovative proposals has been translated into a working prototype although South Africa plans to build a test version of a gas-cooled, pebble-bed reactor before 2010. Moreover, Lochbaum (2001) noted that the reduced cost of the pebble-bed modular reactor has been achieved by eliminating the reactor containment building and hence making the structure an attractive target to terrorists who could ignite the graphite and thus cause the release of large amounts of highly radioactive material from the burning core.

All of the units under construction or on order are just improved versions of standard water-cooled reactors first introduced two generations ago. I see little chance of any vigorous worldwide renaissance of nuclear generation without a strong resurrection of the technique in the United States, but such a shift appears unlikely. To begin with, public distrust of nuclear generation runs deep and it has been accentuated by new disclosures regarding the seriousness of environmental and health consequences of nuclear weapons production, both in the United States and the former Soviet Union, and by the estimates of enormous expenditures needed to clean up the sites of 17 U.S. nuclear weapons plants (see, among many others, Hohenemser,

Goble, and Slovic 1990; GAO 1995b; Makhijani, Hu, and Yih 1995; Zorpette 1996). And the unease might be even greater if the public were to realize that there have been some unpardonable cases of negligence and malpractice in operating the U.S. reactors that did not result in any accident but forced an early retirement (Yankee Row in Massachusetts) or temporary shutdowns for extensive repairs (Lochbaum 2001).

No less important, the country's decision-making system, with its division of power, its propensity for adversarial policymaking and legal confrontations and its maze of competing special interests, is peculiarly vulnerable to irresolution where uncertainty prevails and consensus is absent. Perhaps the only two developments that could change that would be an acceleration of global warming approaching, even surpassing, the upper limits of current estimates for the first half of the twenty-first century (for more see chapter 6), or a resolute quest (as opposed to a number of ineffective plans dating back to the early 1970s; see chapter 3) for a much higher degree of the domestic U.S. energy self-sufficiency brought on by repeated conflagrations in the Middle East that would lead to interruptions of global oil supply.

While acting as risk minimizers, and hence keeping all possibilities open, we should not discard the nuclear option. Beck (1999, p. 135) concluded that "the future of nuclear industry is more in hands of research than in receiving orders for more plants"—but he also admitted that research funding will not be forthcoming without changing public opinion to at least a neutral stance. What will it take to do that? A generation ago Rose (1974, p. 358) noted that "the hazards of burning fossil fuels are substantially higher than those of burning nuclear ones, yet many debates have enticed the uncritical spectator to just the opposite conclusion." How much have the concerns about global warming, rarely noted in 1974 but both widely publicized and increasingly real a quarter-century later, shifted this perception?

And I would argue that even extraordinary advances in public acceptance of nuclear energy based on the future impressive record of operation safety and security may not be enough to usher in a new nuclear era in a world of more rational energy uses and more affordable conversions of renewable energies. And so the only certain conclusion is that a new nuclear era could be fueled comfortably for many generations to come either by deploying various breeder designs or by exploiting thorium, whose resources are much more plentiful (about fivefold) and also more accessible than those of uranium. But, for many reasons, we may never make any resolute decisions to follow either of these routes and nuclear generation of electricity may gradually fade out during the first half of the twenty-first century.

6

Possible Futures

Nothing is more fascinating about the future of complex systems than their essential openness, and this attribute is particularly intriguing when contemplating the future of civilizations. To be sure, there are many confining parameters but the realm of what is possible is eventually always larger than people had imagined because surprises—be they natural catastrophes, technical breakthroughs, or social upheavals—keep reshaping human expectations and ambitions. But, as I clearly demonstrated in my extended critique of long-range forecasts of energy affairs (chapter 3), our record of looking ahead is very poor even when dealing with relatively orderly systems because their complex dynamics have repeatedly made most of the specific predictions obsolete in very short periods of time.

Consequently, I will not offer either any specific quantitative forecasts of global or regional energy needs for the years 2025, 2050, or 2100 or any predictions about the shares of various primary energies that may supply that demand. Instead, I will evaluate what appears to be possible, both in technical and social terms, and suggest what I believe to be desirable by outlining the key goals that are worth pursuing.

But before doing so I must first address the matters of efficient and reduced use of energy and explain why the strategies promoting these objectives are no less essential than the quest for new energy sources and techniques. Anybody even cursorily familiar with the current state of principal energy conversion techniques appreciates the enormous opportunities for using fuels and electricity more efficiently. I will explore the technical potential for these gains in a few key consumption areas and appraise the limits of translating this potential into actual energy savings. Whatever the future gains may be, the historical evidence is clear: higher efficiency of energy conversions leads eventually to higher, rather than lower, energy use, and eventually we will have to accept some limits on the global consumption of fuels and electricity.

Although there may be no apparent economic reasons to do so immediately (there is no reason according to standard economic accounts that neglect many externalities), we will have to do so in order to maintain the integrity of the biosphere that is essential for the perpetuation of any civilization. Doing so will be extraordinarily challenging but realistic assessments indicate that it can be done. Critical ingredients of an eventual success are straightforward: beginning the quest immediately, progressing from small steps to grander solutions, persevering not just for years but for generations—and always keeping in mind that our blunders may accelerate the demise of modern, high-energy civilizations while our successes may extend its life span for centuries, perhaps even for millennia.

Efficient Use of Energy

Few activities are seen by the educated public in a more approving light than the quest for higher energy efficiency. By increasing the ratio of energy services to energy inputs we reduce the initial use of fuels and electricity thereby helping to lower energy prices and to moderate numerous environmental impacts of energy conversions. This chain of benefits is repeatedly cited by many energy and environmental experts in their endorsements. For example, Casten (1998) says that the United States can, and must, double the efficiency of its electricity generation and that this doubling will reduce prices by 30–40% while cutting CO_2 emissions in half, and Ferguson (1999, p. 1) concludes that "energy efficiency is a winner for both the economy and the environment." Let us agree with these statements and take a closer look at what increased efficiency can do. Subdividing this review into two basic categories is helpful for appraising the chances for further efficiency gains.

The first category includes a large variety of improved performances that have resulted from technical fixes and innovations. These measures may embody quite a bit of energy (economists would prefer to say capital) but once marketed they do not demand any changes of habitual lifestyle: flipping switches and turning ignition keys is no different with fluorescent lights (instead of incandescent lightbulbs) or hybrid cars (instead of ICE machines). Historical perspectives make it clear that most of the technical innovations producing these higher efficiencies were not necessarily motivated by the quest for less wasteful energy uses. New, and more efficient, converters and processes are often introduced simply as a part of normal stock turnover after years (many electrical appliances, cars) or decades (turbogenerators, airplanes) of service. The quest for improved productivity is another ubiquitous factor and the truly revolutionary replacement of steam-driven group shaft power by individual electric motors (described in chapter 1) is perhaps the best example of such a substitution.

Ease of control and operation are major factors favoring more efficient household converters: a natural gas-fired furnace linked to a thermostat replacing a manually stoked coal stove is a perfect example. Considerable efficiency improvements can be also realized indirectly by introducing products and services whose direct energy cost (fuel or electricity needed for their operation) remains basically the same but whose manufacture, handling, and disposal requires much less energy. This strategy relies on better design, consideration of life-cycle costs, replacement of more energy-intensive components, and widespread recycling of materials.

The quest for higher conversion efficiencies is ultimately limited by thermodynamic considerations and by stoichiometric minima. Some widely used converters and processes, including efficient chemical syntheses, large electric motors and the best boilers, have come fairly close to these barriers. At the same time, even after generations of technical innovation there are many opportunities for efficiency gains on the order of 30–60% that could be realized within 20–25 years, and potential savings on the order of 10–25% are very common. Jochem (2000) offers good continental and regional summaries of potential efficiency gains, disaggregated by major economic sectors, by the years 2010 or 2020.

Modes, accomplishments and advantages of realizing these savings have been described and evaluated in numerous post-1973 writings: their representative selection should include Ford et al. (1975), Lovins (1977), Gibbons and Chandler (1981), Hu (1983), Rose (1986), Culp (1991), Casten (1998), and Jochem (2000). While it is difficult not to be impressed by the scope of potential savings it is also imperative to realize that there will always be substantial gaps between technical potentials, best possible practices and perfectly optimized processes on one hand and actual quotidian performance on the other.

Consequently, too many reviews and appraisals of higher energy conversion efficiency promise too much, and too many predictions of future savings err on a high side. Nobody has been a more enthusiastic practitioner of this oversell than Amory Lovins and his Rocky Mountain Institute. As already noted in chapter 3 (see fig. 3.8), in 1983 Lovins put the global energy consumption in the year 2000 at just 5.33 Gtoe, more than 40% below the actual total. In 1991 RMI claimed that the potential cost-effective savings could cut the U.S. energy demand by 75% (Shepard 1991). And these measures were supposed to be not just cost-effective but quite inexpensive, with most of the electric efficiency gains to be implemented for under 3 c/kWh (or less than half of the typical generation cost) and large amounts of oil to be saved at a cost of less than $12 per barrel. Less than a decade later Hawken, Lovins, and Lovins (1999, p. 11) claimed in their *Natural Capitalism* that "within one generation nations can achieve a tenfold increase in the efficiency with which they use energy, natural resources and other materials."

My attitude to Lovins's sermons has not changed during the past 25 years: I have always wholeheartedly agreed with many of his conclusions and ideas and parts of his and my writings, although informed by very different backgrounds, are interchangeable (Smil 2001). I share his quest for technical rationality, higher conversion efficiencies, and reduced environmental impacts. I agree with many of his goals—but not with any of his excessive claims and recurrent exaggerations. The tenfold claim is a perfect example of this propensity to ignore the real world. Obviously, it cannot refer to improvements of today's dominant energy conversions because even the least efficient ones among them (e.g., ICEs) average more than 20%, and the most efficient ones (e.g., large boilers and electric motors) turn more than 90% of initial energy inputs into useful outputs. Tenfold increases in efficiency of these conversions are thus unthinkable. Only radically different conversions combined with fundamentally restructured patterns of final demand could yield much larger savings—but such arrangements cannot be accomplished in the course of a single generation.

If the tenfold gain were true, North America's and EU's natural capitalists content with maintaining today's standard of living would have no need for 90% of today's energy needs by the year 2020, and as those countries consume nearly half of the global TPES such a flood of unwanted primary energy and other mineral resources would bring a virtual collapse of all commodity prices. And today's low-income countries would have no need to rush in and snap up these cheap commodities: these countries now claim about one-third of the world's resources, but by converting them with ten times higher efficiency they could easily afford a standard of living equivalent to the current Western mean.

Using average per capita energy consumption as a guide China's standard of living would be lifted at least 30% above today's Japanese mean! And, remember, the arrival of this nirvana was forecast to be less than 20 years away. There is no need to belabor the "tenfold increase" claim any further: it is not just an exaggeration, it is a patently ridiculous one. But readers who did not skip chapter 3 will recognize in these extravagant claims a pattern reborn, the idea of social salvation through efficiency fix. The only difference is that this time it is not nuclear energy that is to liberate the world in a single generation but fuel cells and Lovinsian hypercars that are to usher the age of contentment as they

> will save as much oil as OPEC now sells, decouple road transport from climate and air quality, and provide enough plug-in fuel-cell generating capacity when parked to displace the world's coal and nuclear power plants many times over (Denner and Evans 2001, p. 21).

If true, this book, and anybody else's concerns about the future of global energy supply are superfluous, and indeed useless. Moreover, RMI claims that the hydrogenous salvation (appraised critically in chapter 5) is right around the corner: Hypercar (2002), RMI's spinoff outfit, promises that these wonder machines will become widely available in less than ten years. Back to the reality that also includes the phenomenon quite opposite to Lovinsian exaggerations. Because long-term technical innovations and their rates of diffusion are unpredictable or can be, at best, foreseen in partial and disjointed ways (see chapter 3), many forecasts regarding efficient use of energy have also been, and will continue to be—in contrast to the RMI's habitual exaggerations—too timid.

Abundant opportunities for improving efficiency of virtually every energy conversion undertaken by modern societies make it impractical to provide a systematic review of even major options: given the confines of this section it would have to be nothing but a truncated recital of items. A more useful, and a much more revealing, way is to note first those opportunities where our actions would make the greatest difference. I will also look at the potential impact of a few specific energy-saving techniques distinguished either because of their extraordinary savings or because of the multiplier effect of specifically small but massively diffused technical improvements. But there are large gaps between what is possible and what prevails and so after looking at those impressive potential gains I will go on to describe some key barriers to rational energy use.

The most rewarding technical fixes should proceed by focusing first on two kinds of conversions: on relatively large improvements to be gained in the most inefficient sectors of energy use, and on relatively small savings whose large aggregate impact emerges through the multiplier effect as they are eventually adopted by millions of users. And a key reality of promoting more rational use of energy must be always kept in mind: because of the complexity of modern societies even the single most rewarding technical fix or behavioral change can cut the energy use by only a small margin. Consequently, substantial long-term aggregate reductions of energy use that would yield greatest economic, social, and environmental returns can be achieved only by pursuing a wide variety of options in both categories. Given the low average efficiency of conventional thermal electricity systems every effort should be made to lower both the specific use of fuel as well as transmission and distribution losses.

Global mean is barely above 30% for thermal generation, and plant use and transmission and distribution losses cut the energy content of the actually delivered electricity to just above 25% of the fuel supplied to coal- or hydrocarbon-burning plants. And given these large losses it is self-evident that every effort should be also made to install more efficient electric motors, appliances, and lights and less wasteful electrical heat and electrochemical processes. Examples of possible savings are easy to construct. For example, the sequence of conventional gas-fired generation (boiler-turbogenerator, overall 35% efficiency), wasteful transmission and distribution (10% loss) and inferior electric motors (70% efficiency) will result in just 22% of the initial energy providing a useful service. A new combined-cycle gas turbine (60% efficient), minimized transmission losses (6%), and high-efficiency (90%) motors will transform 50% of the fuel's energy content into a useful service, more than doubling the overall performance.

Such illustrative examples can be criticized as high-performance options whose widespread use in energy systems known for longevity of their infrastructures could come only after many years, or decades, of gradual diffusion. Their value is in showing what is technically (and hence not necessarily economically or socially) possible: they should not be mistaken for universal guideposts but neither should they be seen as unrealistic goals because actual performance of the best systems has reached, and surpassed, such rates. In any case, what is more important than setting a particular target for future energy savings is to be engaged in a process of improvements and innovation whose goals could be repeatedly modified and adjusted as circumstances demand but whose continuation should be assured.

As I already noted in chapter 5, cogeneration and combined-cycle gas turbines make it possible to raise the efficiency of thermal generation by at least 50% above the prevailing mean. Inevitably, more widespread cogeneration would bring substantial cuts in CO_2 emissions (Kaarsberg and Roop 1998). Any effort aimed at diffusing these techniques should receive the highest attention, as should, at the opposite end of the system, the large-scale introduction of more efficient converters of electricity. Every inefficient electric motor, light, and appliance is a suitable candidate for replacement but, compared to replacing much more durable and more costly major appliances or industrial motors, the multiplier effect of relamping is easiest to achieve as its cost is relatively low and as the switch can be achieved very rapidly. And the relamping can make a notable difference because in affluent countries about 10% of all electricity is used for lighting.

Replacement of the current mixture of incandescent lightbulbs and standard fluorescent tubes with the most energy-efficient lights, be they compact fluorescents for indoors or various high-efficacy sources for the outdoors, is thus perhaps the most frequently cited case of large and almost instantaneously achievable electricity savings. Looking at an actual large-scale substitution involving a variety of lights is much more realistic than contrasting the least and the most efficient converter for a particular use. Relamping a single city block along Berkeley's Telegraph Avenue in 2001 saved about 45% of previously used electricity, a cut indicative of 30–50% electricity savings that can be achieved by comprehensive urban relamping (Philips Lighting 2001).

Lighting efficacy has been improving in waves of innovation for more than a century (see fig. 1.14) and the process will continue in decades ahead. Universally desirable substitution of incandescent light for indoor use has become even more desirable, and much more acceptable. With more efficient electronic (as opposed to magnetic) ballasts fluorescent lights are now even better performers (efficacies as high as 90 lumen per watt) able to deliver improved color rendition (i.e., closer matching of daylight spectrum) and a greater selection of color temperatures (ranging from 2,500 K to 7,500 K). Among the greatest improvements already on the market or available as working prototypes are the full-spectrum, light-emitting diodes (LEDs) and sulfur lamps.

LEDs, so well known as little red or green indicator lights in electronics, have already replaced lightbulbs in car brake lights, taillights, and turn signals and their larger and brighter varieties are now also found in many traffic lights and exhibit and advertisement displays. But their greatest impact would come from white-light versions that are much more difficult to produce inexpensively (Craford, Holonyak, and Kish 2001). Mixing red, green, and blue wavelengths is tricky to do with great uniformity and so a better way is to use LED photons to excite phosphors. Consequently, it is not unreasonable to conclude that by the middle of the twenty-first century the average efficacy of the rich world's lighting might be 50% above the current rate. At the same time, high-power, high-efficacy (1 kW, 125 lumens/W) sulfur lamps will make inroads in public lighting.

Ubiquity of electricity means that it is not enough to target just the obvious end uses including space and water heating, lighting, and refrigeration. Miscellaneous electricity uses, an eclectic category ranging from TVs and VCRs through personal computers and their paraphernalia to ceiling fans and hot tubs have been, and will continue to be, a major driver of rising demand. In the United States these markets already account for one-fifth of total electricity use (Sanchez et al. 1999). About 20% of this miscellaneous use is due to standby losses from an increasing array of appliances. The average U.S. household now leaks constantly about 50 W, or 5% of all residential electricity use, mostly due to remote-ready TVs, VCRs, and audio equipment and communication devices (Meier and Huber 1997; Thorne and Suozzo 1997; fig. 6.1). Reducing standby losses of all domestic appliances to less than 1 W per device offers one of the best examples of relatively tiny but cumulatively large energy

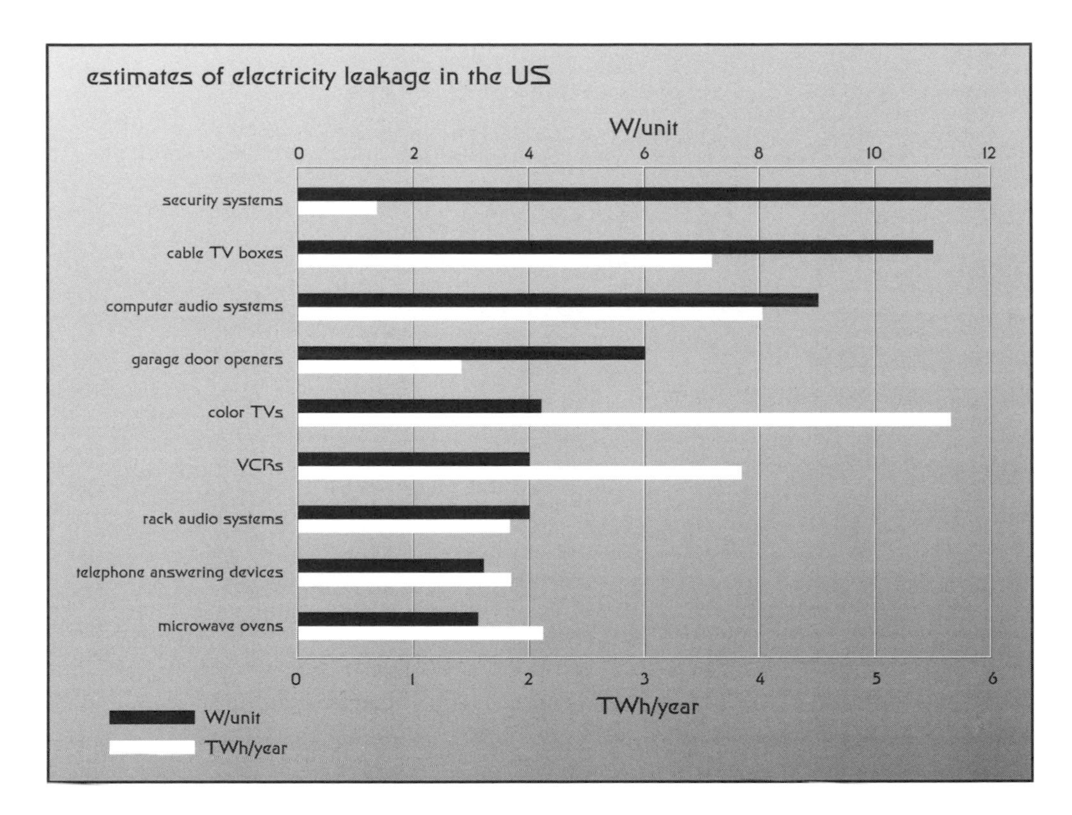

Figure 6.1
Estimates of electricity leaking from commonly owned consumer appliances in the U.S. in 1997. TVs, audio systems, VCRs, and cable boxes accounted for the largest aggregate phantom loads, followed by telephone answering machines and garage door openers. Plotted from data in Thorne and Suozzo (1997) and Meier and Huber (1997).

savings: this painless and technically doable measure (a Nokia TV 7177 already draws less than 0.1 W) would cut the wasteful load by 70%.

Maximum practically achievable lighting efficiencies and minimum standby losses should also be among the strategic energy management concerns in all modernizing countries that are undergoing rapid urbanization and industrialization and that can expect substantially higher ownership of lights and appliances throughout the first half of the twenty-first century. Two other electricity uses that are particularly good candidates for efficiency improvements in low-income countries are room air conditioners, whose ownership is becoming common among the more affluent households in subtropical and tropical cities, and irrigation pumps that will be even more essential for assuring reliable water supply for larger harvests. And buildings in all sunny climates could moderate electricity use without any changes in existing converters by adopting 'cool' roof standards.

Highly absorptive dark roofs will get up to 50 °C warmer than the ambient temperature while the highly reflective 'cool' roofs (be they painted white or made of light-colored materials) will be just 10 °C warmer and if their emissivity is also improved they can cut energy demand for air-conditioning by anywhere between 10–50% (CEC 2001b). New California standards for clay tiles now require a minimum total reflectance of 0.4 and a minimum emissivity of 0.75 and should save, on the average, about 3.75 W/m^2, or more than the power density of common wind or hydro-energy fluxes that can be tapped by turbines. In addition, higher albedo of 'cool' roofs will help to lower the ambient air temperature of densely built-up areas and hence moderate the heat island effect and reduce the formation of photochemical smog.

In North America there can be no dispute about the second most rewarding technical fix next to raising the efficiency of electric systems: improving the performance of private cars whose driving now claims about one-sixth of the continent's TPES. The extraordinarily wasteful performance of that fleet is an obvious target for major efficiency gains (fig. 6.2). This is an apposite place to introduce the second category of efficiency measures, those that do not rely on technical fixes but on measures affecting habits and lifestyles. Limiting high car speeds indubitably raises energy efficiency of driving: the well-known, hump-shaped curve charting the efficiency of cars as a function of speed peaks between 45–55 km/h, and performances are 10–20% less efficient at lower speeds but as much as 40% less efficient at speeds above 100 km/h with the decline being greater for lighter cars. This example is an excellent

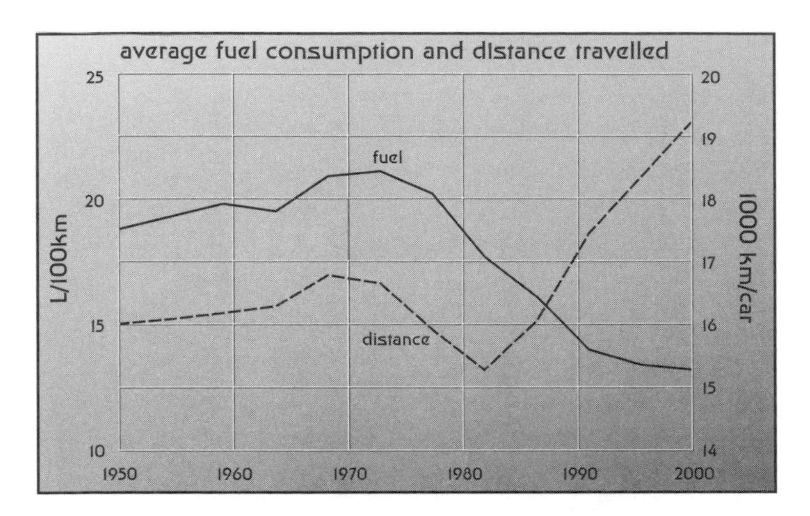

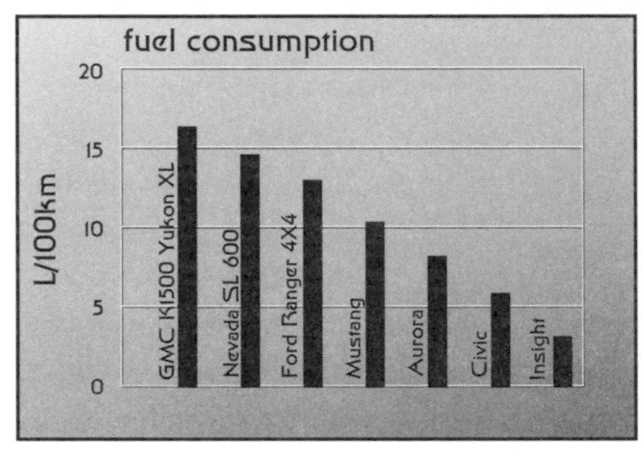

Figure 6.2
Average fuel consumption and distance travelled by the U.S. passenger cars, 1950–2000, and the efficiency range of 2002 models. Disappointingly small performance gain of the 1990s was more than erased by longer drives, and the most efficient popular brand (Honda Civic) consumes only about one-third of gasoline required by GMC K1500 Yukon XL. Plotted from data in EIA (2001a) and DOE (2000c).

illustration of a generally applicable trade-off between efficiency and higher rates of performance. Industrial designers would describe this phenomenon as an increasing input enthalpy with rising rate of production, a proverb would say, more pithily, that haste makes waste (van Gool 1978).

Not surprisingly, making trade-offs between time and efficiency, particularly in populations that feel chronically short of free time, is not a highly favored choice. Saving time is often much preferable to saving energy yet, as Spreng (1978) remarked more than 20 years ago, discussions of efficiency and energy policy commonly ignored (and continue to ignore) the importance of available (free) time in modern lives. If the power that we gained by using fossil energy is, to a large extent, the power to allocate time (Weinberg 1979b), then we cannot expect any rush to a voluntary surrender of this ability. Appeals for switching from private cars to infrequent and substandard public transportation should not be even attempted.

Besides the matter of speed, driving also illustrates how preferences for comfort, convenience, and status rank so often so much higher than energy efficiency. These preferences make many people very reluctant to switch to public transportation even if it is frequent and inviting. For most of those who tasted the freedom (however abridged by heavy traffic) of a private car going back to public transportation appears to be an extreme inconvenience, even a mark of personal failure. This is because, as Boulding (1973, p. 255) so aptly noted, the automobile is "remarkably addictive . . . it is a suit of armor . . . It turns its driver into a knight with the mobility of an aristocrat . . . The pedestrian and the person who rides public transportation are, by comparison, peasants looking up with almost inevitable envy at the knights riding by in their mechanical steeds."

In addition, of course, transportation infrastructure of so many of the rich world's megacities whose post-World War II growth has been driven predominantly with private cars in mind leaves little choice to consumers who would be otherwise stranded not just in suburbia but now also increasingly in exurbia that may be quite distant from the nearest stop of public transportation. As a result, modern societies will deploy a great deal of inventiveness to keep private cars on the road. These adjustments already range from hybrid vehicles to electronically monitored pricing of road space and they will increasingly include various near zero- or zero-emission cars.

Failures or limitations of efficiency gains predicated on changes of our behavior spring from our preferences and feelings. But why do not more people take advantage of widely available technical fixes that would not demand any changes of behavior and that would save them money? The world would be a different place if the decisions were made by perfectly informed consumers driven by rational choices guided by the goal of maximized energy efficiency and minimized economic cost and environmental impacts. Any voluntary effort aimed at reduced energy consumption is predicated on the informed readiness of consumers to adopt less wasteful arrangements or to buy more efficient converters. This would be even more so if the minimization of life-cycle environmental costs were to be the overriding goal. In reality, we have to consider not only the lack of information, misunderstandings, outright ignorance, and pure disinterest regarding the understanding of available choices, but also a peculiar economic calculus attached to energy efficiency, as well as the peculiar pricing of energy.

More expensive but more efficient converters should be purchased once their cost is offset by lower energy bills but distortions in electricity pricing, ignorance, and perception barriers make this a much less frequent choice than theoretical expectation would lead us to believe. Very commonly, consumers are not simply aware of potential savings arising from substitutions and will keep their outdated and inefficient converters as long as those gadgets will work. Most consumers who will buy a particular electricity-saving device will not be aware of any savings as bills are not traced to specific appliances and as intervening purchases of other household items may have actually raised the total electricity bill. People who would benefit most from saving energy are in the lowest income groups and may not be able to afford the cost of more efficient converters. But perhaps the greatest barrier is the now so deeply ingrained preference of modern affluent society that puts initial capital investment far ahead of lifetime costs.

Millions of decisions that will cost more money over the course of next 5 or 30 years are made every day because they need less (of often expensively borrowed) capital right now. Of course, as Spreng (1978) pointed out, there is also a more rational aspect to this as the uncertain perception of the future favors this lower capital outlay and higher operating costs. Owners of a new home in a highly mobile society may leave their dream house in a few years, a period too short to justify the cost of higher energy inputs embodied in its proper construction—but not too short to build large foyers and curved staircases. The lowest possible capital cost of ostentatious display repeatedly wins over maximized energy savings that would require more spending on energy embodied in superior product: flimsiness beats endurance. And so it is not surprising that studies have shown that even well-informed consumers will buy energy-saving converters only if such devices have extremely short payback periods of two years or less, effectively demanding a much higher rate of return on their investment than they used to get on their guaranteed investments or in the rapidly rising stock market (Wells 1992). We will have to wait in order to see to what extent this behavior will change under the conditions of low inflation (even deflation), and hence low interest rates and stagnating (or falling) stock prices.

Consequently, in spite of decades of promoting energy efficiency, and studying the impacts of utility-driven savings programs, tax subsidies, performance regulations and energy costs labelling (such as EPA's Energy Star) much remains uncertain. Utility demand-side management programs have not been easy to evaluate (Fels and Keating 1993) and the resulting uncertainties led to some irreconcilable claims. Some early studies of demand-side management (DSM) programs showed that for a typical U.S. utility these actions will eliminate up to 15% of the electricity demand expected over a period of ten years and that some states could avoid more than half of the growth in their electricity demand during the same period of time (GAO 1991b) and well-run schemes were judged to have the potential to reduce electricity demand by at least 1% a year over a period of 10–20 years (Nadel 1992).

But soon some enthusiasts saw an even bigger role for the DSM as an effective substitute for investment in new stations (a position best summed up by Lovins's concept of negawatt) while critics of the concept found that its implementation is not that inexpensive.

Complexities and intricacies of assessing the real costs and economic success of the DSM undertaken since the 1980s by the U.S. utilities are perhaps best exposed in claims made on behalf of this approach by the RMI (Lovins and Lovins 1991) and in criticism of their conclusions by Joskow and Marron (1992, 1993). RMI supply curves show that utilities could save up to 20% of electricity consumed by lighting at no cost, and up to 70% at a cost less than 4c(1991)/kWh while the curves developed by the EPRI show that savings larger than 30% come at a steeply rising cost (fig. 6.3). Subsequent pointed exchange between the two protagonists (Joskow 1994; Lovins 1994) only served to show how irreconcilable were their positions. And if there are sharp disputes about the cost of energy savings there is no clear understanding how much and how long these savings will persevere.

Socolow's (1985, p. 27) complaint—"we still know pitifully little about the determinants of durability of hardware, and even less about the determinants of durability

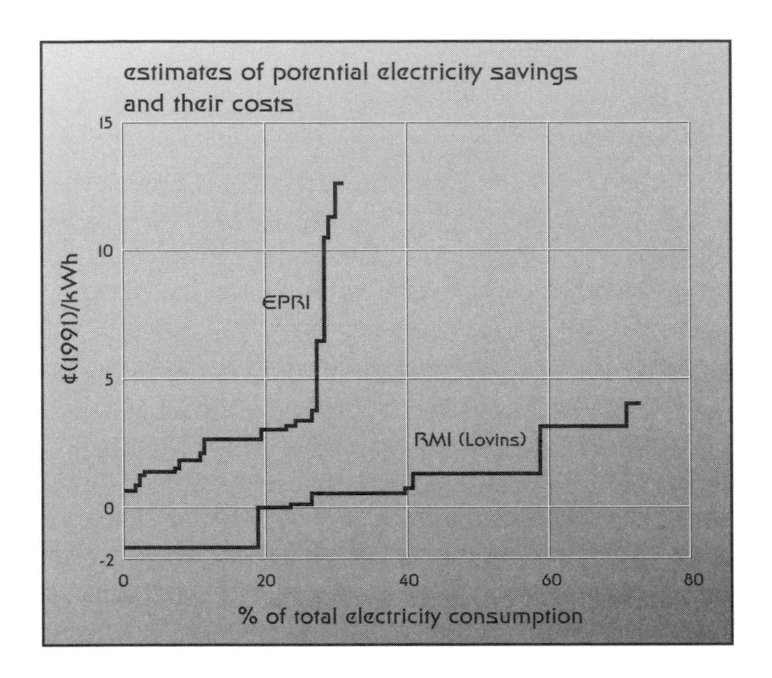

Figure 6.3
Very different estimates of typical costs of electricity savings according to RMI and EPRI. Based on Kahn (1991).

of attitudes and behavior"—remains as valid today as when it was made in 1985. As Vine (1992) concluded, persistence of energy savings is perhaps the single most important unanswered question in DSM (Vine 1992). Installing a more efficient furnace or boiler should guarantee at least a decade of only slightly degraded performance but Nolden (1995) found that in 15 retrofits of boilers in apartment buildings the most recent savings (5–9 years later) expressed as shares of the first years gains ranged from a mere 4% to 156%, a range that makes difficult any generalizations about the persistence of gains and points out the importance of specific on-site management. Nor can we generalize about how much will retrofitted door and window weather-stripping deteriorate (or be removed or replaced) over five or ten years or how many owners of subcompacts will suddenly decide to match their rising incomes by buying SUVs.

And then there are efficiency improvements whose gains are promptly erased by higher system demands. These can arise from physical necessities, secular changes, or individual choices. More efficient electric motors offer a perfect illustration of the first impact (Hoshide 1994). Most of the high efficiency motors will have higher full load speeds than do the standard models—but the power needed by a pump or a fan goes up with the cube of the machine's speed. Consequently, a typical 15-kW, 1,780-rpm motor replacing a 1,750-rpm unit will save 5.2% of electricity but increased load from rotating equipment will consume 5.2% more and there will be no savings. Similarly, a large part of the higher combustion efficiency in subcompact and compact cars has been negated by the more common choice of automatic transmission and by air conditioning becoming standard equipment on many of these models.

Assessments of possible gains of energy efficiency are thus recurrent exercises in contrasts. On one hand there is a multitude of technically mature and hence readily available and indisputably cost-effective options to reduce energy demand. On the other hand there are the quotidian realities of the affluent world dominated by casual waste of resources, growing displays of ostentatious consumption, and a palpable absence of concern about the impact of high energy use on the fate of the biosphere. This gap between what is easily possible and what actually takes place has been a recurrent source of my exasperation. Energy use in our household can be hardly classed as punishingly frugal yet our demand is only a small fraction of the typical rate: this makes me impatient with the argument that anything beyond modest energy savings requires a great deal of inconvenience or even intolerable personal sacrifices. This is particularly true in a society where the only 'sacrifice' asked of the general population after 9/11 was to do more shopping to keep the economy humming.

The car I drive (Honda Civic) needs 60% less gasoline to take me to work than does a 4 × 4 Range Rover favored by some in the show-off crowd. A compact fluorescent light in my lamp consumes 65% less electricity than does the incandescent bulb that preceded it in the same socket. The high-efficiency furnace in my basement converts 94% of natural gas into warmth inside my house, compared to 50% efficiency for an older furnace and −10% efficiency (or even worse during Canadian winter) for heat-wasting, wood-burning fireplaces whose multiple installations are de rigeur in megahouses of nouveaux riches.

Clearly, opportunities for doing things better remain so abundant and so obvious that the actions we have taken so far add up to only a small share of a truly enormous potential. But at this point I must ask a broader and a truly fundamental question that is habitually overlooked in the quest for higher conversion efficiencies: does this effort really save energy? And if not what actions will?

Beyond Higher Efficiency

There is no doubt that relying on devices and machines that convert fuels and electricity with higher efficiency leads to lower energy use and to savings of money at the microeconomic level, that is for individual consumers, households, and companies and even at the mesoeconomic level, for entire industries. But what happens at the national, that is macroeconomic, level? Historical evidence shows unequivocally that secular advances in energy efficiency have not led to any declines of aggregate energy consumption. Stanley Jevons, whose concerns about the exhaustion of British coal we encountered in chapter 4, was the first economist to address the potential of higher efficiency for "completely neutralising the evils of scarce and costly fuel" (Jevons 1865, p. 137).

He was well aware that good engines and furnaces of his day converted only a small part of the consumed coal into useful work but he concluded that

> *It is wholly a confusion of ideas to suppose that the economical use of fuels is equivalent to a diminished consumption. The very contrary is the truth.* As a rule, new modes of economy will lead to an increase of consumption according to a principle recognised in many parallel instances (Jevons 1865, p. 140; the emphasis is in the original).

Jevons cited the example of Watt's low-pressure steam engine and later high-pressure engines whose efficiencies were eventually more than 17 times higher than that of Savery's atmospheric machine but whose diffusion was accompanied by a huge increase in coal consumption (Smil 1994a). His conclusions have been shared, and elaborated, by virtually all economists who have studied the macroeconomic impacts of increased energy efficiency—and they have been disputed by many environmentalists and efficiency advocates. Herring (1998, 2001) provides excellent surveys of these debates. The most resolute counterarguments claim that the future elimination of large, existing conversion inefficiencies and the shift toward increasingly service-based, or supposedly more dematerialized, economies can lead to no less than stunning reductions of energy use at the national level (Lovins 1988; Hawken, Lovins and Lovins 1999).

A more measured assessment concluded the improvement in efficiency is, per se, only a small part of the reason why total energy consumption may have gone up, and that the overall growth of energy use is more related to increasing population, household formation, and rising incomes (Schipper and Grubb 2000). How small is not easy to determine. To begin with, there are disputes about the relative magnitude of the rebound effect whereby savings accruing from more efficient use of energy lead to lower prices and hence eventually to increased consumption, either in the very same category (direct rebound) or for other goods and service (indirect rebound effect). Lovins (1988) argued that at the consumer's level the overall rebound is minimal, a position rejected by Khazzoom (1989). Some studies have shown direct rebound on the order of 20% but indirect macroeconomic consequences are much harder to quantify.

A much admired trend toward the dematerialization of affluent economies has, indeed, produced often impressive *relative* savings, be they measured by the use of mineral commodities per unit of GDP or by the use of energy and materials per specific finished products and delivered services. But this process has been accompanied by rising *absolute* consumption of many energy-intensive materials and by even more rapidly growing demand for services whose energy cost is anything but low. Examples of these trends abound. Among the most notable ones is the relative dematerialization of cars. Lighter engine blocks (aluminum replacing iron) and lighter bodies (plastics and composite materials replacing steel and glass) introduced since the 1970s have significantly reduced the mass/power ratio of passenger cars, and hence also the energy needed to drive them. But these savings would have translated into lower absolute need for materials and considerable fuel savings only if cars of the early twenty-first century remained as powerful, as equipped with energy-consuming accessories and as much driven as were the vehicles during the mid-1970s.

But in today's United States the vehicles of choice are not even classified as cars as all SUVs are put into the light truck category. Their weight is commonly between 2–2.5 t, with the largest ones topping 4 t, compared to 0.9–1.3 t for compact cars; their fuel consumption in city driving (where they are mostly used) commonly surpasses 15 L/100 km (and 20 L/100 km for some), while efficient subcompacts need less than 8 and compacts average around 10 L/km. But these cars, too, have become heavier and more powerful than a generation ago: my 2002 Honda Civic is actually a bit more powerful and a bit heavier than was my Honda Accord 20 years ago. Moreover, the average distance driven per year keeps increasing—in the United States it is now nearly 20,000 km per motor vehicle, up by about 30% between 1980 and 2000 (Ward's Communications 2000)—as the commutes lengthen and as more trips are taken to remote destinations. The net outcome of all of this: America's motor vehicles consumed 35% more energy in 2000 than they did in 1980.

Similarly, power density of energy use in new houses is now lower, but the houses have grown larger, with average size of the new U.S. house is up by more than 50% since the early 1970s and in 2001 it topped 200 m² (USCB 2002). Moreover, in the country's Sunbelt inhabitants of these houses may buy superefficient air conditioners to maintain indoor summer temperatures they would consider too cold in winter. In view of these recent consumption trends Moezzi (1998) is right not to be impressed by the stress that the contemporary American energy policy puts on energy efficiency: such emphasis overlooks the human actions that tend to increase energy consumption in the long-run and may not necessarily save anything even in the short run.

Herring (2001) offers another excellent example of greatly improved efficiency negated by an even faster growing demand. As I already explained in chapter 1, efficacy of lights rose impressively during the twentieth century, and the improvement in British street lighting since the 1920s was about twentyfold, from 10 lumen/W for incandescent bulbs to about 200 lumen/W for low-pressure sodium lamps. However, more roads (less than 50% increase) and a huge rise in the average light intensity (when measured in lumens per km of road it rose more than 400 times) meant that during the same period electricity consumption per average km of British roads has increased twenty-fivefold, entirely negating the enormous advances in efficiency (fig. 6.4).

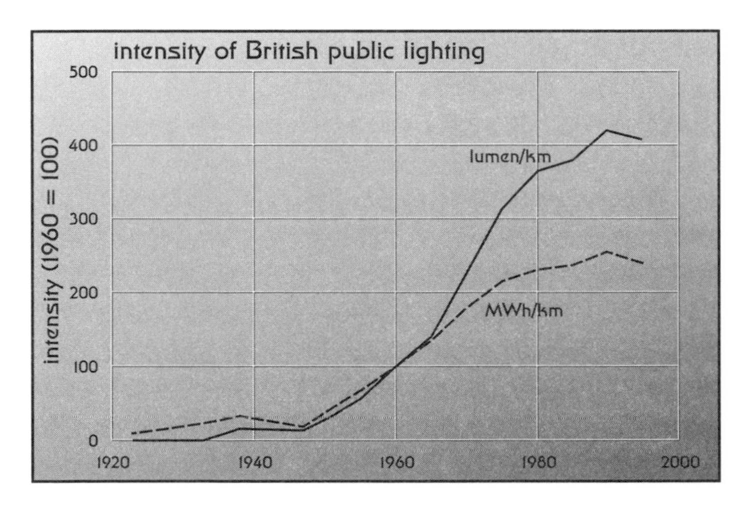

Figure 6.4
Long-term trends in the intensity of British public lighting (Herring 2000) illustrate how higher levels of service reduce, or completely negate, savings arising from higher energy efficiency.

And there is no shortage of examples exposing the illusory dematerialization. Rapid increase in the consumption of paper has coincided with the diffusion of what was to be the paperless office of the electronic age, and short life spans of computers and peripherals have created serious waste disposal problems as these machines contain a variety of plastics and metals that should not be landfilled. And no special analytical skills are needed in order to conclude that if money saved by more efficient furnaces or lights is used to fly for family weekends in Las Vegas, a destination chosen by millions of American families every year, the overall energy use will rise.

Finally, it is easy to perform a revealing exercise on a national level. As we have seen (chapter 2), the average energy intensity of the U.S. economy fell by 34% between the years 1980 and 2000, while the country's population increased by about 22%. If the average per capita GDP remained at the 1980 level then the U.S. TPES in the year 2000 would have been 20% *below* the 1980 level. Even if the average had grown by one-third, the TPES in the year 2000 would have been only about 7% ahead of the 1980 total. In reality, average per capita GDP rose by more than 55% and so in spite of that impressive decline in the energy intensity of the U.S. economy the country's TPES in the year 2000 was about 26% higher (fig. 6.5)! Analogicaly, Ehrlich et al. (1999) examined the total post-1975 material use in the United States, Japan, Germany and the Netherlands and found that in relative terms (per unit of GDP) it fell in all of these nations (by about one-third on the average) but only in the United States did the mass consumed per capita actually decline: in Germany it increased marginally but in Japan and the Netherlands it rose by about 20%.

Historical evidence is thus replete with examples demonstrating that substantial gains in conversion (or material use) efficiencies stimulated increases of fuel and electricity (or additional material) use that were far higher than the savings brought by these innovations. Indeed, the entire history of Western modernization can be seen as a continuing quest for higher efficiencies as generations of engineers have spent their professional lives wringing additional returns from their contraptions and as entire nations, guided by that famously invisible hand, have been relentlessly following the path of reduced waste and higher productivity. But the outcome is indisputable: global energy consumption far higher than the rate of population growth and the need to satisfy not just basic existential needs but also a modicum of comfort and affluence (see chapter 2).

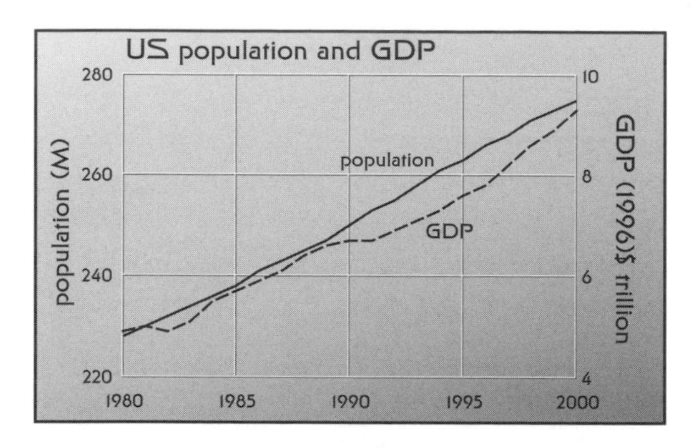

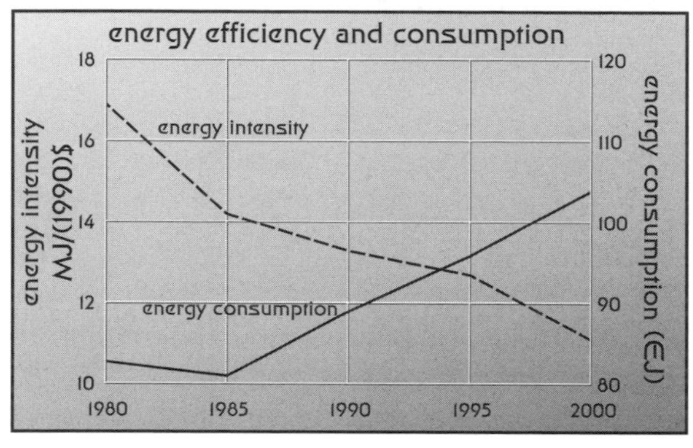

Figure 6.5
U.S. population and GDP growth, energy efficiency, and the rise of primary energy consumption, 1980–2000. Plotted from data in EIA (2001a).

Given the fact that efficiency has become a mantra of modern, globally competitive business whose key goal is to make and sell more, the quest for better performance can be then seen, in Rudin's (1999, p. 1) disdainful view, as a justification "to consume our resources efficiently without limit." And he points out the distinction between relative and absolute savings by noting that "our environment does not respond to miles per gallon; it responds to gallons" (Rudin 1999, p. 2). So if we are to see any actual reductions in overall energy use we need to go beyond increased efficiency of energy conversions. One way to preserve these cuts in energy use would be to tax away the savings accruing from higher efficiency and reinvest them in projects whose low energy-intensity would be combined with demonstrably positive impacts on public welfare and the integrity of the biosphere. Planting of trees and many other activities aimed at restoring natural habitats and preserving biodiversity would be the most obvious choices. But even the proponents of this approach have no illusions about its political feasibility (Costanza and Daly 1992).

A more realistic goal is to promote energy conservation. Of course, in the strict scientific sense the term 'energy conservation' should be avoided because there is no need to perpetuate yet another erroneous usage akin to the already noted interchangeable and incorrect use of terms energy and power. Energy is always conserved: such is the universal physical imperative summed up in the first law of thermodynamics. But the term is too ingrained to ignore it, and it entails any measures aimed at reducing energy use by either voluntary or mandated cuts in quality or rate of energy services.

One of the most iconic entries of the concept of energy conservation into the public consciousness was thanks to cardigan-clad Jimmy Carter imploring the U.S. citizens to lower thermostats and don sweaters during the somber years of the 'energy crisis' of the late 1970s. But it was the speed limit imposed on the U.S. Interstate highways during the same time (the famous double-nickel, 55 mph) that was the most obvious everyday reminder of regulated energy conservation for millions of American drivers. Given the complexity of modern societies regulation would always have a role in energy conservation but the bulk of such savings should be preferably delivered by an enlightened public that chooses to change its behavior and modify its lifestyle. Appeals for this shift have been made by many devoted conservationists. The fact that "improved efficiency coincides with increased use of resources should be enough to make us think in nonbusiness terms . . . Using less energy is a matter of discipline, not fundable political correctness" (Rudin 1999, p. 4).

Seen from this perspective calls for energy conservation are just a part of much broader appeals for moderation (if sacrifice may seem too strong a term), frugality, and cooperation for the sake of a common good that forms the moral foundations of every high civilization. Being content with less or not requiring more in the first place are two precepts that have been a part of both Western and Eastern thought for millennia and that were explicitly voiced by teachers of moral systems as disparate as Christianity and Confucianism. How kindred are these quotes from *The Analects,* in Arthur Waley's translation (Waley 1938), and from Luke (XII:22–34; King James version):

The Master said, He who seeks only coarse food to eat, water to drink and bent arm for pillow will without looking for it find happiness to boot.

And he said unto his disciples, Therefore I say unto you, Be not anxious for your life, what ye shall eat; nor yet for your body, what ye shall put on. For the life is more than the food, and the body than the rainment . . . make for yourselves purses which wax not old, a treasure in the heavens that faileth not . . . for where your treasure is, there will your heart be also.

And I cannot resist to quote one more appeal for conscious dematerialization, Lao Zi's perceptive stanza in R. B. Blakney's translation [Blakney 1955]:

So advantage is had
From whatever is there;
But usefulness arises
From whatever is not.

The two tenets have retained a high degree of moral approbation in affluent countries even as their devotion to religion has weakened considerably. Of course, a mechanistic translation of some very effective past practices would not be the best way to proceed. There is no need to call, for example, for an emulation of what was perhaps the best energy-minimizing arrangement: medieval monastic orders where most of the food, and also most of all clothes and simple wooden and metallic utensils were produced by artisanal labor, where nothing was packaged, everything was recycled and where the inmates had no personal possessions beyond their coarse clothes and a few simple utensils and were content with bleak cells, hard beds, copying of missals, and occasional a capella singing.

What is called for is a moderation of demand so that the affluent Western nations would reduce their extraordinarily high per capita energy consumption not just by 10% or 15% but by at least 25%–35%. Such reductions would call for nothing more than a return to levels that prevailed just a decade or no more than a generation ago. How could one even use the term *sacrifice* in this connection? Did we live so unbearably 10 or 30 years ago that the return to those consumption levels cannot be even publicly contemplated by serious policymakers because they feel, I fear correctly, that the public would find such a suggestion unthinkable and utterly unacceptable? I will return to these fundamental questions later in this chapter.

However modest, or even timid, any calls for living with less may be, they are fundamentally incompatible with the core belief of modern economics that worships not just growth, but sustainable growth at healthy rates (judging by the comments of ubiquitous Chief Economists commenting on the last quarter's GDP this means at least 2–3%/year). Only the widespread scientific illiteracy and innumeracy—all you need to know in this case is how to execute the equation $y = x \cdot e^{rt}$—prevents most of the people from dismissing the idea of sustainable growth at healthy rates as an oxymoronic stupidity whose pursuit is, unfortunately, infinitely more tragic than comic. After all, even cancerous cells stop growing once they have destroyed the invaded tissues.

If we are to prevent the unbounded economic growth doing the same to the Earth's environment then the preservation of the biosphere's integrity must become a high purpose of human behavior. Inevitably, this must entail some limits on human aquisitiveness in order to leave room for the perpetuation of other species, to maintain irreplaceable environmental services without whose provision there could be no evolution and no civilization, and to keep the atmospheric concentrations of greenhouse gases from rising so rapidly and to such an extent that the Earth would experience global tropospheric warming unmatched during the evolution of our species from ancestral hominids.

Energy and the Future of the Biosphere

Growth rates of global population that have prevailed during the twentieth century will not be replicated during the twenty-first century. Global population grew nearly fourfold between 1900 and 2000, from 1.6 to 6.1 billion, but its relative rate of growth has been declining from the peak of about 2% reached during the late 1960s to less than 1.5% by the late 1990s, and its absolute annual increment is now also diminishing (UNO 1998). Consequently, we are now fairly confident that the self-limiting nature of global population growth will prevent yet another doubling during the twenty-first century. There is now a 60% probability that the world's population will not exceed 10 billion people by the year 2100, and about a 15% probability that it will be lower by that time than it is today (Lutz, Sanderson, and Scherbov 2001). But even relatively small population total increases could be associated with further perilous degradation of the biosphere (Smil 2002).

For example, if today's low-income countries were to average at least one-third of energy now used per capita by affluent nations (the fraction was less than one-fifth in the year 2000) and if the rich economies were to limit themselves just to a 20% increase in their energy demand, the global TPES would be roughly 60% above the 2000 level. With fossil fuels still the dominant, albeit a declining, source of that energy the cumulative generation of carbon, sulfur, and nitrogen oxides would be substantially higher during the first half of the twenty-first century than it was during the latter half of the twentieth century, and so would be the consequences for climate, air, and water quality, land use, and biodiversity.

Not surprisingly, not all of the many environmental impacts caused by extraction, transportation and conversion of commercial, and traditional, energies (reviewed in chapter 2) are equally worrisome. Some may cause a severe injury to biota but are highly localized (acid mine drainage, oil spills on beaches); others may not seem to be so destructive when seen in isolation but their impact adds up when viewed in a global perspective. Contribution of energy industries and uses to the global loss of biodiversity is perhaps the best example in the latter category. Land use changes caused by deforestation and conversion of grasslands and wetlands (be it for new agricultural land, pastures, timber production, or new settlements), periodic grassland fires set by pastoralists, and inappropriate farming methods opening the way to excessive soil erosion are the main causes for outright destruction of natural ecosystems or for their considerable impoverishment (Smil 2002).

But energy industries contribute to these degradations mostly because of the surface mining of coal, construction of water reservoirs, rights-of-way for pipelines and HV lines, as well as due to the refining of crude oil and thermal generation of electricity. I estimate that by the end of the twentieth century the aggregate land claim of the global energy infrastructure reached about 290,000 km^2 (an area slightly smaller than that of Italy), with about 60% of this total covered by hydroelectric reservoirs (fig. 6.6). This is less than 2% of the total area of natural ecosystems (mostly forests and grasslands) that has been converted to crop, pasture, industrial, transportation, and urban land during the past 250 years (Smil 2002). But the actual effects of energy infrastructure on biota cannot be measured so simplistically because many energy facilities are located in relatively vulnerable environments and because they, and the associated transportation and transmission corridors, help to impoverish the surviving ecosystems.

Examples of locations in or upwind from highly biodiverse and sensitive environments range from hydrocarbon drilling in wetlands and coastal waters to large concentrations of coal-fired, electricity-generating plants emitting sulfur and nitrogen oxides that are transported to acidification-prone lakes and forests. Deltas of the Mississippi or the Niger, and Caspian Sea or the Venezuela's Lake Maracaibo are the best examples in the first category (Stone 2002). Large coal-fired stations in the Ohio Valley, whose acidifying emissions are carried to the Adirondacks and all the way to the Canadian Maritimes, or the Central European plants, whose emissions have acidified lakes and forests of southern Scandinavia, are the best known cases of the second kind. Energy infrastructures contribute to the loss of biodiversity due to the fragmentation of ecosystems.

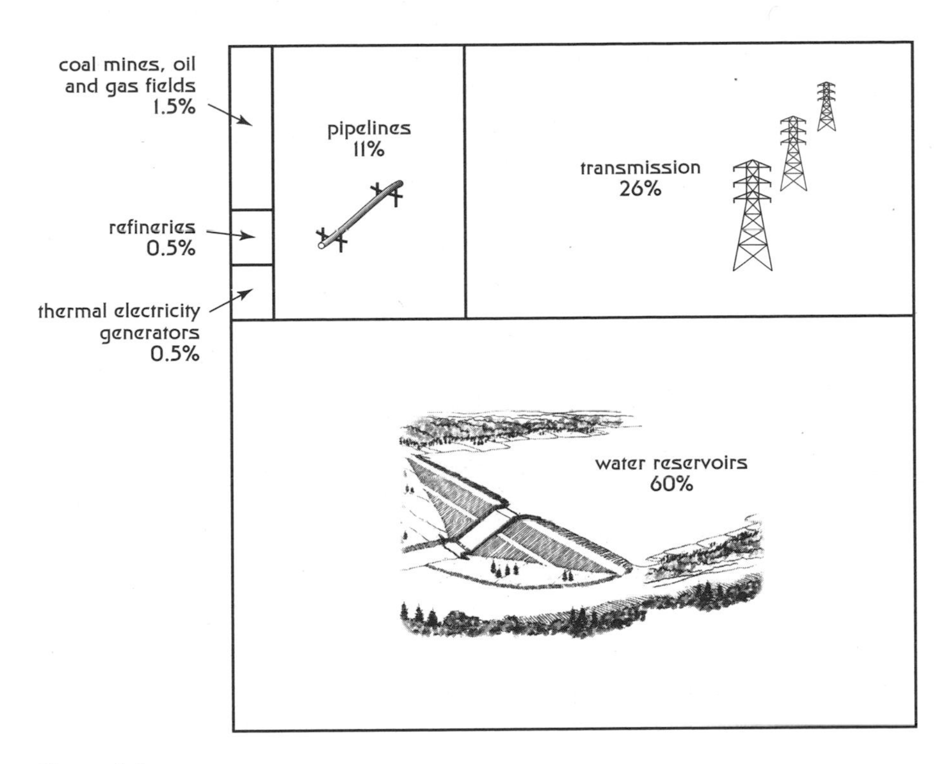

Figure 6.6
Global estimates of land required by primary and secondary energy industries and their supporting infrastructures. Based mostly on data in Smil (1991).

The good news is that during the past 50 years many of these environmental impacts have been impressively reduced in relative terms, i.e., when measured as efficiencies of pollution control equipment, power densities of fuel extraction, or as specific releases of air- and waterborne pollutants per J of extracted or transported energy or per kWh of generated electricity. Combination of higher conversion efficiencies combined with a variety of pollution-control devices has made the combustion of fossil fuels much more acceptable than it was 50 years ago. Commercial techniques can now eliminate all but a few tenths of one percent of all particulate emissions and they can lower the emissions of major gaseous air pollutants by anywhere between 70–95%. New drilling methods (see chapter 4) mean that the footprint of oil and gas extraction can be much reduced as many hydrocarbon-bearing structures can be reached from a single well.

Traditional coal-fired electricity generation would produce typically between 3–5 g SO_2/kWh while plants with FGD release less than 1 g/kWh and combined-cycle gas turbines emit no more than 0.003 g/kWh. Carbon emission cannot be cut by such margins but going from conventional coal-fired power plant to combined-cycle gas turbines entails their reduction by no less than two-thirds (Islas 1999). These gains would look even more impressive over time if the comparisons were done on the basis of fuel used for the generation: while steam turbine efficiencies have levelled off after 1970, performance of combined-cycle gas turbines rose from less than 40 to 60% during the same period of time.

The bad news is that, as stressed previously in this chapter, higher efficiency is not enough and that impacts of energy industries and uses have been increasing in absolute terms. Even more importantly, a much more worrisome change has now been added to many longstanding local and regional concerns as the combustion of fossil fuels became the single largest cause of human interference in the global carbon cycle. The resulting rise of tropospheric CO_2 is the main reason for increasing absorption of outgoing IR radiation, the process whose continuation would result in global warming at a rate unprecedented not only during the history of human civilization but during the evolution of our species. We have to go back about 15,000 years in order to encounter the most recent rise of more than 1-2 °C in average annual temperatures of the Northern hemisphere (Culver and Rawson 2000). And as it took about 5,000 years of postglacial warming to raise the temperature by 4 °C, global warming of more than 2 °C during the twenty-first century would proceed, as already noted, at a rate unprecedented since the emergence of our species about 0.5 Ma ago.

A possible event of this magnitude must be taken seriously because pronounced global warming would affect virtually every process in the entire biosphere. Warming would accelerate water cycles, alter the rates of many transfers within major nutrient cycles, affect the NPP of all biomes, shift the boundaries of ecosystems and heterotrophic species (including disease vectors), and influence the functioning of many indispensable ecosystemic services such as rates of organic decay, soil water storage, or control of pathogens. And so it is imperative to engage in rigorous interdisciplinary examinations of multiple impacts this rapid global warming would have on the biosphere and on our society. At the same time we must be mindful of the fact that assessments of these risks are an imperfect work in progress, and that we are unable to pinpoint the tropospheric temperatures that will prevail 25, 50, or 100 years in the future, and that we can be even less certain about the regional effects and about the net long-term impacts of such a change.

Global energy consumption scenarios prepared for the latest IPCC assessment translate to CO_2 concentrations between 540–970 ppm by the year 2100, or between 46–262% above the 2000 level (Houghton et al. 2001; fig. 6.8). Together with rising levels of other greenhouse gases these concentrations would produce average global forcing between 4–9 W/m^2 and this, in turn, would elevate the average global ground temperature by between 1.4–5.8 °C, with the extremes representing a largely tolerable rise and quite a worrisome climate change (fig. 6.7). All forecasts of GHG emissions, their climatic effects, and the ensuing environmental and economic impacts thus remain highly uncertain and researchers have increasingly tried to assess these

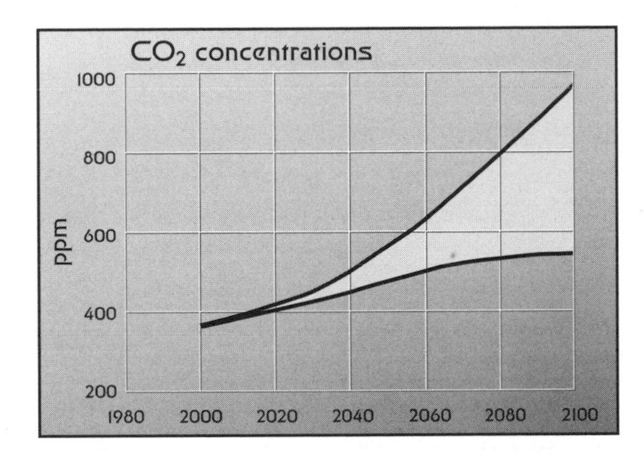

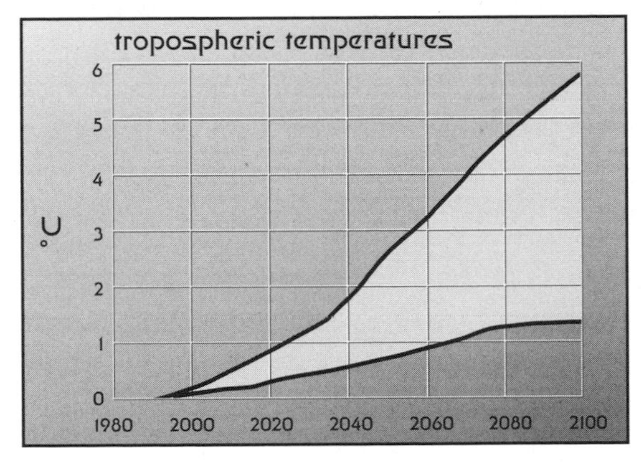

Figure 6.7
Energy consumption scenarios considered by the latest IPCC report (Houghton et al. 2001) result in atmospheric CO_2 levels between 540–970 ppm and tropospheric temperatures 1.4–5 °C higher by the end of the twenty-first century. Based on figures in SRES (2001).

uncertainties and to come up with the most plausible warming ranges and impact scenarios.

After taking into account uncertainties in emissions, climate sensitivity, carbon cycling, ocean mixing and atmospheric aerosols Wigley and Raper (2001) concluded that the probability of warming values outside the latest range of the IPCC assessment, that is above or below the projected range of 1.4–5.8 °C, is very low, and set the 90% probability warming estimate in the absence of any GHG reduction at 1.7–4.9 °C by the year 2100. In contrast, Forest et al. (2002) believe, on the basis of the marginal probability distributions, that the higher limit for the climate sensitivity with the equivalent doubling of CO_2 is as much as 7.7 °C, much higher than the IPCC value. Allen et al. (2000) looked only as far ahead as the decade of 2036–2046 (expected to be, again without any GHG controls, 1–2.5 °C warmer than in the preindustrial times) and concluded that unlike this relatively robust prediction, the final equilibrium warming (after the atmospheric composition stabilizes) remains very uncertain. What is perhaps most important is that detailed projections that are needed to assess regional impacts are unreliable (Reilly et al. 2001).

Nor can we forget the far from perfectly understood complexities of the atmospheric dynamics. One of their consequences is a notable difference between satellite and ground temperature measurements since 1979: the latter record shows warming of 0.3–0.4 °C but the former observations found that the midtroposphere, where models predict a stronger warming, had its temperature unchanged (NRC 2000; Santer et al. 2000). And, just to mention of the more interesting recent findings, satellite studies of cloud coverage over a part of the western Pacific showed that the area of cirrus coverage decreases about 22%/°C increase in the surface temperature of the cloudy region and Lindzen, Chou, and Hou (2001) ascribed this effect to a diminished cirrus detrainment from cumulus convention. If this attribution is correct this would constitute a negative temperature feedback (akin to the eye's iris that adjusts to changing light levels) that would more than cancel all the positive feedbacks in today's more sensitive climate models.

Summary of the latest IPCC report acknowledges the uncertainties of our risk assessments. Although it concludes that "the balance of evidence suggests a discernible human influence on global climate" (IPCC 2001, p. 3), it also notes that our ability to quantify the human influence on global climate is currently limited because the expected signal is still emerging from the noise of natural variability, and because there are uncertainties in key factors. Unfortunately, these inherent uncertainties have made it easy to turn the debate about global climate change into pointless, and endless, arguments about the extent and the rate of future warming and, for economists, about the cost of reducing future GHG emissions.

Protracted negotiations preceding the formulation of the Kyoto Protocol of December 1997 and subsequent debates, controversies, modifications, and opposition to the agreement are a perfect illustration of these distractions. All of this is even more remarkable given the fact that even a fully implemented Kyoto Protocol would have done little to prevent further substantial increases of GHG emissions. The agreement was just a timid start in the right direction, obliging only the affluent countries (so called Annex B Parties) to reduce their overall GHG emissions 5.3% below the 1990 levels by 2008–2012. Low-income countries, including such large GHG producers as China and India, were not asked for any commitments, not even for voluntary undertakings. Depending on the choice of published studies a reader could believe that the implementation of the protocol would have great economic benefits or carry substantial costs: Nordhaus (2001) is a recent example of this now common genre.

In spite of the fact that the protocol offered the Annex B countries a variety of approaches to be used in order to achieve their differentiated emission targets the United States, by far the world's largest GHG producer, refused to implement its quota (calling for 7% reduction) and other large producers, most notably Canada, will not reach their targets. Nobody familiar with recent trends of the TPES has been surprised to see the U.S. carbon emissions more than 10% ahead of their 1990 levels by the year 2000, making any compliance with the protocol's targets most unlikely.

I see all of these never-ending debates about the rate, effects, and costs of global warming not just as being unproductive but as being clearly counterproductive. When faced with major uncertainties concerning the anthropogenic impact on climate, the key determinant of the biosphere's dynamics, the only responsible way is to act as risk minimizers and to begin taking all reasonable steps to reduce GHG emissions. Of course, many economists are correct when arguing that such steps, given the inherent uncertainties, may impose unnecessarily large burdens on societies that might use the resources committed to what may turn out to be a wasteful exercise in many more beneficial ways. This is where the appeal of the no-regret approach comes in. Such a course makes perfect sense even if the coming decades would not see any pronounced anthropogenic global warming, that is if the warming signal were to stay submerged within natural fluctuations or if the rise of GHG concentrations were to result in only a minor and entirely tolerable environmental change.

This is due to the fact that the no-regret strategy of reduced energy consumption would not only cut GHG emissions but by doing so it would also reduce the releases (or the formation) of aerosols, the presence of photochemical smog and acid deposition, water pollution, land degradation, and then loss of biodiversity—and there is no uncertainty about high desirability of reducing every one of these environmental degradations and little doubt about the long-term economic benefits of such actions.

Most aerosols, including the abundant sulfates and nitrates formed by atmospheric oxidation of SO_x and NO_x, cool the surface by reducing the incoming solar radiation. This cooling amounts to as much as 30 W/m^2 over the tropical northern Indian Ocean when the winter monsoon brings the subcontinent's increasing air pollution, and to about 26 W/m^2 over the North Atlantic off the U.S. coast (Satheesh and Ramanathan 2000). In contrast, black carbon emitted during imperfect combustion of biomass and fossil fuels contributes not only to local and regional air pollution and (its smaller fractions, with diameter below 10 μm) to higher incidence of respiratory illnesses but it also helps to warm the troposphere and its effect in this regard may be now surpassing methane's contribution (Jacobson 2001).

Photochemical smog has become a semipermanent presence in every one of the world's major cities, be it Atlanta (together with Houston, the new U.S. smog capital) or Athens, Bangkok or Beijing, Taipei or Toronto (Colbeck and MacKenzi.e., 1994; Mage et al. 1996). Moreover, effects of the urban-generated photochemical smog have been spilling into the surrounding countryside. Crop productivity in Western Europe, Eastern North America, and East Asia has been already affected by high levels of ozone, an aggressive oxidant generated by photochemical reactions (Chameides et al. 1994). If unchecked, ozone damage would be thus weakening agricultural capacity just as food requirements of at least additional two billion people, and higher worldwide demand for animal foods, would be making higher crop yields even more necessary (Smil 2000c). And while the acidifying emissions have been significantly reduced in North America (see fig. 2.10) and Europe, they have been increasing in Asia. Expanding land claims of the world's energy infrastructure have been noted earlier in this section.

Relative decline of coal combustion, rise of hydrocarbons, and a larger share of electricity coming from primary hydro and nuclear generation has resulted in a gradual decline of the average carbon intensity of the global TPES, from just over 24 t C/TJ in 1900 to about 18 t C/TJ in 2000 (Grübler and Nakićenović 1996; Smil 2000a; see fig. 5.26). This translates to an average annual decline of almost 0.3% and at that rate the carbon share would be reduced by about 25% during the twenty-first century. Energy intensity of the world economic product was falling by about 0.7% during the last 30 years of the twentieth century (Smil 2000a), and continuation of this trend would halve the energy content of the world output by the year 2100. Combination of these two trends would mean that compared to the year 2000 the delivery of an average unit of useful energy would generate only about 38% as much CO_2.

Consequently, the overall TPES by the year 2100 could increase no more than 2.6 times in order to maintain the emissions at the 2000 level. This increase implies average global TPES growth rate of almost exactly 1%/year, a rate far below the means for the last 50 and 20 years of the twentieth century (3.2% and 1.75%, respectively). Even with this reduced growth rate of global energy consumption there would be no stabilization of atmospheric CO_2 concentrations at the current level as the cumulative emissions of the gas would amount to about 630 Gt C during the twenty-first century (compared to about 230 Gt C generated from fossil-fuel combustion during the twentieth century). Models show that this influx would raise the tropospheric level to about 450 ppm, or just over 20% above the current level. Needless to say, the much more likely global TPES growth in excess of 1%/year would then require additional steps to reduce GHG emissions or, in the absence of such action, it would raise CO_2 concentrations to well above 450 ppm by the year 2100 (see fig. 6.8).

What is needed is an active no-regret strategy of reducing our dependence on fossil fuels: relying on the eventual impacts of continued secular trends of lower energy intensity and gradual decarbonization to ease all of these burdens is not enough if we are to prevent the more acute warming outcomes and to realize sooner the multiple benefits of reducing smog, acid deposition, and water and land degradation. Yet in spite of the risk that the world could experience historically unprecedented global environmental change there has been virtually no progress toward this rewarding objective. The only processes working in our favor are the two just described secular improvements that have been now under way for more than a century. This is disappointing and regrettable but not surprising. When the concerns about anthropogenic global warming rose to a new prominence during the late 1980s many people were eagerly citing the successful international agreements to control CFCs as an inspiring example that could point the way toward effective GHG control. But that is a very inappropriate comparison.

CFCs were produced in a relatively small number of countries that agreed quickly to phase them out because the two companies whose output dominated the global synthesis of these compounds (Dupont and ICI) had suitable substitutes (HCFCs) at hand (Smil 1994a). In contrast, as was made clear in preceding chapters, our civilization has energized its accomplishments predominantly through its dependence on fossil fuels that now supply nearly 90% of the world's commercial TPES. Three other realities made this dependence even less easy to break. First, CO_2 cannot be managed readily after it was generated by add-on air pollution control techniques that we have used so successfully to reduce (or in some instances almost completely eliminate) the releases of particulate matter, CO, SO_2, and even NO_x: there is too much of it, there is no easy way to capture it, and while there are options to sequester it even their relatively widespread use would have a very limited impact.

Mass comparison with other major gaseous by-products of fuel combustion reveals a difference of two orders of magnitude in annual global production: we now generate annually about 25 Mt N and less than 80 Mt S in combustion gases, but release nearly 6.5 Gt C by burning fossil fuels. Sequestration of the generated CO_2 in abandoned mines and in hydrocarbon wells is conceivable for some nearby point sources but not for the gas produced far away from suitable deposition sites and obviously not for hundreds of millions of vehicle and airplane engines. Proposals for CO_2 dumping into the deep ocean and for geochemical and photosynthetic sequestration on land and in open seas can be seen as interesting mind-stretching exercises or as ideas whose time should never come.

Second, the worldwide use of fossil fuels now rests on an immense global infrastructure of facilities for extraction, transportation, processing, conversion, transmission, and final use whose replacement cost is at least $10 trillion in current monies. Some ingredients of this now so ubiquitous infrastructure were originally put in place more than a century ago and have been repeatedly expanded and updated ever since. Most of its machines and converters have useful life spans ranging between 30–50 years. Consequently, this enormous and long-lasting capital investment could not be readily abandoned even if many nonfossil alternatives would be already persuasively competitive and available in requisite capacities. As I explained in chapter 5, neither is yet true.

The third major complication is also connected with this lack of sufficiently large nonfossil capacities. Renewables capacities added annually at MW or GW rates could make a great difference in low-income countries where per capita energy use of fossil fuels is still very modest, being equivalent to less than 1 kW/capita. Unfortunately, these mostly tropical countries cannot readily afford still very high capital cost of large-scale modular PV, the conversion best suited for their environment. In contrast, in the world's affluent countries the dependence on fossil fuels has risen to such a high level—averaging now in excess of 4 kW/capita—that there is no alternative technique of nonfossil energy conversion that could take over a large share of the supply we now derive from coal and from hydrocarbons in just a few decades.

But long-term prospect for moderating the human impact on the biosphere would change if the world would pursue a grand energy convergence whereby the gradually reduced energy demand in affluent economies would leave more room for the TPES expansion that is necessary to lift the averages in low-income countries above the subsistence level. This grand strategy would accomplish a number of desirable goals. Moderated growth of global energy demand would begin to reduce immediately the growth rate of GHG emissions and of all other environmental impacts associated with extracting and converting more than 6 Gt of carbon a year. Reduced growth of the TPES would also make any contributions of new nonfossil capacities relatively more important, effectively accelerating the annual rate of global decarbonization. And this strategy would be directly addressing one of the potentially most destabilizing realities of the modern world, the persistent gap between the affluent (highly or relatively) minority of about one billion people and the rest of the world.

There are several obvious questions to ask. Is this really desirable? If so, is it doable? If yes, what are the chances of any genuine practical progress toward this goal? And if there will be such a progress, what difference it can realistically make in a generation or two, that is by 2025 or 2050? I will try to answer all of these questions as directly as possible.

What Really Matters

The first question—is the moderation of our transformation of the biosphere desirable?—is the easiest one to answer. We know enough about the true foundations of our civilization to realize that it is not imperiled by our inability to sustain high rates of economic growth for prolonged periods of time but rather by continuing degradation of the environment that weakens its biospheric foundations. Historical record shows that many societies collapsed because of local or regional environmental degradation (Taintner 1988). During the twentieth century this degradative process had become evident even on the planetary scale. This is a matter of high concern because the existence of complex human societies depends on the incessant provision of essential and hence invaluable environmental services.

What ecologists take for granted, most of the economists—transfixed by human actions, and viewing the environment merely as a source of valuable goods—have yet to accept: no matter how complex or affluent, human societies are nothing but open subsystems of the finite biosphere, the Earth's thin veneer of life, which is ultimately run by bacteria, fungi, and green plants (Smil 2002). Advanced civilization, humane and equitable, could do very well without Microsoft and Wal-Mart or without titanium and polyethylene—but, to choose just one of many obvious examples, not without cellulose-decomposing bacteria. Only these organisms can take apart the most massive component of plant matter accounting for about half of all trunks, roots, leaves, and stalks. In their absence more than 100 billion tons of plant tissues that die every year (the mass more than ten times larger than all fossil fuels we now extract in a year) would begin piling up in forests, on grasslands, and on fields.

If ours were a rational society we would be paying much more anxious attention to nature's services than to Dow Jones and NASDAQ indices. Above all, we would not be destroying and damaging with such abandon the stocks of natural capital—intricate assemblages of living organisms in forests, grasslands, wetlands, fertile soils, coastal waters, or coral reefs—that produce the astounding array of environmental services. Our pursuit of high rates of economic growth has resulted in an ever larger share of the Earth's primary productivity being either harvested for human needs or affected by our actions, it has already destroyed large areas of natural ecosystems, polluted, modified much of what remains, and it keeps on impoverishing the global biodiversity that took immense energy flows and long time spans to evolve.

These trends cannot continue unabated for yet another century. We cannot double the share of the photosynthetic production that we already harvest without severely imperilling the availability of irreplaceable environmental goods and weakening the delivery of indispensable ecosystemic services (Daily 1997; Smil 2002). As energy uses are responsible for such a large share of this worrisome anthropogenic transformation it is imperative to at least begin the process of limiting their environmental impacts in general, and the threat of unprecedented global warming in particular.

Nor is the second question—is such a moderation of human impact doable?—difficult to answer. Yes, we can reduce all of these impacts while maintaining acceptable quality of life. If our actions were guided by the two greatest concerns a sapient terrestrial civilization can have—for the integrity of the biosphere and for the dignity of human life—then it would be inescapable to ask the two most fascinating questions in energy studies: what is the maximum global TPES compatible with the perpetuation of vital biospheric services, and what is the minimum per capita energy use needed for a decent quality of life? These questions get asked so rarely not only because they are so extraordinarily difficult to answer but also because they compel us to adopt attitudes incompatible with the reigning economic ethos of growth and because they demand clear moral commitments. Formulation of our goals must be aided by science but everyday exercise of effective commitments and their intergenerational transfer can succeed only when the challenge is understood as a moral obligation.

Needless to say, many uncertainties in our understanding of the structure and the dynamics of the biosphere on one hand, and continuing technical and social innovation on the other, make it both impossible and highly undesirable to offer any rigid quantitative answers to the two great questions. Consequently, I will not tell you, as some bold-but-naïve prescribers have done, that the Earth should support no more than 1.2 billion people, and I will not ask you to begin agitating for a society where everybody gets a certificate for personal energy use worth a certain amount of GJ/year. At the same time, I refuse to treat the two great questions as unanswerable. Our understanding of basic biospheric functions is reasonably advanced, and as loaded as the term 'decent quality of life' may seem to be its common-sense definition is not surely beyond the rational consensus.

Careful readers will remember that I laid the foundation for such a consensus in chapter 2 when I reviewed the links between energy use and different indicators of quality of life. If the energy requirements of a good life were to be quantified on the basis of health alone then the two most sensitive indicators, infant mortality and life expectancy at birth, point to annual maxima of about 110 GJ/capita: virtually no gains accrue beyond that level, and only marginal gains are to be had once the consumption passes 70–80 GJ/capita. Correlation between higher education and energy consumption is very similar: no more than 100 GJ/capita is needed to assure easy access to postsecondary schooling, while primary and secondary requirements are well-satisfied at less than 80 GJ/capita.

Let me remind you at this point that most of the rewarding enrichments of human life—be it personal freedoms and artistic opportunities, or pastimes of physical or mental nature—do not claim large amounts of additional fuels or electricity. Of course, as far as pastimes go there are those high-powered, noisy, polluting, and high-risk worlds of car racing, speedboating, or snowmobiling but most of the leisure activities and hobbies cost only a modest amount of energy embodied in books, music recordings, or table games. Other pastimes require just a small additional food input to provide kinetic energy for scores of sports and outdoor activities. And the activity that has been shown to be most beneficial in preventing cardiovascular disease, by far the most important cause of death in Western populations, is a brisk walk for 30–60 minutes a day most days of the week (Haennel and Lemire 2002)—and it requires moderate energy expenditure of only 4.2 MJ/week, or the food energy equivalent to a single good dinner.

Consequently, this rich evidence leads to the conclusion that the average consumption of between 50–70 GJ/capita provides enough commercial energy to secure general satisfaction of essential physical needs in combination with fairly widespread opportunities for intellectual advancement and with respect for individual freedoms. Moreover, convincing historical evidence demonstrates that the only outcome guaranteed by the increasingly more ostentatious energy use above that level (and particularly in excess of 150 GJ/capita) is a higher intensity of environmental degradation. Remarkably, the global mean of per capita energy consumption at the beginning of the twenty-first century, 58 GJ/year, is almost exactly in the middle of the 50–70 GJ range. This means that equitable sharing of the world's fuels and electricity would supply every inhabitant of this planet with enough energy to lead a fairly healthy, long, and active life enriched by more than a basic level of education and made more meaningful by opportunities for the exercise of individual liberties. Naturally, all of the desirable minima of per capita consumption refer to the now prevailing energy conversion efficiencies, and hence they should be substantially lowered in the future as all common energy uses leave much room for additional improvements.

Secular trends toward higher efficiencies and lower energy intensities show no signs of early saturations. Of course, efficiencies of particular converters may be pushed to their practical, if not thermodynamic, limit but introduction of new techniques may start a new cycle of improvements and a series of such innovations will form an ascending wavelike pattern. As I noted in the preceding section, the global economy has been able to lower the energy intensity of its output by 0.7%/year during the past 30 years, while estimates using more generous values of the world economic product, put the average gain at 1%. A mere continuation of the latter rate would mean that in one generation (by the year 2025) today's global mean of 58 GJ/capita would be able to energize the production of goods and services for which we need now about 75 GJ. Conversely, energy services provided by today's 58 GJ required about 70 GJ/capita of initial inputs during the early 1970s—and that rate was the French mean of the early 1960s and the Japanese mean of the late 1960s.

This leads to an obvious question: would the billions of today's poor people be distressed when a generation from now they could experience the quality of life that was enjoyed by people in Lyon or Kyoto during the 1960s? Their life expectancies were above 70 years and infant mortalities were below 20/1,000 births (UNO 1998). And, in addition to being globally recognized epitomes of sophisticated culture, enviable cuisine, and admirable style, the two countries were also highly innovative. Their trains were, and remain, perhaps the best embodiment of this innovative drive. In 1964 Japan introduced its remarkable *shinkansen* (literally 'new trunk line' but better known as the bullet train) that has operated ever since without a single accident and injury carrying about 130 million people a year at speeds originally exceeding 200 and now even 300 km/hour and with delays averaging a mere 36 s/ train (CJRC 2002). And in France of the 1960s the state railway corporation (SNCF), stimulated by the Japanese example, launched its high-speed train program that produced *train de grand vitesse,* Europe's fastest means of land transport (TGV 2002).

And so the answer is obvious: for more than 90% of people that will be alive in today's low-income countries in the year 2025 it would be an immense improvement to experience the quality of life that was reached in France and Japan during the 1960s, a categorical gain that would elevate them from a barely adequate subsistence to what I would label perhaps most accurately as incipient affluence. A few hundreds millions of high income urbanites, ruling elites, and *narcotraficantes* of tomorrow's low-income world aside, that part of the grand convergence would be an easy sell. In contrast, lowering the rich world's mean seems to be an utterly unrealistic proposition. But I will ask any European reader born before 1950 or shortly afterwards and hence having good recollection of the 1960s, this simple question: what was so unbearable about life in that decade? What is so precious that we have gained since that time through our much increased energy use that we seem to be unwilling even to contemplate a return to those levels of fuel and electricity consumption? How fascinating it would be to collect a truly representative sample of honest answers!

But there is no doubt that the large intervening increase of average energy consumption has made any substantial reductions more difficult as they would require cuts far above marginal adjustments of 10–15%. For example, both France and Japan now average over 170 GJ/capita and so they would have to cut their consumption energy by about two-thirds. And the retreat assumes gargantuan proportions in the United States and Canada where an eventual sharing the global energy consumption mean of some 70 GJ/capita would mean giving up at least four-fifths of today's enormous energy use. An unprecedented catastrophic development aside, there is no realistic possibility of such a reduction within a single generation. A perceptive reader senses that I am merely gauging the reality against the potential, not charting a serious course of possible action.

The enormous global inequity that has been put in place by the economic development of the last two centuries cannot be undone in a generation but these simple yet factually unassailable comparisons demonstrate that an impressively high quality of life could be enjoyed worldwide with virtually unchanged global energy consumption. And the exercise is also useful as a benchmark for exploring the options that would significantly reduce the gap between the rich and poor worlds and establish a more secure global civilization. Continuation of existing trends (a mere theoretical exercise to be sure) shows the two means becoming equal in about 300 years. The need for a much faster narrowing of this gap is self-evident. More than a quarter-century ago Starr (1973) was concerned that by the year 2000 the average global per capita energy use will have risen from only about one-fifth of the U.S. average at that time to about one-third of that mean. In reality, in the year 2000 the global mean of 58 GJ/capita in 2000 was less than one-fifth (18% to be exact) of the U.S. average, and so the progress has been backwards.

But what strategy, what approach, would be most helpful in shifting the global energy system in the desirable direction of lessened environmental impacts and higher consumption equity in general, and reduced burning of fossil fuels in affluent countries in particular? Which one of the scores of technical fixes and new conversion techniques reviewed in this book offers the best potential for gradual but eventually truly radical changes of the world's energy use? Which set of policies would take us with the greatest expediency and with the least amount of social friction to desired goals of higher efficiency, better environmental protection, and enhanced human dignity?

What Does, and Does Not, Help

What does not help is easier to list, and a critical worldwide scan makes for a very sobering conclusion. States where public understanding and political commitment combines with technical and economic means and with decent governance to create conditions necessary for the formulation and pursuit of rational energy policies are in a small minority. At least two-thirds, and more realistically some three-quarters of the world's nations do not qualify. Either their overall capacities to be effective are quite inadequate, or even absent, or there are so many other constant crises, profound concerns, chronic complications, and never-ending entanglements claiming their limited resources that any thoughts that the leadership of these countries can contemplate conversion efficiency or accelerated decarbonization are truly bizarre. A listing of major barriers, in ascending order of total populations affected, makes this point clear.

How naïve to expect all kinds of dysfunctional states to become suddenly engaged in an effective quest for a more rational energy future. They include those that have seen nothing but decades of brutal civil wars (Sudan, Angola), experienced recurrent domestic genocides (Rwanda, Liberia), transborder conflicts (Central Africa) or various combinations of all of the above (Ethiopia, Eritrea, Somalia), are beset by sweeping AIDS epidemics (more than a score of sub-Saharan countries could be listed in this category), subverted to different degrees by *narcotraficantes* (not just Colombia but also Mexico, Bolivia, and Afghanistan), ruled by small dictatorial or military cliques (Congo, Myanmar, and North Korea), governed by extraordinarily corrupt bureaucracies (here the circle widens to include such major actors as China, India and Indonesia), or are trying to cope with growing populations living overwhelmingly in poverty (from Bangladesh to Vietnam).

These realities are the obvious ammunition for many science- and policy-based preachers of impending doom, but the end is not near. Evolution has built in a great deal of resilience into biota and history offers also plenty of inspiring examples of hope, recovery, and the rise of new human enterprises. Historical perspective also points out one of the most constant and most frustrating realities: we could do much better—in fact, given our current potential for technical fixes and social adjustments, almost miraculously so—but if we will do so remains always an open question. How does a democratic society, where the demand for energy is driven by myriads of individual actions and whose basic policies are subject to vigorous (often debilitating) contests of sometimes apparently irreconcilable opinions, come up with an effective course of action? More specifically, how do we begin the transition from the worship of unlimited growth to the sensibility of moderate consumption? Precedents may not offer any good guidance but they are revealing.

The only case of clearly encountered limits on energy use is that of the average per capita food consumption in affluent countries. This is not a supply limit but a limit on what should be actually consumed in order to maintain a healthy and productive life—and the way the populations and governments of affluent countries have, so far, dealt with this matter is hardly encouraging. These countries now have available at retail level more than 3,000 kcal/day per capita (in the United States this rate is about 3,700 kcal/day) while their actual average daily intake is only about 2,000 kcal/capita (fig. 6.8). This disparity creates enormous food waste amounting to between 1,100 kcal/day in Canada to 1,700 kcal/day in the United States—but

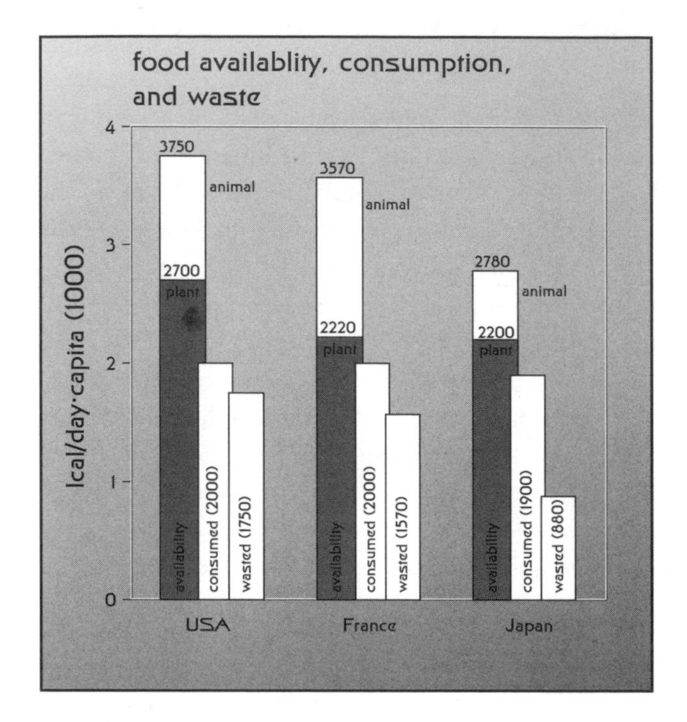

Figure 6.8
Per capita food availability, actual consumption, and apparent waste in the United States, France, and Japan. Plotted from data in FAO (2001) and Smil (2000c).

because the average requirements in the increasingly sedentary societies are below the average intakes the incidence of overeating is rampant, and obesity has been on a steep rise. In the United States its prevalence was fairly stable between 1960 and 1980 (at about 25% of adult population) and it increased by 8% during the 1980s (Kuczmarski et al. 1994).

By the early 1990s the mean weight gain of 3.6 kg had made every third U.S. adult overweight, with the highest increases among men over 50 and women between 30–39 and 50–59 years of age (Flegal 1996). Even more alarmingly, when looking at actual weights associated with the lowest mortality three-quarters of adults are above the optimal body mass. Obesity rates are still lower in Europe but there, too, they are on the rise. A rational response to this veritable epidemic of obesity that brings higher incidence of cardiovascular diseases and diabetes would be to begin limiting the overall food output destined for domestic consumption and to remove at least some of the existing production subsidies in order to raise food prices. Yet, in spite of the stunning diffusion of obesity, North American and European governments continue to promote higher food production, to finance research leading toward higher crop and animal yields, and to subsidize farmers in order to keep food prices at levels unprecedented in modern history (amounting to less than 15% of disposable income for North America's large middle class).

On the other hand, we have adjusted quite effectively to other energy challenges. Market response to the energy crises of the 1970s has been, after the initial hesitation, very effective. Some new public policies brought better than initially hoped-for results, and some that were counterproductive were deservedly abandoned: U.S. cars, as inefficient as they are, consume, on the average, only half as much fuel per unit distance as they did in 1973, and we are not spending billions on exploiting shale oil. But these successes are not enough to subscribe to Julian Simon's message of "joy and celebration" and to believe that

> There is no physical or economic reason why human resourcefulness and enterprise cannot forever continue to respond to impending shortages . . . and leave us with the bonus of lower costs and less scarcity in the long run. The bonus applies to such desirable resources as better health, more wilderness, cheaper energy, and a cleaner environment (Simon 1996, p. 588).

As with every complex system, no simplistic precepts will do, no single-line-of-attack solution will work. Moreover, as I stressed repeatedly, infrastructural problems preclude any rapid technical and resource shifts even if there were an unprecedented political will and a social compact to act. What *The Economist* (December 15, 2001, p. 9) editorialized in relation to the Western addiction to oil applies to any effective energy policy in general: "Gradualism is the key to doing this intelligently. The time to start is now." So these are the essential ingredients of any practical, effective approach: working without delay and with persistent commitment. This leaves little room for counterproductive ideological policy wars and their associated rhetoric, no matter in what special-interest camps they originate.

Demonization of nuclear power is not helpful in a world that, for better or worse, gets nearly one-fifth of its electricity from fissioning uranium and where many countries would find themselves in a precarious situation in regard to their electricity supply if they were to opt for a rapid closure of their reactors. Nor does it help to advocate the nuclear path as the best way to avoid the future consequences of rising GHG emissions: there are, clearly, many other solutions deserving serious attention. Elevation of still merely marginal, and often outright immature, renewable conversions to nearly exclusive pillars of the global TPES of the twenty-first century is not just naïve, it is profoundly irresponsible. On the other hand, a perfunctory dismissal of these conversions as inherently limited and permanently overrated choices is equally obtuse and shortsighted.

Consequently, what does not help is to adopt a factional true-believer attitude and cite as revealed truths one of the fairly extreme evaluations that pretends to end all disputes by staking out some unyielding positions. Unfortunately, most of the studies that compare the cost of renewable energy conversions with costs and burdens of fossil-fuel-based systems fall into this category and hence they cannot guide our actions. Instead of being hedged and honestly perplexed by the complexity of the task they gravitate toward one of the two incompatible conclusions. How I wish these confident analysts would have taken note of Alvin Weinberg's (1978, p. 158) honest admission:

> Having been guilty in the past of predicting that nuclear electricity would cost ten times less than it now costs, I am profoundly skeptical of any estimates of the costs of an energy system to be deployed a decade or more in the future.

The first group of these comparative analyses stresses that some of the new renewable conversions already are, and others will shortly become, competitive with fossil-fueled generations (e.g., Swezey and Wan 1995; Turner 1999; Herzog et al. 1999; UCS 2001; and just about everything produced by Amory Lovins and the RMI). Wind-generated electricity is repeatedly singled out as being already competitive with even the cheapest fossil-fueled generation and its price is predicted to drop by nearly half before 2010. Naturally, these analyses conclude that renewables can cover a large share of national and global TPES within the next generation and do so by saving huge capital investments, cutting the extraction and combustion of fossil fuels, and reducing the rate of global warming.

A subset of these analyses, assessments of the economic cost of significantly reducing GHG emissions, shows that renewable conversions should be the cornerstone of least-cost climatic stabilization and that abating global warming would have no net economic cost but would save hundreds of billions of dollars every year without any drastic changes to the current U.S. lifestyles. Lovins and Lovins (1991) put the overall savings at U.S.$200 billion (1990)/year and stressed that they can be achieved without any grand plans and government interventions by relying purely on free choice and free enterprise (Lovins and Lovins 1991). In the same year Lovins (1991) published another version of this claim under a playful title "Abating global warming for fun and profit."

And, to give just one more specific example, UCS (2001) believes that renewables can meet at least 20% of the U.S. electricity needs by 2020 while saving nearly half of $1 trillion, avoiding construction of nearly 1,000 new electricity-generating plants and reducing CO_2 emissions by two-thirds compared to the business-as-usual scenario. Not surprisingly, Turner (1999, p. 689) urges us that "we should embark upon this path with all due speed." In a sharp contrast, the second set of appraisals concludes that the renewables are neither cheap nor exceptionally 'green,' that their uneven distribution may do little for local or regional energy supply and that subsidies for their development are a waste of money (CEED 1995; Bradley 1997, 2001; Scheede 2001).

As far as the cost of cutting GHG is concerned, Nordhaus (1991) put it more than a decade ago at U.S.$200 billion (1990)/year for a 20% emissions cut to the late 1980s level. More recently, Nordhaus and Boyer (2000) used better integrated models of climate change and global economy to conclude that except for geoengineering—including such controversial planetary-scale interventions as the injection of particulate matter into the atmosphere or the stimulation of oceans as carbon sink and assumed to cost $10/t C (NAS 1992; Keith 2000)—none of the examined policies had major economic benefits. Limiting global emissions to the 1990 level would entail a loss of $1.125 trillion in present value and an even more ambitious goal of reducing GHG output to 80% of the 1990 level would cost the global economy $3.4 trillion. No Lovinsian fun and profit in these models!

As I have many points of disagreement with either of these extreme views I will, yet again, make few friends in these two opposite camps by advocating a very critical scrutiny of any supposedly totipotent solution. I have no respect for all those exaggerated claims made on behalf of immature techniques that have not been through the only test that matters: millions of units providing efficient, economical (when judged by any reasonable standard), reliable, and environmentally acceptable service for a couple of decades. Except for hydrogeneration using turbines of all sizes, none of the renewable conversions belongs to this category. I am also astonished by the continued acceptance of pronouncements made by experts whose forecasting record has been patently and repeatedly poor: remember (chapter 3), Lovins (1976) told us that by now one-third of the U.S. TPES will come from decentralized renewables, a forecast that missed its target by more than 90%!

On the other hand, I have little sympathy for the critics of new renewable conversions who are upset by subsidies received by wind farms or by the PV industry. The last thing I want to see is taxpayers' money wasted on such dubious schemes as corn-derived ethanol, cultivation of kelp forests for biomass or building of large-scale OTEC prototypes, but subsidies, overt or hidden, for new energy conversions are nothing new. Sieferle (2001) gives a number of interesting examples of how the Prussian state promoted and subsidized the use of coal during the late eighteenth century. And, as already noted in chapter 2, nothing in recent history equals the largesse made available to the U.S. nuclear industry that received between 1947 and 1998 about U.S.$ 145 billion (1998) or 96% of all government subsidies for energy R&D, with U.S.$5 billion (1998) going to renewables. As the NIRS (1999) summed it up, nuclear subsidies cost the average household more than $U.S.1,400 (1998) compared to U.S.$11 for (1998) wind-generated electricity. More recently, non-R&D federal tax credits have been heavily tilted in favor of fossil fuels, with nearly two-thirds of them going recently to the natural gas industry in order to encourage hydrocarbon production from nonconventional sources, and with all renewables receiving only 1% of all of these breaks (EIA 1999b).

Broader perspectives also help to make a case for judicious subsidies. Affluent economies are willing to subsidize European, North American, and Japanese farmers to the tune of more than $1 billion dollars a day in order to add to an already excessive supply of commodities whose glutonous consumption brings widespread obesity and a higher incidence of several civilizational diseases. Why then should not we be ready to subsidize the Great Plains farmers from Texas to North Dakota to tend windfarms rather than to keep producing superfluous crops whose growing accelerates soil erosion, adds more nitrogen leached from fertilizers to streams and aquifers, and reduces biodiversity? Analogically, why not subsidize some California and Arizona farmers to cover their fields with PV cells and generate electricity rather than to subsidize the cost of irrigation water used in their fields to grow crops that could be easily grown in rainier areas?

And although I do not belong to habitual worshippers of the market, and hence have no great illusions about its perfect operation, I do not accept the argument that it continues to ignore untold opportunities to realize large returns from energy savings, or that it persistently fails to internalize any important externalities. Many industries have pursued energy savings quite aggressively and, as already noted (in chapter 2), many of the major costs of coal-fired electricity generation have been internalized to a surprisingly high degree. If the external costs of coal-fired generation remain really as high as some renewable proponents claim, then, as Singer (2001, p. 2) acidly remarks, "the Clean Air Act has been a colossal failure and EPA has failed miserably in doing its job" and the competitiveness of wind-generated electricity would be assured even without any subsidies. On the other hand, it is not credible to claim that only marginal costs remain to be internalized. Absence of any links between the price of gasoline and the cost of military and political stabilization of the Middle East, or the toll of respiratory illness caused and aggravated by photochemical smog created by car emissions are perhaps the two most obvious examples of missing internalization.

At the same time, even the best available figures regarding the specific cost of today's conversion techniques must not be uncritically accepted. This skeptical attitude is doubly justified when looking at comparisons of highly dissimilar processes (e.g., wind vs. nuclear; hydroelectricity vs. PV cells) and it must be the dominant mode of any evaluations of future cost claims. While learning curves undoubtedly exist, and while specific costs of conversion processes tend to decline with time, there is no universal law that dictates the rate of such declines or their persistence. Costs of nuclear fission (contrary to predictions published during the 1950s and 1960s) did not fall so low that the technique could displace all other forms of electricity generation; the cost of electric cars remains uncompetitive with the ICE-powered machines after more than a century of predicting a bright future for electric vehicles; and the costs of assorted renewable conversion did not decline as rapidly during the last quarter of the twentieth century as was envisaged by advocates of the soft-energy path.

And how can I take seriously the conclusion made by Nordhaus and Boyer (2000) that geoengineering is the only profitable strategy to combat global warming? No such deliberately biosphere-shaping operation has ever been tried on a planetary scale and all one has to do is to plug into their model an entirely fictional estimate, made by the NAS (1992), that such actions may cost $10/t of carbon. But, avoiding dogmatism once more, I am not saying that we should simply dismiss the idea of planetary-scale engineering as hubristic and utterly impractical. Every one of the geoengineering solutions recently devised to deal with the human interference in the carbon cycle—be it the sequestration of liquid CO_2 in the deep ocean, nitrogen fertilization of the ocean's euphotic layer to remove the atmospheric carbon in a sinking dead phytomass, or installation of extraterrestrial shields to reduce the insolation (Jones and Otaegui 1997; Schneider 2001)—faces immense implementation challenges and none of them may prove to be practical.

I agree with Keith (2000, p. 280) who argues that "we would be wise to begin with a renewed commitment to reduce our interference in natural systems rather than to act by balancing one interference with another." But while I believe that effective purposeful planetary management is only the second-best choice and that it is, in any case, still beyond our technical capacity and social acceptance I would also argue that, to a significant degree, we are already engaged, unwittingly and unsystematically, in such a process (Smil 2002). Consequently, any thoughtful, systematic exploration of geoengineering possibilities and opportunities should be welcome rather than dismissed, even though none of these ideas may become a reality during this century. Just remember that it has taken more than 150 years to elevate fuel cells from a laboratory concept to one of the most promising energy converters.

Finally, and most fundamentally, even the most carefully constructed quantifications of existing or future costs of energy conversions can go only so far as there is no satisfactory method to come up with a meaningful cost–benefit ratio of the ultimate prize, the long-term preservation of the biosphere's integrity. This is because we have no meaningful way to value the nature's worth and recent quantifications at the global level are perfect examples of the futility of such estimates. Costanza et al. (1997) ended up with an average annual value of $33 trillion while Pimentel et al. (1997) put the total just short of $3 trillion, an order of magnitude difference. But this huge discrepancy is only a part of the problem. What do such figures really mean? That we can replicate the biosphere by investing that much, or that we can provide equivalent environmental services by spending that amount of money for alternative arrangements? Either idea is, obviously, ridiculous as no amount of money can replicate or substitute for nearly four billion years of the biosphere's evolution.

Realities and a Wish List

I made it clear (particularly in chapter 3) that long-term forecasts of energy affairs are of little value. Exploratory scenarios are much more preferable but as they have become quite fashionable during the 1990s I see no point of adding to this new industry: cognoscenti of that genre can get their fill by turning to scores of alternatives from WEC (1993), IIASA-WEC (1995), Morita and Lee (1998), SRES (2001), Nakićenović (2000) and EIA (2001). My preference for looking ahead is to outline a set of fairly unexceptionable realities that will shape the future developments and complement it by a short wishlist of actions we should take, attitudes we should think about, and commitments we should accept and pursue, as individuals and as societies, in order to move toward those grand twin objectives of life with dignity and the biosphere with assured integrity.

During the first two decades of the twenty-first century the long-established energy sources and conversion techniques will continue to dominate the markets. Coal will remain a key fuel for thermal generation of electricity and large plants will carry the bulk of the global base-load. Hydrocarbons will keep providing more than half of the global TPES, with a great deal of effort and investment going into further development of conventional resources that are either remote or difficult to access (or both) and with increasing attention paid to nonconventional deposits and to enhanced rates of recovery. There is nothing inevitable about the global output of conventional oil peaking before the year 2010 or shortly afterwards. The oil era's duration may be determined more by the demand for the fuel than by its availability. In any case, an early peak of oil extraction should be no reason for panic or regrets, and it would be unlikely to lead to such desperate technical fixes as large-scale coal liquefaction and extraction of oil shales. Natural gas, more efficient conversions and nonfossil sources will ease the transition.

The transition from societies energized overwhelmingly by fossil fuels to a global system based predominantly on conversions of renewable energies will take most of the twenty-first century. A very long road lies ahead of the renewables. At the beginning of the twenty-first century commercially exploited renewable energy sources supplied less than 30 EJ or between 7–8% of the world's TPES, but most of this total was made up of woody biomass (16 EJ) and hydroelectricity (9 EJ) with the new renewables—direct solar energy conversions and electricity generated by wind, waves, and the Earth's heat—adding merely 2 EJ (Turkenburg 2000; fig. 6.9). That is roughly 0.5% of the global TPES but, as with many other techniques in early stages of their diffusion, all of these conversions are growing at high rates and many countries have ambitious expansion plans, aiming at 15–20% of all electricity in the years 2020–2025 to be generated from renewable flows.

Of course, the pace of adopting new renewable conversions may be eventually determined more by the perceived and actual concerns about the global environmental change in general, and climate warming in particular, rather than by any supply shortages. But even with an unexpectedly speedy transition from fossil fuels to renewables the world would still experience a further increase of tropospheric GHG concentrations and the growing risk of nontrivial climate warming. This concern, and the lasting uncertainty about the eventual net global impact of climate change, will remain one of the key formative factors of energy decision-making during the twenty-first century.

Nor will many other environmental concerns disappear. Wider use of better techniques and continuing decarbonization of the global TPES will continue to moderate the environmental degradation caused by energy industries and uses. Still, none of

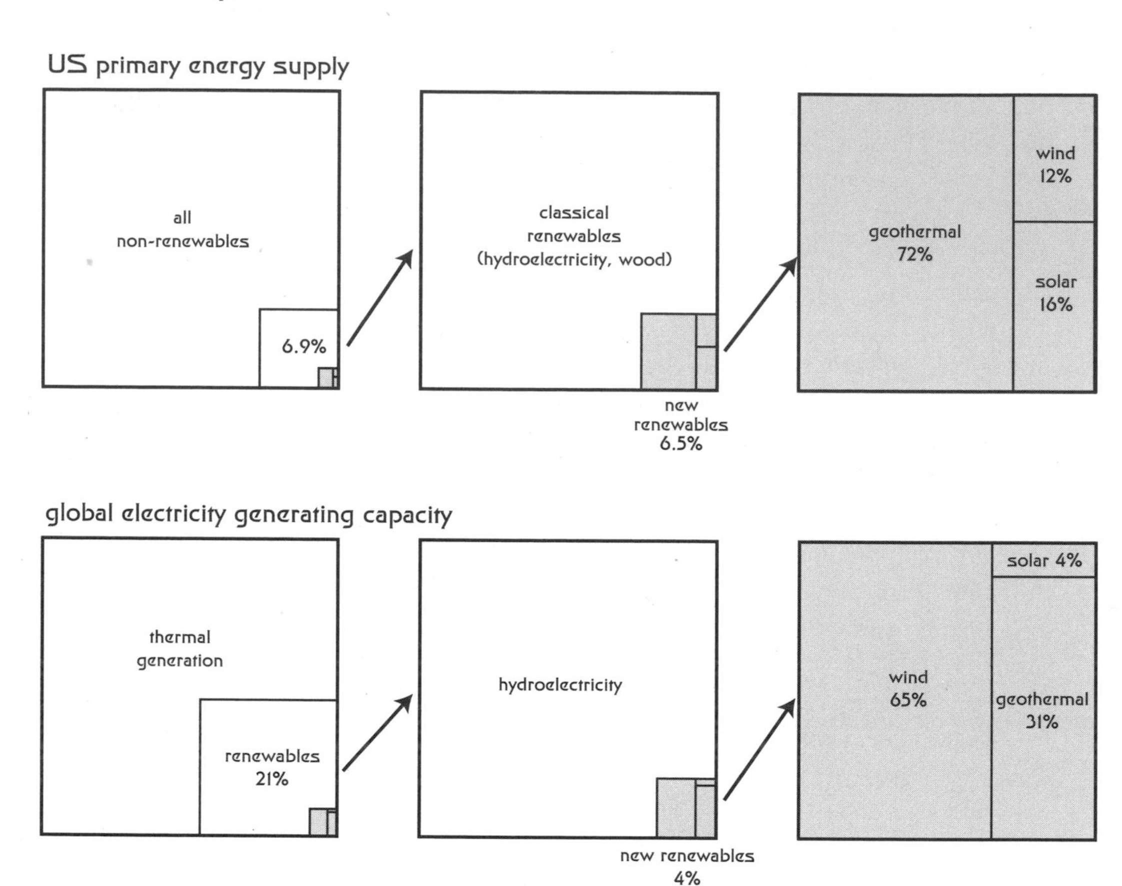

Figure 6.9
A long road ahead for the renewables: large-scale hydro generation aside, they still account for only minuscule shares of the U.S. and global TPES or electricity generation. Based on data in EIA (2001a), UNO (2001), and international statistics for specific renewable conversions.

the many local and regional impacts that are caused (entirely or partially) by our use of fossil fuels and electricity and that figure prominently among today's environmental worries (be it acid deposition and photochemical smog or loss of biodiversity or land-use changes) will be reduced to a nonconsequential degree during the next generation. And given the late-twentieth-century history of an unexpected emergence of several major environmental concerns (persistent pesticides, acid deposition, and depletion of stratospheric ozone) we should expect new, as yet unidentified, risks to assume a sudden prominence.

In contrast, the one concern that can be safely set aside is the availability of energy resources: reserves of conventional fossil fuels are sufficient to meet any conceivable demand during the next two generations, and nonconventional oil and gas resources can extend the duration of hydrocarbon era into the second half of the twenty-first century. Beyond the fossil fuels the world can tap several enormous renewable flows: direct solar radiation and wind energy in the accessible layer of the troposphere are both several orders of magnitude larger than the current global TPES and they can be supplemented by hydroenergy and geothermal flows.

There is little doubt that several nonfossil conversions, most notably wind- and PV-generated electricity, will become much more important during the coming decades, not just in absolute terms but also as shares of relevant aggregates. In contrast, mass production of liquid fuels from biomass has (and should have) low probability and the emergence of the much-touted hydrogen economy will almost certainly take much longer than is suggested by many uncritical promoters of the switch. Forecasting specific shares of these new renewable conversions at specific times is futile: we know most, if not all, practical ingredients of the energy system of the next 20–30 years, but we cannot pinpoint their future rates of growth, substitution speed, and resulting supply mix. The only indisputable certainty is the continuing rise of electricity's importance reflected by a higher share of fossil fuels consumed in this indirect way as well as by a determined pursuit of more efficient and less costly means of renewably generated electricity.

The only unassailable statements regarding the global TPES are that it should grow faster than the expected rate of population increase during the first half of the twenty-first century and that it should continue to grow even if there were an early stabilization of the global population. This growth is necessary in order to bring about the badly needed improvements in the quality of life to the world's low-income populations and to narrow the intolerably huge consumption gap between rich and poor countries. But even the highest practical rates of growth of energy consumption in today's low-income countries will not be able to close the gap between the two worlds of energy use.

Availability of useful energy will be increasing faster than the TPES due to continuing technical innovations and management improvements that will result in substantial gains in conversion efficiency and in lower transportation and transmission losses. Continuation of the historical trend should see gains on the order of 0.5–1% a year but faster gains are both desirable and possible. High-efficiency natural gas furnaces aside, just about every major household energy convertor in affluent countries should be 50% more efficient well before the middle of the twenty-first century, and many industrial and commercial processes should do even better. Efficiency gains as large, and even larger, should be promoted in every modernizing country.

As a result, the energy intensity of the global economy will continue to fall but the quality of the supplied energy will rise as a larger share of it will be consumed as electricity and natural gas. More electricity will be generated by conversions of renewable flows but, in accord with historically long transitions times, the shift from today's predominantly fossil-fuel-based system to one dominated by conversions of direct and indirect solar flows will not be completed before 2050. The greatest uncertainties concern the future contributions of nuclear generation. Even in the short run (until 2020) it is not certain if it will retain an important, albeit diminished, share of aggregate electricity supply, or if it will shrink to less than 10% of the total output. Its long-run fortunes are anybody's guess.

So is the future success and diffusion of technical breakthroughs that could make a fundamental difference in the world's energy supply and that are today either still in the experimental stage or are just taking the first steps toward commercialization. Given electricity's fundamental role in modern society superconductivity is perhaps the most alluring example. Until 1986 the phenomenon was possible only up to 23 K but in that year the limit was lifted a bit to 36 K and by the century's end there were several compounds able to superconduct above 77 K, the boiling point of nitrogen, with the record at 164 K (Grant 2001). Although it is challenging to fashion the copper oxide-based ceramics into wires, these advances make commercial applications of high-temperature superconducting (HTS) cables cooled by cheap liquid nitrogen possible.

The first uses are not motivated by eliminating transmission losses but by the opportunity to supply rising urban demand by carrying several times more power by merely retrofitting the existing links with the same thickness of a HTS cable. Detroit's Edison Frisbee substation was the first facility to lay several hundred meters of such cables in order to supply about 30,000 residential customers. By 2001 HTS cables were still 8–10 times more expensive than copper but some forecasts envisage that they may take half of the underground market in a decade. Among the recently reported advances are the discovery of C_{60} crystals, easier to make into electronic components, superconducting at 117 K (Service 2001), and the progress in making relatively long wires from magnesium diboride (MgB_2), a compound that is inexpensive and simple to make. The compound superconducts only up to 39 K but because it is so much easier to turn into wires it may find some commercial applications sooner than most oxide-based materials (Service 2002).

Examination of historical trends shows that many critical transitions took place even in spite of deliberate policies, or that planning and assorted ways of government intervention have not played critical roles in the diffusion of new conversions or new consumption patterns. On the other hand, many energy-related developments would not have taken place without such interventions. These are the challenges of public policy that never go away. What are the best choices to bring about many desired outcomes? How to promote efficiency and innovation without ending up with higher overall consumption? How to reduce energy use among those who are grossly overconsuming fuels and electricity and boost it among those who can only dream about dignified lives?

When facing so many uncertainties and when unable to foresee not just inevitable surprises but also the course of many key trends, we should pursue any effective means that bring us closer to those goals. That means adopting the ying–yang approach to reality: acting as complexifying minimalists rather than as simplifying maximalists, being determined but flexible, eclectic but discriminating. The first contrast means favoring a multitude of approaches rather than relying on any single (and purportedly perfect) solution and championing minimal inputs compatible with the highest achievable useful services. The other contrasts mean that there should be no place for a priori ideological purity that would not tolerate particular solutions, no categorical exclusions of certain ingredients (such as 'renewable future cannot be even partially nuclear' or 'there is no good large dam'), and no inflexible insistence on what is best ('distributed generation is the only way' or 'hydrogen economy is a must').

This will require, most commonly, a tolerant but discriminating approach to taxes and subsidies. Taxes (on imported oil, on gasoline, and on coal-generated electricity) are a common tool, perhaps second only to subsidies (for ethanol, wind-generated electricity, and home insulation) in shaping energy markets. A recent favorite is tax the emissions of carbon in order to stimulate the introduction of carbon-free transportation techniques and hence also to weaken the OPEC's future influence. Both of these tools can be wasteful, regressive, and basically ineffective—but when carefully designed and judiciously implemented they can be useful and rewarding. For example, international comparisons show that high purchase and ownership taxes are particularly effective in restraining car size, while taxes on weight, power, or displacement seem to have a smaller effect (Schipper et al. 1993). Carbon taxes have been widely debated as a tool for moderating the rate of global warming and their smart application (i.e., investing the money in GHG- and pollution-reducing activities and increasing the society's resilience rather than to disappear into the general revenue) could be very effective.

But I believe that one kind of government intervention —a targeted quest for massively expanded supply of a particular kind of energy—demands a great deal of skepticism and an extraordinarily careful scrutiny. These 'solutions' to perceived shortages of energy supply have been favored for decades by many interventionist advocates but looking back, there is a great deal to be thankful that so many of these proposals did not fly. To give just one notable example, during the late 1970s many experts favored the creation of a large-scale, government-subsidized oil shale industry and, as noted in chapter 3, massive public funding began to flow to create such an inefficient and environmentally objectionable entity. And even a decade later Abelson (1987) called for regulations requiring the incorporation of a certain share of synthetic liquids in gasoline in order to create a vigorous and competitive synthetic fuels program. Perhaps only obvious crises could justify that approach; otherwise I would keep this option at the very bottom of my energy wishlist.

Highest on my list—and, alas, the least realistic, at least in short- to midterm—would be a radical change in attitudes regarding the material consumption and the stewardship of the biosphere. Ever since the publication of *Our Common Future* (WCED 1987) these aspirations have been expressed in countless calls for sustainable development. Its definition as development that "meets the needs of the present without compromising the ability of future generations to meet their own needs" (WCED 1987, p. 8) and that promotes "harmony among human beings and between humanity and nature" (WCED 1987, p. 65) is fuzzy (how many future generations, what kind of needs?), even naïve (what kind of harmony?). If the term is to be valid on a civilizational scale, i.e., across a span of 1,000 years, then it is obvious that our fossil-fuel-based civilization is inherently unsustainable if its full benefits were to be extended to 7–9 billion people. And it is also clear that retaining its high-energy comforts and extending them to billions of people who still can only dream about them by relying entirely on renewable energy flows will be a monumental task that can be accomplished only after many generations of determined effort.

How fast can we push in that direction? How much easier can we make the transition by lowering the energy use in affluent countries? What are the realistic parameters of moderation? A classic of this genre gives a rather uncompromising answer. Georgescu-Roegen's attempts to examine modern economics in the light of physical realities ignored by the mainstream practitioners of the dismal science led him not just to reject the standard model of economic growth but to conclude that the most desirable state of affairs is not a stationary-state economy favored by many anti-growth advocates but a declining one (Georgescu-Roegen 1971, 1975).

His proposals to foster what he called a minimal bioeconomic program were headed by prohibition of all instruments of war, not just war itself. In the second place he put the aid to underdeveloped nations so they could attain as quickly possible a good (but not luxurious) standard of life. The other key steps included the reduction of global population to a level entirely supportable by organic farming, concerned efforts to minimize conversion inefficiencies, to eliminate the morbid craving for extravagant gadgetry and the habit of discarding items that can still perform useful service, a goal to be helped by designing more durable goods. Although unwavering in following the logic of the entropic nature of the economic process, Georgescu-Roegen was doubtful if humanity will adhere to any effort that implies a reduction of its addiction to exosomatic comfort. Inevitably, this led him to conclude that the modern civilization may have a brief, exciting, and extravagant existence rather than conforming to conditions needed for long, uneventful survival, a task better left to microorganisms that have no spiritual ambitions.

"First things first" might be the best way to argue with that sobering verdict. To begin with, I would be overjoyed to see the worship of moderate growth coupled with an unwavering commitment to invest in smart, that is appropriately targeted, protection of key biospheric goods and services. Two formidable obstacles are in the way: a disproportionate amount of our attention continues to go into increasing the supply rather than moderating the demand, and modern economists, zealous worshippers of growth, have no experience with running a steady-state economy and an overwhelming majority of them would probably even refuse to think about its possible modalities. Yet there is little doubt that many of these moderating steps can be taken without materially affecting the high quality of life and at a very acceptable cost (or even with profit). I do not think I exaggerate when I see this to be primarily an issue of attitude rather than of a distinct and painful choice.

A single impressive example illustrates the opportunities. James et al. (1999) calculated that safeguarding most of the world's existing biodiversity would cost annually about $16.6 billion in addition to the inadequate $6 billion that is being spent currently. This boost would secure adequate budgets for national parks and various nature reserves that cover nearly 5% of the world area and it would extend their coverage to 10% of area in every major region. The sum of $23 billion is equal to only about 0.1% of the combined GDP of the world's affluent economies—or to the annual global sales of cinema tickets, and it is a small fraction of environmentally harmful subsidies going to growing surplus food.

Examples of this kind show that *shaping the future energy use in the affluent world is primarily a moral issue, not a technical or economic matter. So is the narrowing of the intolerable quality-of-life gap between the rich and the poor world.* This divide usually refers to the huge difference between the high- and low-income economies but it also exists within all modernizing countries where the absence of decent governance, inequity of opportunities, and widespread corruption create gaps that are unnecessarily wide. Economic development of the last quarter of the twentieth century has made a great deal of difference (China's quadrupling of GDP in a single generation is certainly the best example of welcome progress) and Collier and Dollar (2001) concluded that if the trends of the 1990s persist, poverty in the low-income countries will be cut by about one-half by 2015. The two best things the rich countries can do to accelerate this process is to allow fair access at fair prices to exports from modernizing countries and to transfer to them the energy efficient techniques so the rising energy use can be accommodated with high efficiency.

All of the items on my wishlist (and it could be expanded considerably) are a matter of sensible choice: there are no insurmountable obstacles preventing their adoption, there is no excruciating price to pay for their pursuit; in fact there is a very high probability of substantial long-term rewards. But they would require some rather fundamental departures from many prevailing practices and attitudes. Any hope-inspiring, long-term perspective must come to this conclusion. That is why the most comprehensive multiauthored assessment of world energy prospects also argues that the best course available to us will not succeed without significant policy and behavioral changes (Goldemberg 2000). Making these changes seems forbidding and unrealistic when the task is seen through the prism of ingrained practices but retrospectives show repeatedly that many shifts that were considered too radical or too demanding were eventually accomplished with a surprising ease. But, of course, the success is not foreordained, as some attitudes seem immune to change.

And so we may succeed (at least relatively)—or we may fail. Any diligent student of long-term historical trends must appreciate unpredictably cyclic nature of human affairs and hence cannot pick with high confidence tomorrow's winners on the basis of today's performance (Taintner 1988; Mallmann and Lemarchand 1998). Russia's history offers the most fascinating twentieth-century example of these swings. A weak Soviet state that emerged from the Russian defeat in WWI was transformed a generation later by its victories during WWII into a new superpower, and for 40 years afterwards the seemingly relentless rise of the Soviet Union inspired mortal fears among the Western decision makers. Their policies were guided by one overarching goal, to prevent the Soviet global takeover, to deny them a draw in the cold war. The Soviet regime was indisputably evil as well as powerful in so many ways but once it fell it also became clear how fundamentally hollow and weak it had been for decades, how (fortunately) exaggerated were our fears, including the CIA's famously huge overestimates of the country's GDP (see, for example, CIA 1983).

China's course during the last 300 years is the best example of these dramatic waves unfolding on a multicentennial scale. Under Qianlong (1711–1799) the Qing empire was the world's richest large economy with average standard of living surpassing that of European powers (Frank 1998). Slow decline was then followed by more than a century of disintegration, wars, and suffering, from the Opium War (1839–1842) to Mao Zedong's death in 1976, 15 years after the world's worst famine, caused by his criminal decisions, cost 30 million lives between 1959 and 1961 (Smil 1999c). And then Deng Xiaoping's rise to power in 1979 ushered in a rapid emergence of China as the world's third-largest economy of the 1990s and the country is now clearly aspiring to return, once again, to a great power status.

Needless to say, these waves have a profound effect on energy use. Just before the beginning of World War II, Russia produced only about one-thirteenth of the U.S. output of primary energy (Schurr and Netschert 1960; CSU 1967), but just before its collapse the Soviet Union was producing about 10% more of the primary energy than did the United States and it led the world in both crude oil and natural gas extraction. In 1950 China consumed less than 3% of the U.S. energy demand, half a century later it rivaled the country in coal extraction and its TPES rose to one-third of the U.S. total (BP 2001). And the link, not surprisingly, goes also in the opposite direction: Watt (1989) concluded that the exploitation of energy resources has itself been an important factor in producing long waves in the U.S. economy.

And so it may be that after centuries of technical, economic, and political ascendance the West, now increasingly dependent on energy imports, may falter during the twenty-first century. And perhaps even the resurgent East will not be able to meet easily the concatenated demands imposed by global environmental change, a transition to a new energy foundation, radical demographic shifts, and political instability. Perhaps the evolutionary imperative of our species is to ascend a ladder of ever-increasing energy throughputs, never to consider seriously any voluntary consumption limits and stay on this irrational course until it will be too late to salvage the irreplaceable underpinnings of biospheric services that will be degraded and destroyed by our progressing use of energy and materials.

In 1900 the world's TPES was about 35 EJ, more than one-third of it from biomass fuels. By 1950 the total rose about 2.5 times to almost 90 EJ—but during the second half of the twentieth century it went up more than 4.5 times to about 400 EJ (Smil 1994a; BP 2001). A reprise of this growth (nearly 12-fold expansion) during the twenty-first century would bring the global TPES close to 5,000 EJ by the year 2100 and even with the world of 10 billion people that would prorate annually to 500 GJ/capita, more than 40% above the current North American mean (fig. 6.10). That is not going to happen—but how far will we push in that direction regardless of the increasing signs of the biospheric distress? Where are the precedents for substantial self-imposed limits on consumption in any modern democratic society or for successful legislated curbs on demand that have reduced *total* (as opposed to specific) energy or material flows in the long run?

But serious crises induced by excessive consumption may be no harbinger of a civilizational demise: just the opposite may be true. New anthropological perspectives indicate that our species, unlike all other organisms, including our much less encephalized hominin ancestors, has evolved not to adapt to specific conditions and tasks but to cope with change (Potts 2001). This ability does not guarantee steady progress but it is an essential quality for dealing with crises. So perhaps the best developments we should wish for would be the denial of the Persian Gulf oil or indubitable evidence of imminent and near-catastrophic global warming. Such crises would challenge our ability to deal with shifts that endanger the very foundations of our existence as an advanced civilization but given the level of our understanding and our capacity for change we might be able to overcome them no less successfully than our ancestors did the enormous climatic and ecosystemic seesaws of the last 200,000 years. Unless, of course, both the magnitude and the pace of these changes proves to be too overwhelming.

None of us knows what lies ahead. What we know is that our uses of energy that define and sustain our physical well-being and allow for an unprecedented exercise of our mental capacities will be the key ingredients in shaping that unknown future. "When we try to pick out anything by itself, we find it hitched to everything else in the Universe," wrote John Muir (1911, p. 326). These links are particularly strong for energy flows and conversions. Tug at any human use of energy and you will find

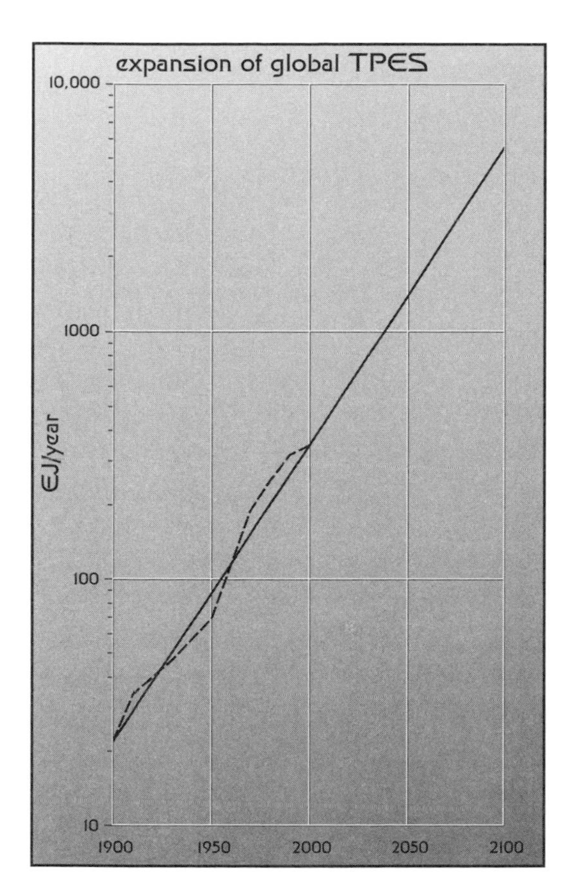

Figure 6.10
As this graph shows, the impressive record of the post-1900 expansion of the TPES cannot be repeated during the twenty-first century. A new road must be found and followed, but there are no easy precepts to follow.

its effects cascading throughout the society, spilling into the environment and coming back to us. As we were building the edifice of the first high-energy society many things got unraveled in the process but one key reality made the task easier: during the twentieth century we were largely on a comfortable, and a fairly predictable, energy path of a mature fossil-fueled civilization. Things are different now: the world's energy use is at the epochal crossroads. The new century cannot be an energetic replica of the old one and reshaping the old practices and putting in place new energy foundations is bound to redefine our connection to the universe.

Units and Abbreviations

a	year
b and bbl	barrel
°C	degree Celsius
ce	coal equivalent
dwt	dead weight ton
g	gram
h	hour
ha	hcctare
hp	horsepower
Hz	Hertz
J	Joule
K	temperature, in Kelvin
L	liter
m	meter
mpg	miles per gallon
oe	oil equivalent
Pa	Pascal
ppm	parts per million
Rmb	yuan renminbi
s	second
t	tonne (metric ton)
V	volt
W	watt
¥	yen

Prefixes

μ	micro	10^{-6}
m	milli	10^{-3}
h	hecto	10^{2}
k	kilo	10^{3}
M	mega	10^{6}
G	giga	10^{9}
T	tera	10^{12}
P	peta	10^{15}
E	exa	10^{18}

Acronyms

AC	Alternating Current
AFC	Alkaline Fuel Cell
AFBC	Atmospheric Fluidized Bed Combustion
AWEA	American Wind Energy Association
BP	British Petroleum
CAFE	Corporate Automobile Fuel Efficiency
CCT	Clean Coal Technology
CDIAC	Carbon Dioxide Information and Analysis Centre
CEC	California Energy Commission
CEED	Center for Energy and Economic Development (United States)
CIA	Central Intelligence Agency (United States)
CJRC	Central Japan Railway Company
CSE	Centre for Science and Environment (United States)
CSP	Central Solar Power
DC	Direct Current
DMFC	Direct Methanol Fuel Cell
DOE	Department of Energy (United States)
DSM	Demand Side Management
DWIA	Danish Wind Industry Association
EC	European Commission
EI	Energy Intensity
EIA	Energy Information Administration (DOE)

EPA	Environmental Protection Agency (United States)
EPRI	Electric Power Research Institute (United States)
ESA	Electricity Storage Association (United States)
EU	European Union
FAO	Food and Agriculture Organization (UNO)
FBC	Fluidized Bed Combustion
FGD	Flue Gas Desulfurization
FPC	Federal Power Commission (United States)
GAO	General Accounting Office (United States)
GDP	Gross Domestic Product
GHG	Greenhouse Gas
GNP	Gross National Product
HDI	Human Development Index
HTGR	High Temperature Gas Reactor
IAEA	International Atomic Energy Agency
ICAO	International Civil Aviation Organization
ICE	Internal Combustion Engine
ICOLD	International Commission on Large Dams
IEA	International Energy Agency
IEE	Institute of Energy Economics (Japan)
IGA	International Geothermal Association
IGCC	Integrated Gasification Combined Cycle
IHA	International Hydropower Association
IIASA	International Institute for Applied Systems Analysis
IISI	International Iron and Steel Institute
IPCC	Intergovernmental Panel on Climatic Change
JAMSTEC	Japan Marine Science and Technology Center
LBL	Lawrence Berkeley Laboratory
LED	Light Emitting Diode
LIMPET	Land Installed Marine Powered Energy Transformer
LMFBR	Liquid Metal Fast Breeder Reactor

LNG	Liquefied Natural Gas
LSP	Lunar Solar Power
MCFC	Molten Carbonate Fuel Cell
MSHA	Mine Safety and Health Administration (United States)
NAS	National Academy of Sciences (United States)
NAPAP	National Acid Precipitation Assessment Program (United States)
NBS	National Bureau of Statistics (China)
NCES	National Center for Education Statistics (United States)
NCP	National Center for Photovoltaics (United States)
NIRS	Nuclear Information and Resource Service (United States)
NPP	Net Primary Productivity
NRC	National Research Council (United States)
NREL	National Renewable Energy Laboratory (United States)
NYMEX	New York Mercantile Exchange
OECD	Organization for Economic Cooperation and Development
OPEC	Organization of Petroleum Exporting Countries
ORNL	Oak Ridge National Laboratory
OSM	Office of Surface Mining (United States)
OTA	Office of Technology Assessment (United States)
OTEC	Ocean Thermal Energy Conversion
PAFC	Phosphoric Acid Fuel Cell
PDVSA	Petroleos de Venezuela SA
PEMFC	Proton Exchange Membrane Fuel Cell
PFBC	Pressurized Fluidized Bed Combustion
PG&E	Pacific Gas & Electric
PPP	Purchasing Power Parity
PV	Photovoltaics
PWR	Pressurized Water Reactor
RMI	Rocky Mountain Institute
R/P	Reserve/Production
RWEDP	Regional Wood Energy Development Programme in Asia

SFC	Synthetic Fuels Corporation (United States)
SPS	Solar Power System
SRES	Special Report on Emission Scenarios
SSF	Simultaneous Saccharification and Fermentation
SUV	Sport Utility Vehicle
TGV	Train de Grand Vitesse
TIGR	The Institute for Genomic Research
TPES	Total Primary Energy Supply
TRC	Texas Railroad Commission
UCS	Union of Concerned Scientists (United States)
UKAEA	United Kingdom Atomic Energy Agency
UNDP	United Nations Development Programme
UNO	United Nations Organization
USCB	U.S. Census Bureau
USGS	U.S. Geological Survey
WAES	Workshop on Alternative Energy Strategies
WCD	World Commission on Dams
WCED	World Commission on Environment and Development
WCI	World Coal Institute
WEC	World Energy Conference
WHO	World Health Organization
WNA	World Nuclear Association
WTO	World Trade Organization
ZEV	Zero Emissions Vehicle

References

ABC News. 2001. That sinking feeling. <http://abcnews.go.com/sections/politics/DailyNews/poll010604.html>.

Abelson, P. 1987. Energy futures. *American Scientist* 75:584–593.

Adelman, M. 1990. OPEC at thirty years: What have we learned? *Annual Review of Energy* 15:1–22.

Adelman, M. 1992. Oil resource wealth of the Middle East. *Energy Studies Review* 4(1):7–21.

Adelman, M. 1997. My education in mineral (especially oil) economics. *Annual Review of Energy and the Environment* 22:13–46.

AeroVironment 2001. *Unmanned Aerial Vehicles*. Monrovia, CA: AeroVironment. <http://www.aerovironment.com/area-aircraft/unmanned.html>.

Alewell, C., et al. 2000. Environmental chemistry: Is acidification still an ecological threat? *Nature* 407:856–857.

Alexander's Gas & Oil Connections. 1998. World offshore energy output expected to rise strongly. <http://www.gasandoil.com/goc/reports/rex84414.htm>.

Allen, M. R., et al. 2000. Quantifying the uncertainty in forecasts of anthropogenic climate change. *Nature* 407:617–620.

Al Muhairy, A., and E. A. Farid. 1993. Horizontal drilling improves recovery in Abu Dhabi. *Oil & Gas Journal* 91(35):54–56.

Alpert, S. B. 1991. Clean coal technology and advanced coal-based power plants. *Annual Review of Energy and the Environment* 16:1–23.

Anderson, B. 2001. The energy for battle. *The Spectator* (December 8): 20–21.

Anderson, J. H. 1987. Ocean thermal power comes of age. York, PA: Sea Solar Power.

Anderson, R. N. 1998. Oil production in the 21st century. *Scientific American* 278(3):86–91.

Appleby, A. J., and F. R. Foulkes. 1989. *Fuel Cell Handbook*. New York: Plenum.

Aringhoff, R. 2001. Concentrating solar power in the emerging marketplace. Paper presented at the Solar Forum 2001, Washington, DC, April. <http://www.eren.doe.gov/troughnet/pdfs/rainer_aringhoff_proj_devel.pdf>.

Ariza, L. M. 2000. Burning times for hot fusion. *Scientific American* 282(3):19–20.

Artsimovich, L. 1972. The road to controlled nuclear fusion. *Nature* 239:18–22.

Ash, R. F., and Y. Y. Kueh, eds. 1996. *The Chinese Economy under Deng Xiaoping*. Oxford: Clarendon Press.

Ashley, S. 1992. Turbines catch their second wind. *Mechanical Engineering* 114(1):56–59.

Atkins, S. E. 2000. *Historical Encyclopedia of Atomic Energy*. Westport, CT: Greenwood Press.

Aurbach, D., et al. 2000. Prototype systems for rechargeable magnesium batteries. *Nature* 407:724–726.

Ausubel, J. H. 1996. Can technology spare the Earth? *American Scientist* 84:166–178.

AWEA (American Wind Energy Association). 2001a. Global Wind Energy Market Report. Washington, DC: AWEA. <http://www.awea.org/faq/global2000.html>.

AWEA (American Wind Energy Association). 2001b. Wind Energy Projects throughout the United States. Washington, DC: AWEA. <http://www.awea.org/projects/index.html>.

Bainbridge, G. R. 1974. Nuclear options. *Nature* 249:733–737.

Ballard Power Systems. 2002. XCELLSIS: The Fuel Cell Engine Company. Burnaby, BC: Ballard Power Systems. <http://www.ballard.com/viewreport.asp>.

Bankes, S. 1993. Exploratory modeling for policy analysis. *Operations Research* 41:435–449.

Barczak, T. M. 1992. *The History and Future of Longwall Mining in the United States*. Washington, DC: US Bureau of Mines.

Basile, P. S., ed. 1976. *Energy Demand Studies: Major Consuming Countries*. Cambridge, MA: The MIT Press.

Battelle Memorial Institute. 1969. *A Review and Comparison of Selected United States Energy Forecasts*. Washington, DC: USGPO.

Beck, P. W. 1999. Nuclear energy in the twenty-first century: Examination of a contentious subject. *Annual Review of Energy and the Environment* 24:113–137.

Becker, J. 2000. Henan dam fails to find customers. *South China Morning Post*, October 19, 2000, p. 3.

Bethe, H. 1977. The need for nuclear power. *Bulletin of the Atomic Scientists* 33(3):59–63.

Betts, K. S. 2000. The wind at the end of the tunnel. *Environmental Science & Technology* 34:306A–312A.

Björk, S., and W. Granéli. 1978. *Energivass*. Lund: Limnologiska Institutionen.

Blakney, R. B. 1955. *The Way of Life* (Translation of Laozi's *Daodejing*). New York: New American Library.

Blair, B. G. 2001. *Threat scenarios*. Washington, DC: Center for Defense Information. <http://www.cdi.org/terrorism/nuclear.cfm>.

Blue Energy Canada. 2001. Advantages of the Blue Energy Power System. Blue Energy Canada: Vancouver. <http://www.bluenergy.com/advantages.html>.

Blyth Offshore Wind Farm. 2001. Blythe Offshore—The U.K.'s first offshore wind farm. Hexham, UK, AMEC Border Wind. <http://www.offshorewindfarms.co.uk/sites/bowl.html>.

Bockris, J. O. 1980. *Energy Options.* Sydney: Australia New Zealand Book Company.

Boeing. 2001. Boeing 777 program milestones. 777-300 technical characteristics. <http://www.boeing.com/commercial/>.

Boulding, K. E. 1973. The social system and the energy crisis. *Science* 184:255–257.

Bowers, B. 1998. *Lengthening the Day: A History of Lighting Technology.* Oxford: Oxford University Press.

BP (British Petroleum). 1979. *Oil Crisis Again.* London: BP.

BP (British Petroleum). 2001. *BP Statistical Review of World Energy 2001.* London: BP. <http://www.bp.com/worldenergy>.

Bradley, R. L. Jr. 1997. *Renewable Energy: Not Cheap, Not "Green."* Washington, DC: Cato Institute.

Bradley, R. L. Jr. 2001. Green energy. In *Macmillan Encyclopedia of Energy,* ed. J. Zumerchik, Vol. 2, 598–601. New York: Macmillan.

Brantly, J. E. 1971. *History of Oil Well Drilling.* Houston, TX: Gulf Publishing.

Braun, G. W., and D. R. Smith. 1992. Commercial wind power: Recent experience in the United States. *Annual Review of Energy and the Environment* 17:97–121.

Bruce, A. W. 1952. *The Steam Locomotive in America.* New York: W. W. Norton.

BTM Consult. 1999. Summary. Ringkøbing: BTM Consult. <http://www.btm.dk/Articles/fed-global/fed-global.htm>.

California Energy Commission (CEC). 1999. *High Efficiency Central Gas Furnaces.* <http://www.energy.ca.gov/efficiency/appliances>.

Campbell, C. J. 1991. *The Golden Century of Oil 1950–2050.* Dordrecht: Kluwer.

Campbell, C. J. 1996. The Twenty-First Century: The World's Endowment of Conventional Oil and its Depletion. <http://www.hubbertpeak.com/campbell/camfull.htm>.

Campbell, C. J. 1997. *The Coming Oil Crisis.* Brentwood, UK: Multi-Science Publishing and Petroconsultants.

Campbell, C. J., and J. Laherrère. 1998. The end of cheap oil. *Scientific American* 278(3): 78–83.

Capstone MicroTurbines. 2002. CEC Certifies Capstone MicroTurbines to stringent California grid interconnect standards. <http://www.microturbine.com/whatsnew/pressrelease.asp?article=163>.

Caro, R. A. 1982. *The Years of Lyndon Johnson: The Path to Power.* New York: Knopf.

Casten, T. R. 1998. *Turning off the Heat: Why America Must Double Energy Efficiency to Save Money and Reduce Global Warming.* Amherst, NY: Prometheus Books.

Catalytica Energy Systems. 2001. XONON™ Combustion System. Mountain View, CA: Catalytica Energy Systems. <http://www.catalyticaenergy.com/xonon/how_it_works3.html>.

Cavender, J. H., D. S. Kircher, and A. J. Hoffman. 1973. *Nationwide Air Pollutant Emission Trends 1940–1970.* Research Triangle Park, NC: EPA.

CDIAC (Carbon Dioxide Information and Analysis Center). 2001. Current greenhouse gas concentrations. Oak Ridge, TN: CDIAC. <http://cdiac.esd.ornl.gov/pns/current_ghg.html>.

CEC (California Energy Commission). 2001a. Energy Statistics. Sacramento, CA: California Energy Commission. <http://www.energy.ca.gov/>.

CEC (California Energy Commission). 2001b. *Cool Savings with Cool Roofs.* Sacramento, CA: California Energy Commission. <http://www.consumerenergycenter.org/coolroof>.

CEED (Center for Energy and Economic Development). 1995. *Energy Choices in a Competitive Era: The Role of Renewable and Traditional Energy Resources in America's Electric Generation Mix.* Alexandria, VA: CEED.

Chambers, A. 2000. Wind powers spins into contention. *Power Engineering* 104(2):15–18.

Chameides, W. L., et al. 1994. Growth of continental-scale metro-agro-plexes, regional ozone pollution, and world food production. *Science* 264:74–77.

Chanute, O. 1904. Aerial navigation. *Popular Science Monthly* (March 1904):393.

Chapin, D. M., et al. 2002. Nuclear power plants and their fuel as terrorist targets. *Science* 297:1997–1999.

Chatterjee, A. 1994. *Beyond the Blast Furnace.* Boca Raton, FL: CRC Press.

Chesshire, J. H., and A. J. Surrey. 1975. World energy resources and the limitations of computer modelling. *Long Range Planning* 8:60.

China E-News. 2001. Forecasts of primary energy consumption in China. <http://www.pnl.gov/china/chect2.htm>.

Chynoweth, D. P., and R. Isaacson, eds. 1987. *Anaerobic Digestion of Biogas.* New York: Elsevier.

CIA. 1983. *Soviet Gross National Product in Current Prices, 1960–1980.* Washington, DC: CIA.

CJRC (Central Japan Railway Company). 2002. Databook 2001. Tokyo: CJRC. <http://www.jr-central.co.jp>.

Clark, E. L. 1986. Cogeneration—Efficient energy source. *Annual Review of Energy* 11:275–294.

Claude, G. 1930. Power from the tropical seas. *Mechanical Engineering* 52:1039–1044.

Colbeck I. 1994. *Air Pollution by Photochemical Oxidants.* New York: Elsevier.

Colbeck, I., and A. R. MacKenzie. 1994. *Air Pollution by Photochemical Oxidants.* Amsterdam: Elsevier.

Collier, P., and D. Dollar. 2001. Can the world cut poverty in half? How policy reform and effective aid can meet international development goals. *World Development* 29:1787–1802.

Coming Global Oil Crisis, The. 2001. <http://www.oilcrisis.com/>.

Colombo, U., and U. Farinelli. 1992. Progress in fusion energy. *Annual Review of Energy and the Environment* 17:123–159.

Committee for the Compilation of Materials on Damage Caused by the Atomic Bombs in Hiroshima and Nagasaki. 1981. *Hiroshima and Nagasaki.* New York: Basic Books.

Constant, E. W. 1981. *The Origins of Turbojet Revolution.* Baltimore, MD: Johns Hopkins University Press.

Cook, J. H., J. Beyea, and K. H. Keeler. 1991. Potential impacts of biomass production in the United States on biological diversity. *Annual Review of Energy and the Environment* 16: 401–431.

Cooper, G. A. 1994. Directional drilling. *Scientific American* 270(5):82–87.

Corten, G. P., and H. F. Veldkamp. 2001. Insects can halve wind-turbine power. *Nature* 412: 41–42.

Costanza, R., and H. Daly. 1992. Natural capitalism and sustainable development. *Conservation Biology* 6:37–46.

Costanza, R., et al. 1997. The value of the world's ecosystem services and natural capital. *Nature* 387:253–260.

Cowan, R. 1990. Nuclear power reactors: a study in technological lock-in. *Journal of Economic History* 50:541–567.

Craford, M. G., N. Holonyak, and F. A. Kish. 2001. In pursuit of the ultimate lamp. *Scientific American* 284(2):63–67.

Craig, P. P. 1999. High-level nuclear waste: The status of Yucca Mountain. *Annual Review of Energy and the Environment* 24:461–486.

Craig, P. P., A. Gadgil, and J. G. Koomey. 2002. What can history teach us? A retrospective examination of long-term energy forecasts for the United States. *Annual Review of Energy and the Environment* 27:83–118.

Crany, R., et al. 1948. *The Challenge of Atomic Energy.* New York: Columbia University Press.

Creagan, R. J. 1973. Boon to society: The LMFBR. *Power Engineering* 77(2):12–16.

Creutz, E. C. 1972. How soon fusion? *Science* 175:43.

Criqui, P. 1991. After the Gulf crisis: The third oil shock is yet to come. *Energy Studies Review* 3(3):205–216.

Criswell, D. R., and R. G. Thompson. 1996. Data envelopment analysis of space and terrestrially-based large scale commercial power systems for Earth: A prototype analysis of their relative economic advantages. *Solar Energy* 56:119–131.

CSE (Centre for Science and Environment). 2001. Green alternatives. *Down to Earth* 10(7): 56.

CSU (Central'noye statisticheskoie upravleniye). 1967. *Strana sovetov za 50 let.* Moscow: CSU.

Cullen, R. 1993. The true cost of coal. *The Atlantic Monthly* 272(6):38–52.

Culp, A.W. 1991. *Principles of Energy Conservation.* New York: McGraw-Hill.

Culver, S. J., and P. F. Rawson, ed. 2000. *Biotic Response to Global Change: The Last 145 Million Years.* Cambridge: Cambridge University Press.

Dai, Q., ed. 1994. *Yangtze! Yangtze!* London: Earthscan.

Daily, G. C., ed. 1997. *Nature's Services: Societal Dependence on Natural Ecosystems.* Washington, DC: Island Press.

Daly, H. E., and J. B. Cobb, Jr. 1989. *For the Common Good.* Boston: Beacon Press.

Darmstadter, J. 1997. *Productivity Change in U.S. Coal Mining.* Washington, DC: Resources for the Future.

David, P. A. 1991. The hero and the herd in technological history: Reflections on Thomas Edison and the battle of the systems. In *Favorites of Fortune,* eds. P. Higonnet et al., 72–119. Cambridge, MA: Harvard University Press.

de Beer, J., E. Worrell, and K. Blok. 1998. Future technologies for energy-efficient iron and steel making. *Annual Review of Energy and the Environment* 23:123–205.

De Gregori, T. R. 1987. Resources are not; they become: An institutional theory. *Journal of Economic Issues* 21(3):1241–1263.

DeCarolis, J. F., and D. Keith. 2001. The real cost of wind energy. *Science* 294:1000–1001.

Deffeyes, K. S. 2001. *Hubbert's Peak: The Impending World Oil Shortage.* Princeton, NJ: Princeton University Press.

Denner, J., and T. Evans. 2001. Hypercar makes its move. *RMI Solutions* 17(1):4–5, 21.

Denning, R. S. 1985. The Three Mile Island unit's core: A post-mortem examination. *Annual Review of Energy* 10:35–52.

Derickson, A. 1998. *Black Lung: Anatomy of a Public Health Disaster.* Ithaca, NY: Cornell University Press.

Devine, R. S. 1995. The trouble with dams. *The Atlantic Monthly* 276(2):64–74.

Devine, W. D. 1983. From shafts to wires: Historical perspective on electrification. *The Journal of Economic History* 63:347–372.

DeZeeuw, J. W. 1978. Peat and the Dutch Golden Age. *AAG Bijdragen* 21:3–31.

Dickens, G. R., et al. 1997. Direct measurement of in situ methane quantities in a large gas-hydrate reservoir. *Nature* 385:426–428.

Dickinson, H. W. 1967. *James Watt: Craftsman and Engineer.* New York: A. M. Kelly.

Diener, E., E. Suh, and S. Oishi. 1997. Recent findings on subjective well-being. *Indian Journal of Clinical Psychology* 24:25–41. <http://s.psych.uiuc.edu/~ediener/hottopic/paper1.html>.

Dillon, W. 1992. Gas (methane) hydrates—A new frontier. <http://marine.usgs.gov/fact-sheets-hydrates/title.html>.

DiPardo, J. 2000. *Outlook for Biomass Ethanol Production and Demand.* Washington, DC: EIA. <http://www.eia.doe.gov/oiaf/analysispaper/biomass.html>.

Dixon, D. A. 2000. A growing problem. *International Water Power & Dam Construction* 52(5):23–25.

Dockery, D. W., and C. A. Pope. 1994. Acute respiratory effects of particulate air pollution. *Annual Review of Public Health* 15:107–132.

DOE (Department of Energy). 2000a. *Electric Power Annual.* Washington, DC: DOE.

DOE (Department of Energy). 2000b. *Clean Coal Technology Demonstration Program.* Washington, DC: DOE.

DOE (Department of Energy). 2000c. *Model Year 2001 Fuel Economy Guide.* Washington, DC: DOE. <http://www.fueleconomy.gov>.

DOE (Department of Energy). 2001a. Price–Anderson Act. <http://www.gc.doe.gov/Price-Anderson/default.html>.

DOE (Department of Energy). 2001b. Fluidized Bed Coal Combustion. <http://www.fe.doe.gov/coal_power/fluidizedbed/index.shtml>.

Douma, A., and G. D. Stewart. 1981. Annapolis Straflo turbine will demonstrate Bay of Fundy tidal power concept. *Modern Power Systems* 1981(1):53–65.

Dracker, R., and P. De Laquill. 1996. Progress commercializing solar-electric power systems. *Annual Review of Energy and the Environment* 21:371–402.

Duncan, R. C. 1997. US national security threatened by a new Alliance of Muslim Petroleum Exporting Countries. Letter to President William J. Clinton, May 13, 1997.

Duncan, R. C. 2000. The peak of world oil production and the road to the Olduvai Gorge. *Pardee Keynote Symposia,* Geological Society of America Summit 2000, Reno, Nevada, November 13, 2000.

Duncan, R. C., and W. Youngquist. 1999. Encircling the peak of world oil production. *Natural Resources Research* 8:219–232.

Dunn, S. 1999. King coal's weakening grip on power. *World Watch* September–October: 10–19.

Dunn, S. 2001. *Hydrogen Futures: Toward a Sustainable Energy System.* Washington, DC : Worldwatch Institute.

DWIA (Danish Wind Industry Association). 2001. Danish Wind Energy. <http://www.windpower.org>.

Economist, The. 1986. Nuclear's charm. *The Economist,* March 29, 1986:11–12.

EC (The European Commission). 2001a. *ExternE—Externalities of Energy.* Brussels: EC. <http://externe.jrc.es/infosys.html>.

EC (The European Commission). 2001b. Small-scale hydro: Future potential. Brussels: EC. <http://europa.eu.int/comm/energy_transport/atals/htmlu/hydfpot.html>.

EC (The European Commission). 2001c. *Photovoltaics in 2010.* Brussels: EC. <http://www.agores.org/Publications/PV2010.htm>.

Edison, T.A. 1889. The dangers of electric lighting. *North Review,* November 1889:630.

Ehrlich, P. R. 1990. *The Population Explosion.* New York: Simon and Schuster.

Ehrlich, P. R., and J. Holdren, eds. 1988. *Cassandra Conference: Resources and the Human Predicament.* College Station, TX: Texas A & M University Press.

Ehrlich, P. R., et al. 1999. Knowledge and the environment. *Ecological Economics* 30:267–284.

EIA (Energy Information Administration). 1996. *International Energy Outlook.* Washington, DC: EIA. <http://www.eia.doe.gov/oiaf/ieo96/oil.html#head1>.

EIA (Energy Information Administration). 1999a. *A Look at Residential Energy Consumption in 1997.* Washington, DC: EIA. <http://www.eia.doe.gov/pub/pdf/consumption/063297.pdf>.

EIA (Energy Information Administration). 1999b. *Federal Financial Interventions and Subsidies in Energy Markets 1999: Energy Transformation and End Use.* Washington, DC: EIA. <http://www.eia.doe.gov/oiaf/servicerpt/subsidy1/index.html>.

EIA (Energy Information Administration). 2001a. *Annual Review 2001.* Washington, DC: EIA. <http://www.eia.doe.gov/emeu/aer/contents.html>.

EIA (Energy Information Administration). 2001b. *Saudi Arabia.* <http://www.eia.doe.gov/emeu/cabs/saudi.html>.

EIA (Energy Information Administration). 2001c. Photovoltaic cell and module shipments by type, price, and trade, 1982–1999. <http://www.eia.doe.gov/emeu/aer/txt/tab1005.htm>.

EIA (Energy Information Administration). 2001d. *International Energy Outlook 2001.* Washington, DC: EIA. <http://www.eia.doe.gov/oiaf/ieo>.

EIA (Energy Information Administration). 2001e. *Annual Energy Outlook 2001.* Washington, DC: EIA. <http://www.eia.doe.gov/oiaf/archive/aeo01/>.

EIA (Energy Information Administration). 2001f. *World Oil Market and Oil Price Chronologies: 1970–2000.* Washington, DC: EIA. <http://www.eia.doe.gov/emeu/cabs/chron.html>.

EIA (Energy Information Administration). 2001g. *Caspian Sea Region.* Washington, DC: EIA. <http://www.eia.doe.gov/emeu/cabs/caspian.html>.

EIA (Energy Information Administration). 2001h. *North Sea Fact Sheet.* Washington, DC: EIA. <http://www.eia.doe.gov/emeu/cabs/northsea.html>.

EIA (Energy Information Administration). 2001i. *U.S. Natural Gas Markets: Recent Trends and Prospects for the Future.* Washington, DC: U.S. Department of Energy. <http://www.eia.doe.gov/oiaf/servicerpt/naturalgas/>.

Einstein, A. 1905. Über einen die Erzeugung and Verwandlung des Lichtes betreffendedn heuristischen Gesichtspunkt. *Annalen der Physik* 17:132.

Elliott, D. L., et al. 1987. *Wind Energy Resource Atlas of the United States.* Golden, CO: Solar Energy Research Institute.

Elliott, D. L., and M. N. Schwartz. 1993. *Wind Energy Potential in the United States.* Richland, WA: Pacific Northwest Laboratory. <http://www.nrel.gov/wind/potential.html>.

Elliott, R. N., and M. Spurr. 1999. *Combined Heat and Power: Capturing Wasted Energy.* Washington, DC: ACEE. <http://www.acee.org/pubs/ie983.htm>.

Eltony, M. N. 1996. On the future role of Gulf oil in meeting world energy demand. *Energy Studies Review* 8(10):57–63.

EPA (Environmental Protection Agency). 2000. *Latest Findings on National Air Quality: 1999 Status and Trends.* Research Triangle Park, NC: EPA. <http://www.epa.gov/oar/aqtrnd00/brochure/brochure.pdf>.

EPA (Environmental Protection Agency). 2001. Energy Star. <http://www.energystar.gov>.

ESA (Electricity Storage Association). 2001. Large scale electricity storage technologies. <http://www.energystorage.org/technology/pumped_hydro.htm>.

ESA (Energy Storage Association). 2002. *Energy Storage Association.* Morgan Hill, CA: ESA. <http://www.energystorage.org/default.htm>.

ExternE. 2001. Summary of Results for Air Pollutants. <http://externe.jrc.es/All-EU+Summary.htm>.

Fallows, J. M. 1989. *More Like Us: Making America Great Again.* Boston: Houghton Mifflin.

Falsetti, J. S., and W. E. Preston. 2001. Gasification offers clean power from coal. *Power* 145(2):50–51.

FAO (Food and Agriculture Organization). 1999. State of the World's Forests. Rome: FAO.

FAO (Food and Agriculture Organization). 2000. *The State of Food Insecurity in the World 2000.* Rome: FAO. <http://www.fao.org/DOCREP/X8200/X8200E00.HTM>.

FAO (Food and Agriculture Organization). 2001. FAOSTAT Statistics Database. <http://apps.fao.org>.

Farber, D., and J. Weeks. 2001. A graceful exit? Decommissioning nuclear power reactors. Environment 43:8–21.

Farey, J. 1827. *A Treatise on the Steam Engine.* London: Longman, Rees, Orme.

Farrell, A., and D. W. Keith. 2001. Hydrogen as a transportation fuel. *Environment* 43:43–45.

Fecher, F., and S. Perelman. 1992. Productivity growth and technical efficiency in OECD industrial activities. In *Industrial Efficiency in Six Nations,* eds. R.E. Caves, et al., 459–488. Cambridge, MA: The MIT Press.

Federal Energy Administration. 1974. *Project Independence.* Washington, DC: FEA.

Feinstein, C. 1972. *National Income, Expenditure and Output of the United Kingdom 1855–1965.* Cambridge: Cambridge University Press.

Felix, F. 1974. Energy independence: Goal for the '80s. *Electrical World* (March 1): 1–4.

Fels, M. F., and K. M. Keating. 1993. Measurement of energy savings from demand-side management programs in US electric utilities. *Annual Review of Energy and the Environment* 18:57–88.

Ferguson, R. 1999. The value of energy efficiency in competitive power markets. *Clean Power Journal* (Summer 1999):1–2.

Ferris, B. G. 1969. Chronic low-level air pollution. *Environmental Research* 2:79–87.

Firebaugh, M. W., ed. 1980. *Acceptable Nuclear Futures: The Second Nuclear Era.* Oak Ridge, TN: Institute of Energy Analysis.

Flavin, C. 1999. Bull market in wind energy. *WorldWatch* (March/April 1999):24–27.

Flegal, K. M. 1996. Trends in body weight and overweight in the U.S. population. *Nutrition Reviews* 54:S97–S100.

Flink, J. J. 1988. *The Automobile Age.* Cambridge, MA: The MIT Press.

Flower, A. R. 1978. World oil production. *Scientific American* 283(3):42–49.

Fluck, R. C., ed. 1992. *Energy in Farm Production.* Amsterdam: Elsevier.

Flynn, J., and P. Slovic. 1995. Yucca Mountain: A crisis for policy. *Annual Review of Energy and the Environment* 20:83–118.

Ford, H. 1929. *My Life and Work.* New York: Doubleday.

Ford, K. W. et al., eds. 1975. *Efficient Use of Energy.* New York: American Institute of Physics.

Forest, C. E., et al. 2002. Quantifying uncertainties in climate system properties with the use of recent climate observations. *Science* 295:113–117.

Forrester, J. W. 1971. *World Dynamics.* Cambridge, MA: Wright-Allen Press.

Fowler, T. K. 1997. *The Fusion Quest.* Baltimore: Johns Hopkins University Press.

FPC (Federal Power Commission). 1964. *National Power Survey.* Washington, DC: FPC.

Francis, E. J., and J. Seelinger. 1977. Forecast market and shipbuilding program for OTEC/industrial plant-ships in tropical oceans. In *Proceedings of 1977 Annual Meeting of American Section of the International Solar Energy Society,* 24–28.

Frank, A. G. 1998. *ReOrient: Global Economy in the Asian Age.* Berkeley, CA: University of California Press.

Frederickson, J. K., and T. C. Onstott. 1996. Microbes deep inside the Earth. *Scientific American* 275(4):68–93.

Freedom House. 2001. *Freedom in the World: The Annual Survey of Political Rights and Civil Liberties 2000–2001.* Freedom House: New York. <http://www.freedomhouse.org/research>.

Freer, R. 2000. Holding back the years. *International Water Power and Dam Construction* 52(7):40–41.

Fridley, D., ed. 2001. *China Energy Databook.* Berkeley, CA: Lawrence Berkeley Laboratory.

Friedrich, R., and A. Voss. 1993. External costs of electricity generation. *Energy Policy* 21: 14–122.

Friends of the Narmada River. 2001. Large dams on the Narmada River. <http://www.narmada.org/>.

Fuel Cells. 2000. *Fuel Cells: The Online Fuel Cell Information Center.* <http://www.fuelcells.org/>.

Fukuyama, F. 1991. *The End of History and the Last Man.* New York: Avon.

Fuller, R. B. 1981. *Critical Path.* New York: St. Martin's Press.

Galloway, J. N., et al. 1995. Nitrogen fixation: Anthropogenic enhancement-environmental response. *Global Biogeochemical Cycles* 9:235–252.

Gammage, R. B., and B. A. Berven, eds. 1996. *Indoor Air Pollution and Human Health.* Boca Raton, FL: CRC Press.

GAO (General Accounting Office). 1989. *What Can Be Done to Revive the Nuclear Option?* Washington, DC: GAO.

GAO (General Accounting Office). 1991a. *Persian Gulf Allied Burden Sharing Efforts.* Washington, DC: GAO.

GAO (General Accounting Office). 1991b. *Utility Demand-Side Management Can Reduce Electricity Use.* Washington, DC: GAO.

GAO (General Accounting Office). 1993. *Energy Security and Policy.* Washington, DC: GAO.

GAO (General Accounting Office). 1995a. *Global Warming: Limitations of General Circulation Models and Costs of Modeling Efforts.* Washington, DC: GAO.

GAO (General Accounting Office). 1995b. *Nuclear Safety Concerns with Nuclear Facilities and Other Sources of Radiation in the Former Soviet Union.* Washington, DC: GAO.

GAO (General Accounting Office). 2001. *Nuclear Waste: Technical, Schedule and Cost Uncertainties of the Yucca Mountain Repository Project.* Washington, DC: GAO.

Garrett, S. L., and S. Backhaus. 2000. The power of sound. *American Scientist* 88:516–523.

Gawell, K., M. Reed, and P. M. Wright. 1999. Geothermal energy, the potential for clean power from the Earth. Washington, DC: Geothermal Energy Association. <http://www.geotherm.org/PotentialReport.htm>.

Geller, H. S. 1985. Ethanol fuel from sugar cane in Brazil. *Annual Review of Energy* 10:135–164.

George, R. L. 1998. Mining for oil. *Scientific American* 278(3):84–85.

Georgescu-Roegen, N. 1971. *The Entropy Law and the Economic Process.* Cambridge, MA: Harvard University Press.

Georgescu-Roegen, N. 1975. Energy and economic myths. *Ecologist* 5:164–174, 242–252.

Giampietro, M., S. Ulgiati, and D. Pimentel. 1997. Feasibility of large-scale biofuel production. *BioScience* 47:587–600.

Gibbons, J. H., and W. U. Chandler. 1981. *Energy: The Conservation Revolution.* New York: Plenum Press.

Giovando, C. 1999. Big Spring makes wind energy more than a pretty, 'green' face. *Power* 143(4):40–42.

Giovando, C. 2001. Markets grow for coal-combustion byproducts. *Power* 145(1):35–43.

Glaser, P. E. 1968. Power from the Sun: Its future. *Science* 162:957–961.

Glover, T. O., et al. 1970. *Unit Train Transportation of Coal: Technology and Description of Nine Representative Operations.* Washington, DC: U.S. Bureau of Mines.

Godbold, D. L., and A. Hütterman. 1994. *Effects of Acid Precipitation on Forest Processes.* New York: Wiley-Liss.

Goela, J. S. 1979. Wind power through kites. *Mechanical Engineering* 101(6):42–43.

Goeller, H. E., and A. M. Weinberg. 1976. The age of substitutability. *Science* 191:683–689.

Goeller, H. E., and A. Zucker. 1984. Infinite resources: The ultimate strategy. *Science* 223:456–462.

Goetzberger, A., J. Knobloch, and B. Voss. 1998. *Crystalline Silicon Solar Cells.* Chichester: Wiley.

Gold, B., et al., eds. 1984. *Technological Progress and Industrial Leadership: The Growth of the U.S. Steel Industry, 1900–1970.* Lexington, MA: Lexington Books.

Gold, T. 1992. The deep, hot biosphere. *Proceedings of the National Academy of Sciences USA* 89:6045–6049.

Gold, T. 1993. *The Origin of Methane (and Oil) in the Crust of the Earth.* Washington, DC: USGS.

Gold, T. 1996. Can there be two independent sources of commercial hydrocarbon deposits? <http://www.people.cornell.edu/pages/tg21/origins.html>.

Gold, T. 1999. *The Deep Hot Biosphere.* New York: Copernicus.

Goldemberg, J. 1996. The evolution of ethanol costs in Brazil. *Energy Policy* 24:1127–1128.

Goldemberg, J., ed. 2000. *World Energy Assessment: Energy and the Challenge of Sustainability*. New York: UNDP.

Goldstein, H. 2002. Waste not, pollute not. *IEEE Spectrum* 39(1):72–77.

Gordon, R. L. 1991. Depoliticizing energy: The lessons of Desert Storm. *Earth and Mineral Sciences* 60(3):55–58.

Gordon, R. L. 1994. Energy, exhaustion, environmentalism, and etatism. *The Energy Journal* 15:1–16.

Gorokhov, V., et al. 1999. Supercritical power plants hike efficiency, gain world market share. *Power* 103(10):36–42.

Grainger, A. 1988. Estimating areas of degraded tropical lands requiring replenishment of forest cover. *International Tree Crops Journal* 5:31–61.

Grant, P. 2001. Up on the C_{60} elevator. *Nature* 413:264–265.

Grätzel, M. 2001. Photoelectrochemical cells. *Nature* 414:338–344.

Greene, D. L. 1992. Energy-efficiency improvement potential of commercial aircraft. *Annual Review of Energy and the Environment* 17:537–573.

Greenwald, J. 1986. Awash in an ocean of oil. *Time* (February 3, 1986):53.

Grossling, B. 1976. *Window on Oil: A Survey of World Petroleum Resources*. London: Financial Times Business Information.

Grubb, M. J., and N. I. Meyer. 1993. Wind energy: resources, systems, and regional strategies. In *Renewable Energy: Sources for Fuel and Electricity*, eds. T. B. Johansson et al., 157–212. Washington, DC: Island Press.

Grübler, A., and N. Nakićenović. 1996. Decarbonizing the global energy supply. *Technological Forecasting and Social Change* 53:97–110.

Grübler, A., and N. Nakićenović. 2001. Identifying dangers in an uncertain climate. *Nature* 412:15.

Grübler, A., et al. 1996. Global energy perspectives: a summary of the joint study by the International Institute for Applied Systems Analysis and World Energy Council. *Technological Forecasting and Social Change* 51:237–264.

Gurney, J. 1997. Migration or replenishment in the Gulf. *Petroleum Review* May 1997:200–203.

Gutman, P. S. 1994. Involuntary resettlement in hydropower projects. *Annual Review of Energy and the Environment* 19:189–210.

Haennel, R. G., and F. Lemire. 2002. Physical activity to prevent cardiovascular disease. *Canadian Family Physician* 48:65–71.

Häfele, W., et al. 1981. *Energy in a Finite World: A Global System Analysis*. Cambridge, MA: Ballinger.

Hagen, A. W. 1975. *Thermal Energy from Sea*. Park Ridge, NJ: Noyes Data Corporation.

Haile, S. M., et al. 2001. Solid acids as fuel cell electrolytes. *Nature* 410:910–913.

Hall, D. O. 1997. Biomass energy in industrialized countries: A view of the future. *Forest Ecology and Management* 91:17–45.

Hall, E. J. 1984. *Radiation and Life.* New York: Pergamon.

Hall, J. V., et al. 1992. Valuing the benefits of clean air. *Science* 255:812–816.

Hammond, O. H., and R. E. Baron. 1976. Synthetic fuels: prices, prospects, and prior art. *American Scientist* 64:407–417.

Hammons, T. J. 1992. Remote renewable energy resources. *IEEE Power Engineering Review* 12(6):3–5.

Hannam, J., et al., eds. 2000. *International Encyclopedia of Women's Suffrage.* Santa Barbara, CA: ABC–CLIO.

Hansen, J., et al. 2000. Global warming in the twenty-first century: An alternative scenario. *Proceedings of the National Academy of Sciences USA* 97:9875–9880.

Hart, D. D., et al. 2002. Dam removal: Challenges and opportunities for ecological research and river restoration. *BioScience* 52:669–681.

Hart, J. 1988. *Consider a Spherical Cow: A Course in Environmental Problem Solving.* Mill Valley, CA: University Science Books.

Hart, J. 2001. *Consider a Cylindrical Cow: More Adventures in Environmental Problem Solving.* Sausalito, CA: University Science Books.

Hatfield, C. B. 1997. Oil back on the global agenda. *Nature* 387:121.

Hawken, P., A. Lovins, and L. H. Lovins. 1999. *Natural Capitalism.* Boston, MA: Little, Brown.

Hayes, B. 2001. The computer and the dynamo. *American Scientist* 89:390–394.

Henzel, D. S., et al. 1982. *Handbook for Flue Gas Desulfurization Scrubbing with Limestone.* Park Ridge, NJ: Noyes Data Corporation.

Herring, H. 1998. *Does Energy Efficiency Save Energy: The Economists Debate.* Milton Keynes: The Open University. <http://www.tec.open.ac.uk/eeru/staff/horace/hh3.htm>.

Herring, H. 2001. Why energy efficiency is not enough. In *Advances in Energy Studies,* ed. S. Ulgiati, 349–359. Padova: SGE.

Herzog, A. V., et al. 1999. Renewable energy: A viable choice. *Environment* 43:8–20.

Hicks, J., and G. Allen. 1999. *A Century of Change: Trends in U.K. Statistics since 1900.* London: House of Commons Library. <http://www.parliament.uk/commons/lib/research/rp99/rp99-111.pdf>.

Hirsch, F. 1976. *Social Limits to Growth.* Cambridge, MA: Harvard University Press.

Hirschenhoffer, J. H., et al. 1998. *Fuel Cells Handbook.* Morgantown, WV: Parsons Corporation.

Hirst, E. 2001. Interactions of wind farms with bulk-power operations and markets. Alexandria, VA: Sustainable FERC Energy policy. <http://www.ehirst.com/PDF/WindIntegration.pdf>.

Hobbs, P. V., and L. F. Radke. 1992. Airborne studies of the smoke from the Kuwait oil fires. *Science* 256:987–991.

Hoffmann, P. 2001. *Tomorrow's Energy: Hydrogen, Fuel Cells, and the Prospects for a Cleaner Planet.* Cambridge, MA: The MIT Press.

Hofmann-Wellenhof, B., et al. 1997. *Global Positioning System: Theory and Practice.* New York: Springer-Verlag.

Hohenemser, C. 1988. The accident at Chernobyl: Health and environmental consequences and the implications for risk management. *Annual Review of Energy* 13:383–428.

Hohenemser, C., R. L. Goble, and P. Slovic. 1990. Institutional aspects of the future development of nuclear power. *Annual Review of Energy* 15:173–200.

Hohmeyer, O. 1989. *Social Costs of Energy Consumption.* Berlin: Springer-Verlag.

Hohmeyer, O., and R. L. Ottinger, eds. 1991. *External Environmental Costs of Electric Power.* Berlin: Springer-Verlag.

Holdr, G. D., V. A. Kamath, and S. P. Godbole. 1984. The potential of natural gas hydrates as an energy resource. *Annual Review of Energy* 9:427–445.

Holdren, J. P. 1992. Radioactive-waste management in the United States: Evolving policy prospects and dilemmas. *Annual Review of Energy and the Environment* 17:235–259.

Holdren, J. P., and K. R. Smith. 2000. Energy, the environment, and health. In *World Energy Assessment,* ed. J. Goldemberg, 61–110. New York: UNDP.

Holley, I. B. 1964. *Buying Aircraft: Material Procurement for the Army Air Forces.* Washington, DC: Department of the Army.

Hoshide, R. K. 1994. Electric motor do's and don'ts. *Energy Engineering* 91:6–24.

Houghton, J. T., et al., eds. 1990. *Climate Change: The IPCC Scientific Assessment.* Cambridge: Cambridge University Press.

Houghton, J. T., et al., eds. 1996. *Climate Change 1995: The Science of Climate Change.* Cambridge: Cambridge University Press.

Houghton, J. T., et al., eds. 2001. *Climate Change 2001: The Scientific Basis.* New York: Cambridge University Press.

Houthakker, H. S. 1997. A permanent decline in oil production? *Nature* 388:618.

Hu, S. D. 1983. *Handbook of Industrial Energy Conservation.* New York: Van Nostrand Reinhold.

Hubbard, H. M. 1989. Photovoltaics today and tomorrow. *Science* 244:297–304.

Hubbard, H. M. 1991. The real cost of energy. *Scientific American* 264(4):36–42.

Hubbert, M. K. 1956. Nuclear energy and fossil fuels. In *American Petroleum Institute, Drilling and Production Practice,* 7–25. Washington, DC: API.

Hubbert, M. K. 1969. Energy resources. In *Committee on Resources and Man, Resources and Man,* 157–242. San Francisco: W. H. Freeman.

Huber, P., and M. Mills. 1999. Dig more coal—The PCs are coming. *Forbes* 163(11):70–72.

Hudson J. L., and G. T. Rochelle, eds. 1982. *Flue Gas Desulfurization.* Washington, DC: American Chemical Society.

Hunt, J. M. 1979. *Petroleum Geochemistry and Geology.* San Francisco: W. H. Freeman.

Hydrate.org. 2001. Simple summary of hydrates. <http://hydrate.org/intro.cfm>.

Hydro-Québec. 2000. Hydroelectricity, Clean Energy. Montréal: Hydro-Québec. <http://www.hydroquebec.com>.

Hypercar Inc. 2002. Tomorrow. <http://www.hypercar.com/pages/when2.html>.

IAEA (International Atomic Energy Agency). 2001a. *Status of Nuclear Power Plants Worldwide in 2000.* Vienna: IAEA. <http://www.iaea.org/cgi-bin/db.page.pl/pris.main.htm>.

IAEA (International Atomic Energy Agency). 2001b. Latest news related to PRIS and the status of nuclear power plants. <http://www.iaea.org/cgi=bin/db.page.pl/pris.main.htm>.

IAEA (International Atomic Energy Agency). 2001c. *Sustainable Development & Nuclear Power.* Vienna: IAEA. <http://www.iaea.or.at/worldatom/Press/Booklets/Development>.

ICAO (International Civil Aviation Organization). 2000. *The ICAO Annual Report.* Montreal: ICAO. <http://www.icao.org>.

ICOLD (International Commission on Large Dams). 1998. *World Register of Dams.* Paris: ICOLD.

IEA (International Energy Agency). 1999. *Coal in the Energy Supply of China.* Paris: IEA.

IEA (International Energy Agency). 2001. *Key World Energy Statistics.* Paris: IEA.

IEE (The Institute of Energy Economics). 2000. *Handbook of Energy & Economic Statistics in Japan.* Tokyo: The Energy Conservation Center.

IGA (International Geothermal Association). 1998. Geothermal power plants on-line in 1998. Pisa: IGA. <http://www.demon.co.uk/geosci/wrtab.html>.

IHA (International Hydropower Association). 2000. *Hydropower and the World's Energy Future.* Sutton: IHA.

IISI (International Iron and Steel Institute). 2001. *Trends & Statistics.* Brussels: IISI. <http://www.worldsteel.org/>.

Imbrecht, C. R. 1995. California's electrifying future. Sacramento, CA: California Energy Commission. <http://home.earthlink.net/~bdewey/EV_californiaev.html>.

Ingram, L. O., et al. 1987. Genetic engineering of ethanol production in *Escherichia coli. Applied Environmental Microbiology* 53:2420–2425.

International Fusion Research Council. 1979. *Controlled Thermonuclear Fusion: Status Report.* Vienna: IAEA.

IPCC (Intergovernmental Panel on Climatic Change). 2001. Summary for Policymakers: The Science of Climate Change—IIPCC Working Group I. Geneva: IPCC. <http://www.ipcc.ch>.

Irving, P. M., ed. 1991. *Acidic Deposition: State of Science and Technology.* Washington, DC: U.S. National Acid Precipitation Assessment Program.

Isaacs, J. D., and W. R. Schmitt. 1980. Ocean energy: Forms and prospects. *Science* 207:265–273.

Isaacson, M. 1998. A damning decision. *International Water Power and Dam Construction* 50(4):16–17.

Islas, J. 1999. The gas turbine: A new technological paradigm in electricity generation. *Technological Forecasting and Social Change* 60:129–148.

Ittekkot, V. et al. 2000. Hydrological alterations and marine biogeochemistry: A silicate issue? *BioScience* 50:776–782.

Ivanhoe, L. F. 1995. Future world oil supplies: There is a finite limit. *World Oil* 216(10): 77–79.

Ivanhoe, L. F. 1997. Get ready for another oil shock! *The Futurist* 31(1): 20–27.

Jacobson, M. Z. 2001. Strong radiative heating due to the mixing state of black carbon in atmospheric aerosols. *Nature* 409:695–697.

Jacobson, M. Z., and G. M. Masters. 2001a. Exploiting wind versus coal. Science 293:1348.

Jacobson, M. Z., and G. M. Masters. 2001b. Response. *Science* 294:1001–1003.

James, A. N. et al. 1999. Balancing the Earth's accounts. *Nature* 401:323–324.

JAMSTEC (Japan Marine Science and Technology Center). 1998. Offshore floating wave energy device Mighty Whale. Yokosuka: JAMSTEC. <http://www.jamstec.go.jp/jamstec/MTD/Whale>.

Jansen, M. B. 2000. *The Making of Modern Japan.* Cambridge, MA: Belknap Press.

Jensen, M. W., and M. Ross. 2000. The ultimate challenge: developing an infrastructure for fuel cell vehicles. *Environment* 42(7):10.

Jevons, W. S. 1865. *The Coal Question: An Inquiry Concerning the Progress of the Nation, and the Probable Exhaustion of our Coal Mines.* London: Macmillan.

Jochem, E. 2000. Energy end-use efficiency. In *World Energy Assessment,* ed. J. Goldemberg, 173–217. New York: UNDP.

Jones, I. S. F., and D. Otaegui. 1997. Photosynthetic greenhouse gas mitigation by ocean nourishment. *Energy Conversion & Management* 38S:367–372.

Josephson, M. 1959. *Edison: A Biography.* New York: Wiley.

Joskow, P. L. 1994. More from the guru of energy efficiency: "There must be a pony!" *The Electricity Journal* 7(4):50–61.

Joskow, P. L., and D. B. Marron. 1992. What does a negawatt really cost: Evidence from utility conservation programs. *Energy* 13:41–74.

Joskow, P. L., and D. B. Marron. 1993. What does utility-subsidized energy efficiency really cost? *Science* 260:281, 370.

Kaarsberg, T. M., and J. M. Roop. 1998. Combined heat and power: How much carbon and energy can it save for manufacturers? Colorado Springs, CO: 33rd Intersociety Engineering Conference on Energy Conversion. <http://www.aceee.org/chp/Summerstudy99/assessment.pdf>.

Kahn, A. 1991. An economically rational approach to least-cost planning. *Electricity Journal* 14(6):11–20.

Kaijser, A. 1992. Redirecting power: Swedish nuclear power policies in historical perspective. *Annual Review of Energy and the Environment* 17:437–462.

Kalhammer, F. R. 1979. Energy-storage systems. *Scientific American* 241(6):56–65.

Kalt, J. P., and R. S. Stillman. 1980. The role of governmental incentives in energy production: An historical overview. *Annual Review of Energy* 5:1–32.

Kammen, D. M. 1995. Cookstoves for the developing world. *Scientific American* 273(1):72–75.

Kazimi, M. S., and N. E. Todreas. 1999. Nuclear power economic performance: Challenges and opportunities. *Annual Review of Energy and the Environment* 24:139–171.

Keeble, J. 1999. *Out of the Channel: The Exxon Valdez Oil Spill in Prince William Sound.* Cheney, WA: Eastern Washington University Press.

Keeling, C. D. 1998. Reward and penalties of monitoring the Earth. *Annual Review of Energy and the Environment* 23:25–82.

Keeney, D. R., and T. H. DeLuca. 1992. Biomass as an energy source for the Midwestern U.S. *American Journal of Alternative Agriculture* 7:137–143.

Keith, D. W. 2000. Geoengineering the climate: History and prospect. *Annual Review of Energy and the Environment* 25:245–284.

Kenney, J. F. 1996. Considerations about recent predictions of impending shortages of petroleum evaluated from the perspective of modern petroleum science. *Energy World Special Edition on the Future of Petroleum:* 16–18.

Khazzoom, J. D. 1989. Energy savings from more efficient appliances: A rejoinder. *Energy Journal* 10:157–166.

Kheshgi, H. S., R. C. Prince, and G. Marland. 2000. The potential of biomass fuels in the context of global climate change: Focus on transportation fuels. *Annual Review of Energy and the Environment* 25:199–244.

Kingston, W. 1994. A way ahead for ocean wave energy? *Energy Studies Review* 6:85–88.

Klass, D. L. 1998. *Biomass for Renewable Energy, Fuels, and Chemicals.* San Diego, CA: Academic Press.

Kleinberg, R. L., and P. G. Brewer. 2001. Probing gas hydrate deposits. *American Scientist* 89:244–251.

Kloss, E. 1963. *Der Luftkrieg über Deutschland, 1939–1945.* München: DTV.

Koomey, J., et al. 1999. Memo to Skip Leitner of EPA: Initial comments on "The Internet Begins with Coal." Berkeley, CA: Lawrence Berkeley Laboratory. <http://enduse.LBL.gov/Projects/InfoTech.html>.

Krebs-Leidecker, M. 1977. Synthetic fuels from coal. In *Congressional Research Service, Project Interdependence: U.S. and World Energy Outlook through 1990,* 327–328. Washington, DC: USGPO.

Krupnick, A. J., and P. R. Portney. 1991. Controlling urban air pollution: A benefit–cost assessment. *Science* 252:522–528.

Kuczmarski, R. J., et al. 1994. Increasing prevalence of overweight among U.S. adults. *Journal of American Medical Association* 272:205–211.

Kudryavtsev, N. A. 1959. *Oil, Gas, and Solid Bitumens in Igneous and Metamorphic Rocks.* Leningrad: State Fuel Technical Press.

Kvenvolden, K. A. 1993. Gas hydrates—Geological perspective and global change. *Reviews of Geophysics* 31:173–187.

Laherrère, J. 1995. World oil reserves—Which number to believe? *OPEC Bulletin* 26(2): 9–13.

Laherrère, J. 1996. Discovery and production trends. *OPEC Bulletin* 27(2):7–11.

Laherrère, J. H. 1997. Oil markets over the next two decades: Surplus or shortage? <http://www.hubbertpeak.com/laherrere/supply.htm>.

Laherrère, J. H. 2000. Global natural gas perspectives. <http://www.hubbertpeak.com/laherrere/perspective/>.

Laherrère, J. H. 2001. *Estimates of Oil Reserves.* Laxenburg: IIASA. <http://www.oilcrisis.com/laherrere>.

Lake, J. A., R. G. Bennett, and J. F. Kotek. 2002. Next-generation nuclear power. *Scientific American* 286(1):72–81.

Lamont Doherty Earth Observatory. 2001. Lamont 4D Technology. <http://www.ldeo.columbia.edu/4d4/>.

Landis, G. A. 1997. A supersynchronous solar power satellite. In *Proceedings SPS '97 Conference,* 327–328. Montreal: Canadian Aeronautics and Space Institute.

Landsberg, H. H. 1986. The death of synfuels. *Resources* 82:7–8.

Larson, E. D. 1993. Technology for electricity and fuels from biomass. *Annual Review of Energy and the Environment* 18:567–630.

Lazaroff, C. 2001. California mandates electric cars. *Environment News Service.* <http://ens.lycos.com/ens/jan2001/2001L-01-30-06.html>.

LBL (Lawrence Berkeley Laboratory). 2001. Measuring furnace efficiency: "AFUE." <http://hes.lbl.gov/aceee/afue.html>.

Lee, R., et al. 1995. *Estimating Externalities of Electric Fuel Cycles.* Washington, DC: McGraw-Hill/Utility Data Institute.

Lesser, I. O. 1991. *Oil, the Persian Gulf, and Grand Strategy.* Santa Monica, CA: Rand Corporation.

Lester, R. K., et al. 1983. *Nuclear Power Plant Innovation for the 1990s: A Preliminary Assessment.* Cambridge, MA: Department of Nuclear Engineering, MIT.

Leyland, B. 1990. Large dams: Implications of immortality. *International Water Power & Dam Construction* 42(2):34–37.

L'Haridon, S., et al. 1995. Hot subterranean biosphere in a continental oil reservoir. *Nature* 377:223–224.

Lieber, R. J. 1983. *The Oil Decade: Conflict and Cooperation in the West.* New York: Praeger.

Lilienthal, D. E. 1944. *TVA—Democracy on the March.* New York: Harper & Brothers.

Lilienthal, D. E. 1980. *Atomic Energy: A New Start.* New York: Harper & Row.

Linden, H. R. 1996. Electrification will enable sustained prosperity. *Engineering* 100(10): 26–28.

Lindzen, R. S., M. Chou, and A. Y. Hou. 2001. Does the Earth have an adaptive iris? *Bulletin of the American Meteorological Society* 82:417–342.

Lipfert, F. W., and S. C. Morris. 1991. Air pollution benefit–cost assessment. *Science* 253: 606.

Lochbaum, D. 2001. Nuclear power's future. Boston, MA: Union of Concerned Scientists. <http://www.ucsusa.org/energy/view_futurenucpower.html>.

Löfstedt, R. E. 2001. Playing politics with energy policy: The phase-out of nuclear power in Sweden. *Environment* 43(4):20–33.

Lorenz, E. N. 1976. *The Nature and Theory of the General Circulation of the Atmosphere.* Geneva: WMO.

Lovins, A. B. 1976. Energy strategy: The road not taken. *Foreign Affairs* 55(1):65–96.

Lovins, A. B. 1977. *Soft Energy Paths: Toward a Durable Peace.* Cambridge, MA: Friends of the Earth and Ballinger.

Lovins, A. B. 1980. Economically efficient energy futures. In *Interactions of Energy and Climate,* eds. W. Bach, et al., 1–31. Boston: D. Reidel.

Lovins, A. 1988. Energy savings resulting from the adoption of more efficient appliances: Another view. *Energy Journal* 9:155–162.

Lovins, A. B. 1992. The soft path—Fifteen years later. *Rocky Mountain Institute Newsletter* 8(1):9.

Lovins, A. B. 1991. Abating global warming for fun and profit. In *The Global Environment,* eds. K. Takeuchi and M. Yoshino, 214–219. New York: Springer-Verlag.

Lovins, A. B. 1994. Apples, oranges, and horned toads. *The Electricity Journal* 7(4):29–49.

Lovins, A. B. 1998. Is oil running out? *Science* 282:47.

Lovins, A. B., and H. L. Lovins. 1991. Least-cost climatic stabilization. *Annual Review of Energy and the Environment* 16:433–531.

Lowrie, A., and M. D. Max. 1999. The extraordinary promise and challenge of gas hydrates. *World Oil* 220(9):49–55.

Lunt, R. R., and J. D. Cunic. 2000. *Profiles in Flue Gas Desulfurization.* New York: American Institute of Chemical Engineers.

Luo, Z. 1998. Biomass energy consumption in China. *Wood Energy News* 13(3):3–4.

Lutz, W., W. Sanderson, and S. Scherbov. 2001. The end of world population growth. *Nature* 412:543–545.

Mabro, R. 1992. OPEC and the price of oil. *The Energy Journal* 13:1–17.

Macilwain, C. 2001. Out of sight, out of mind? *Nature* 412:850–852.

Maddison, A. 1985. Alternative estimates of the real product of India, 1900–1946. *The Indian Economic and Social History Review* 22:201–210.

Maddison, A. 1995. *Monitoring World Economy 1820–1992.* Paris: OECD.

Madsen, B. T. 1997. 4000 MW of offshore wind power by 2030. *Windstats Newsletter* 10(3): 1–3.

Mage, D., et al. 1996. Urban air pollution in megacities of the world. *Atmospheric Environment* 30:681–686.

Mahfoud, R. F., and J. N. Beck. 1995. Why the Middle East fields may produce oil forever. *Offshore* April 1995: 58–64, 106.

Maier, M. H. 1991. *The Data Game.* Armonk, NY: M. E. Sharpe.

Majumdar, S. K., E. W. Miller, and F. J. Brenner, eds. 1998. *Ecology of Wetlands and Associated Systems.* Easton, PA: Pennsylvania Academy of Science.

Makansi, J. 2000. Formidable barriers face users of new coal systems. *Power* 144(1):40–46.

Makhijani, A., H. Hu, and K. Yih, eds. 1995. *Nuclear Wastelands: A Global Guide to Nuclear Weapons Production and Its Health and Environmental Effects.* Cambridge, MA: The MIT Press.

Mallmann, C. A., and G. A. Lemarchand. 1998. Generational explanation of long-term "billow-like" dynamics of societal processes. *Technological Forecasting and Social Change* 59:1–30.

Manabe, S. 1997. Early development in the study of greenhouse warming: The emergence of climate models. *Ambio* 26:47–51.

Manibog, F. R. 1984. Improved cooking stoves in developing countries: Problems and opportunities. *Annual Review of Energy* 9:199–227.

Manne, A. S., and L. Schrattenholzer. 1983. *International Energy Workshop: A Summary of the 1983 Poll Responses.* Laxenburg: IIASA.

March, P. A., and R. K. Fisher. 1999. It's not easy being green: Environmental technologies enhance conventional hydropower's role in sustainable development. *Annual Review of Energy and the Environment* 24:173–188.

Marchetti, C. 1977. Primary energy substitution models: On the interaction between energy and society. *Technological Forecasting and Social Change* 10:345–356.

Marchetti, C. 1978. *Genetic Engineering and the Energy System: How to Make Ends Meet.* Laxenburg: IIASA.

Marchetti, C. 1987. The future of natural gas. *Technological Forecasting and Social Change* 31:155–171.

Marchetti, C., and N. Nakićenović. 1979. *The Dynamics of Energy Systems and the Logistic Substitution Model.* Laxenburg: IIASA.

MarketPlace Cement. 2001. Global Cement Information System. Bad Zwischenhahn, Germany: MarketPlace Construction. <http://www.global-cement.dk>.

Markvart, T., ed. 2000. *Solar Electricity.* New York: Wiley.

Marland, G., et al. 1999. *National CO_2 Emissions from Fossil-Fuel Burning, Cement Manufacture, and Gas Flaring: 1751–1996.* Oak Ridge, TN: Carbon Dioxide Information Analysis Center. <http://cdiac.esd.ornl.gov/ftp/ndp030/nations96.ems>.

Marland, G., et al. 2000. Global, Regional, and National CO_2 Emission Estimates for Fossil Fuel Burning, Cement Production and Gas Flaring. Oak Ridge, TN: ORNL. <http://cdiac.esd.ornl.gov>.

Martin, D. H. 1998. *Federal Nuclear Subsidies: Time to Call a Halt.* Ottawa: Campaign for Nuclear Phaseout. <http://www.cnp.ca/issues/nuclear-subsidies.html>.

Martin, P.-E. 1995. The external costs of electricity generation: Lessons from the U.S. experience. *Energy Studies Review* 7:232–246.

Masters, C. D., et al. 1987. World resources of coal, oil and natural gas. In *Proceedings of the 12th World Petroleum Congress,* 3–33. New York: Wiley.

Masters, C. D., E. D. Attanasi, and D. H. Root. 1994. *World Petroleum Assessment and Analysis.* Reston, VA: USGS. <http://energy.er.usgs.gov/products/papers/WPC/14/index.htm>.

Masters, C. D., D. H. Root, and E. D. Attanasi. 1990. World oil and gas resources—Future production realities. *Annual Review of Energy* 15:23–51.

Mattera, P. 1985. *Off the Books: The Rise of the Underground Economy.* London: Pluto Press.

Maycock, P. D. 1999. *PV Technology, Performance, Cost: 1975–2010.* Warrenton, VA: Photovoltaic Energy Systems.

McCully, P. 2002. *Silenced Rivers: The Ecology and Politics of Large Dams.* London: Zed Books.

McDonald, A. 1999. Combating acid deposition and climate change. *Environment* 41:4–11, 34–41.

McGowan, J. G., and S. R. Connors. 2000. Windpower: A turn of the century review. *Annual Review of Energy and the Environment* 25:147–197.

McInnes, W., et al., eds. 1913. *The Coal Resources of the World.* Toronto: Morang & Company.

McIntyre, R. S. 2001. *The Hidden Entitlements.* Washington, DC: Citizens for Tax Justice. <http://www.ctj.org/hid_ent/>.

McKelvey, V. E. 1973. Mineral resource estimates and public policy. In *United States Mineral Resources,* eds. D. A. Brobst and W. P. Pratt, 9–19. Washington, DC: USGS.

McLarnon, F. R., and E. J. Cairns. Energy storage. 1989. *Annual Review of Energy* 14:241–271.

McShane, C. 1997. *The Automobile: A Chronology.* New York: Greenwood Press.

Meadows, D., and D. Meadows. 1992. *Beyond the Limits.* London: Earthscan.

Meadows, D. H., et al. 1972. *The Limits to Growth.* New York: Universe Books.

Meier, A., and W. Huber. 1997. Results from the investigations of leaking electricity in the USA. Berkeley, CA: Lawrence Berkeley Laboratory. <http://EandE.lbl.gov/EAP/BEA/Projects/Leaking/Results>.

Meier, R. L. 1956. *Science and Economic Development.* Cambridge, MA: The MIT Press.

Meitner, L., and O. R. Frisch. 1939. Disintegration of uranium by neutrons: A new type of nuclear reaction. *Nature* 143:239–240.

Mellow, I. D. 2000. Over a barrel. *The Economist* January 22, 2000:6.

Melville, G. W. 1901. The engineer and the problem of aerial navigation. *North American Review* December 1901, p. 825.

Merriam, M. F. 1977. Wind energy for human needs. *Technology Review* 79(3):29–39.

Methanex. 2000. Methanol fuel cell car: "Fit for practical use." Vancouver, BC: Methanex. <http://www.methanex.com>.

Methanol Institute. 2002. *Methanol.* Washington, DC: Methanol Institute. <http://www.methanol.org/methanol>.

Meyers, S., and L. Schipper. 1992. World energy use in the 1970s and 1980s: Exploring the changes. *Annual Review of Energy and the Environment* 17:463–505.

Mills, M. 1999. *The Internet Begins with Coal.* Washington, DC: Greening Earth Society.

MIT (Massachusetts Institute of Technology). 2001. Device could aid production of electricity. *MIT News,* November 27, 2001.

Mitsui Corporation. 2001. Primary aluminum supply and demand balance in Japan. <http://www.mitsui.co.jp/alm/Key_Data/html/J_Supply-Demand.htm>.

Mock, J. E., J. W. Tester, and P.M. Wright. 1997. Geothermal energy from the Earth: Its potential impact as an environmentally sustainable resource. *Annual Review of Energy and the Environment* 22:305–356.

Moezzi, M. 1998. *The Predicament of Efficiency.* Berkeley, CA: Lawrence Berkeley National Laboratory.

Moody, J. D. 1978. The world hydrocarbon resource base and related problems. In *Australia's Mineral Resource Assessment and Potential,* eds. G. M. Philip and K. L. Williams, 63–69. Sydney: University of Sydney.

Moore, D. W., and F. Newport. 1995. People throughout the world largely satisfied with personal lives. *The Gallup Poll Monthly* 357:2–7.

Moreira, J. R., and J. Goldemberg. 1999. The alcohol program. *Energy Policy* 27:229–245.

Morgan, N. 1995. 3D popularity leads to 4D vision. *Petroleum Economist* 62(2):8–9.

Morita, T., and H. Lee. 1998. IPCC Scenario Database. <http://www-cger.nies.go.jp/cger-e/db/ipcc.html>.

Moxon, S. 2000. Fighting for recognition. *International Water Power & Dam Construction* 52(6):44–45.

MSHA (Mine Safety and Health Administration). 2000. Injury trends in mining. Washington, DC: MSHA. <http://www.msha.gov>.

Muir, J. 1911. *My First Summer in the Sierra.* Boston: Houghton Mifflin.

Murphy, P. M. 1974. *Incentives for the Development of the Fast Breeder Reactor.* Stamford, CT: General Electric.

Nadel, S. 1992. Utility demand-side management experience and potential—A critical review. *Annual Review of Energy and the Environment* 17:507–535.

Nader, L., and S. Beckerman. 1978. Energy as it relates to the quality and style of life. *Annual Review of Energy* 3:1–28.

Naidu, B. S. K. 2000. An encroachment on hydro. *International Water Power & Dam Construction* 52(5):20–22.

Nakićenović, N. 2000. Energy scenarios. In *World Energy Assessment,* ed. J. Goldemberg, 333–336. New York: UNDP.

NAPAP (National Acid Precipitation Assessment Program). 1991. *1990 Integrated Assessment Report.* Washington, DC: NAPAP Office of the Director.

NAS (National Academy of Science). 1992. *Policy Implications of Greenhouse Warming: Mitigation, Adaptation, and the Science Base.* Washington, DC: NAS.

NASA (National Aeronautics and Space Administration). 1989. Report of the NASA Lunar Energy Enterprise Case Study Task Force. Washington, DC: NASA.

Nathwani, J. S., E. Siddall, and N. C. Lind. 1992. *Energy for 300 Years: Benefits and Risks.* Waterloo, ON: University of Waterloo.

National Foreign Assessment Center. 1979. *The World Market in the Years Ahead.* Washington, DC: CIA.

NBS (National Bureau of Statistics). 2001. *China Statistical Yearbook.* Beijing: China Statistics Press. <http://www.stats.gov.cn/english/>.

NCES (National Center for Education Statistics). 2001. *Digest of Education Statistics, 2000.* Washington, DC: NCES.

NCP (National Center for Photovoltaics). 2001. *PV Roadmap.* Golden, CO: NREL. <http://www.nrel.gov/ncpv/vision.html>.

Nehring, R. 1978. *Giant Oil Fields and World Oil Resources.* Santa Monica, CA: Rand Corporation.

Newcomb, T. P., and R. T. Spurr. 1989. *A Technical History of the Motor Car.* Bristol: Adam Hilger.

Nikkei Net Interactive. 2001. Nikkei Index. <http://www.nikkei.nni.nikkei.co.jp>.

Nilsson, C., R. Jansson, and U. Zinko. 1997. Long-term responses of river-margin vegetation to water-level regulation. *Science* 276:798–800.

NIRS (Nuclear Information and Resource Service). 1999. Background on nuclear power and Kyoto Protocol. <http://www.nirs.org/globalization/CDM-Nukesnirsbackground.htm>.

Nolden, S. 1995. The key to persistence. *Home Energy Magazine Online* September/October 1995 <http://www.homeenergy.org/archive>.

Norbeck, J. M., et al. 1996. *Hydrogen Fuel for Surface Transportation.* Warrendale, PA: Society of Automotive Engineers.

Nordhaus, W. D. 1973. World dynamics: Measurement without data. *The Economic Journal* 83:1156–1183.

Nordhaus, W. D. 1991. To slow or not to slow: The economics of the greenhouse effect. *The Economic Journal* 101: 920–937.

Nordhaus, W. D. 2001. Global warming economics. *Science* 294:1283–1284.

Nordhaus, W. D., and J. Boyer. 2000. *Warming the World: Economic Models of Global Warming.* Cambridge, MA: The MIT Press.

Normile, D. 2001. Japan looks for bright answers to energy needs. *Science* 294:1273.

Northeast-Midwest Institute. 2000. *Overcoming Barriers to the Deployment of Combined Heat and Power.* Washington, DC: Northeast-Midwest Institute. <http://www.nemw.org/energy_linx.htm>.

Novick, S. 1976. *The Electric War: The Fight over Nuclear Power.* San Francisco, CA: Sierra Club Books.

NRC (National Research Council). 1981. *Surface Mining: Soil, Coal, and Society.* Washington, DC: National Academy Press.

NRC (National Research Council). 1995. *Coal: Energy for the Future.* Washington, DC: National Academy Press.

NRC (National Research Council). 1998. *Research Priorities for Airborne Particulate Matter.* Washington, DC: National Academy Press.

NRC (National Research Council). 2000. *Reconciling Observations of Global Temperature Change*. Washington, DC: National Academy Press.

NREL (National Renewable Energy Laboratory). 2001a. *Concentrating Solar Power: Energy from Mirrors*. Washington, DC: DOE.

NREL (National Renewable Energy Laboratory). 2001b. Thin-film partnership program: copper indium-diselenide. <http://www.nrel.gov/ncpv/costeam.html>.

NREL (National Renewable Energy Laboratory). 2001c. Ocean thermal energy conversion. <http://www.nrel.gov/otec/>.

Nye, D. E. 1990. *Electrifying America*. Cambridge, MA: The MIT Press.

Odell, P. R. 1984. The oil crisis: Its nature and implications for developing countries. In *The Oil Prospect,* eds. D. C. Ion, P. R. Odell, and B. Mossavar-Rahmani, 33. Ottawa: The Energy Research Group.

Odell, P. R. 1992. Global and regional energy supplies. *Energy Policy* 20(4):284–296.

Odell, P. R. 1999. *Fossil Fuel Resources in the 21st Century*. London: Financial Times Energy.

Odell, P. R. 2001. Gas is the perfect fuel for Europe. <http://www.statoil.com >.

Odell, P. R., and K. Rosing. 1983. *The Future of Oil: World Oil Resources and Use*. London: Kogan Page, London.

OECD (Organization for Economic Cooperation and Development). 1994. *Natural Gas Transportation: Organization and Regulation*. Paris: OECD.

OECD (Organization for Economic Cooperation and Development). 2000. *Nuclear Energy in a Sustainable Development Perspective*. Paris: OECD.

OECD (Organization for Economic Cooperation and Development). 2001. *Main Economic Indicators*. Paris: OECD. <http://www1.oecd.org/std/meiinv.pdf>.

OECD Oil Committee. 1973. *Oil: The Present Situation and Future Prospects*. Paris: OECD.

Ogden, J. M. 1999. Prospects for building a hydrogen energy infrastructure. *Annual Review of Energy and the Environment* 24:227–279.

Ohkawa, K., and H. Rosovsky. 1973. *Japanese Economic Growth*. Stanford, CA: Stanford University Press.

Øhlenschlaeger, K. 1997. The trend toward larger wind turbines. *Windstats Newsletter* 10(4): 4–6.

Oil and Gas Journal Special Report. 1987. New data lift world oil reserves by 27%. *Oil & Gas Journal* 85(52):33–37.

Olds, F. C. 1972. The fast breeder: Schedule lengthens, cost escalates. *Power Engineering* 76(6):33–35.

Olesen, G. B. 2000. *Wind Power for Western Europe*. Copenhagen: INFORSE. <http://www.orgve.dk/inforse-europe/windfor2.htm>.

Oppenheim, P. 1991. *The New Masters: Can the West Match Japan?* London: Business Books.

Oreskes, N. 1999. *The Rejection of Continental Drift*. New York: Oxford University Press.

ORNL (Oak Ridge National Laboratory). 1968. *Nuclear Energy Centers: Industrial and Agro-Industrial Complexes*. Oak Ridge, TN: ORNL.

OSM (Office of Surface Mining). 2001a. *Tonnage Reported for Fiscal Year 2000.* Washington, DC: OSM, US Department of Interior. <http://www.osmre.gov/coal2000.htm>.

OSM (Office of Surface Mining). 2001b. *Abandoned Mine Land Program.* Washington, DC: OSM. <http://www.osmre.gov/aml>.

OTA (Office of Technology Assessment). 1978. *Renewable Ocean Energy Sources.* Washington, DC: OTA.

OTA (Office of Technology Assessment). 1987. *Starpower: The US and the International Quest for Fusion Energy.* Washington, DC: OTA.

OTA (Office of Technology Assessment). 1992. *Fueling Development.* Washington, DC: OTA.

Overbye, T. J. 2000. Reengineering the electric grid. *American Scientist* 88:220–229.

Owenby, J., et al. 2001. *Climatography of the U.S. No. 81—Supplement # 3.* Washington, DC: NOAA. <http://lwf.ncdc.noaa.gov/oa/documentlibrary/clim81supp3/clim81.html>.

Pacific Gas & Electric Company. 2001. *A Concise Guide to the California Energy Crisis.* <http://www.pge.com>.

PacifiCorp Power. 2001. *World's Largest Wind Plant to Energize the West.* Portland, OR: PacifiCorp Power. <http://www.statelinewind.com/rel_01.09.01.html>.

Parkinson, C. L. 1997. *Earth from Above: Using Color-coded Satellite Images to Examine the Global Environment.* Sausalito, CA: University Science Books.

Paris, L. 1992. Grand Inga case. *IEEE Power Engineering Review* 12(6):13–17.

Partl, R. 1977. *Power from Glaciers: The Hydropower Potential of Greenland's Glacial Waters.* Laxenburg: IIASA.

Pasqualetti, M. J., P. Gipe, and R. W. Righter. 2002. *Wind Power in View: Energy Landscapes in a Crowded World.* San Diego, CA: Academic Press.

Paul, I. 2001. *Supercritical Coal Fired Power Plants.* Washington, DC: World Bank. <http://www.worldbank.org/html/fpd/em/supercritical/supercritical.htm>.

PDVSA (Petroleos de Venezuela SA). 2001. PDVSA Orimulsion. <http://www.pdvsa.com/orimulsion/>.

Pedersen, K. 1993. The deep subterranean biosphere. *Earth-Science Reviews* 34:243–260.

Penney, T. R., and D. Bharathan. 1987. Power from the sea. *Scientific American* 256(1):86–92.

Periana, R. A., et al. 1998. Platinum catalysts for the high-yield oxidation of methane to a methanol derivative. *Science* 280:560–564.

Perlin, J. 1999. *From Space to Earth: The Story of Solar Electricity.* Ann Arbor, MI: Aatec.

Perry, A. M. 1982. Carbon dioxide production scenarios. In *Carbon Dioxide Review: 1982,* ed. W. C. Clark, 337–363. New York: Oxford University Press.

PES (Photovoltaic Energy Systems). 2001. Photovoltaic news. <http://www.pvenergy.com/index.shtml>.

Philips Lighting. 2001. The Light Site North America. <http://www.lighting.philips.com/nam/press/2001/062601a.shtml>.

Pimentel, D. 1991. Ethanol fuels: Energy security, economics, and the environment. *Journal of Agricultural and Environmental Economics* 4:1–13.

Pimentel, D. et al. 1997. Economic and environmental benefits of biodiversity. *BioScience* 47: 747–757.

Porfir'yev, V. B. 1959. *The Problem of Migration of Petroleum and the Formation of Accumulations of Oil and Gas.* Moscow: Gostoptekhizdat.

Porfir'yev, V. B. 1974. Inorganic origin of petroleum. *AAPG Bulletin* 58:3–33.

Poten & Partners. 1993. *World Trade in Natural Gas and LNG, 1985–2010: Trades and Prices, Pipelines, Ships, Terminals.* New York: Poten & Partners.

Potts, R. 2001. Complexity and adaptability in human evolution. Paper presented at the AAAS conference "Development of the Human Species and its Adaptation to the Environment," July 7–8, 2001, Cambridge, MA.

Pratt, W. E. 1944. Our petroleum resources. *American Scientist* 32(2):120–128.

Pratt & Whitney. 2001. *PW400 (112 inch).* East Hartford, CT: Pratt & Whitney. <http://www.pratt-whitney.com/3a/products_pw4000112.html>.

PV Power Resource Site. 2001. PV History. <http://www.pvpower.com>.

Ratcliffe, K. 1985. *Liquid Gold Ships: History of the Tanker (1859–1984).* London: Lloyds.

Reagan, R. 1987. *Remarks at the Brandenburg Gate, West Berlin, Germany, June 12, 1987.* <http://www/reaganfoundation.org/reagan/speeches/wall.asp>.

Reilly, J., et al. 2001. Uncertainty and climate change assessments. *Science* 293:430–433.

Reshetnikov, A. I., N. N. Paramonova, and A. A. Shashkov. 2000. An evaluation of historical methane emissions from the Soviet gas industry. *Journal of Geophysical Research* 105:3517–3529.

Revelle, R., and H. E. Suess. 1957. Carbon dioxide exchange between atmosphere and ocean and the question of an increase of atmospheric CO_2 during the past decades. *Tellus* 9:18–27.

Richards, R. 1981. Spanish solar chimney nears completion. *Modern Power Systems* 1981(12):21–23.

RMI (Rocky Mountain Institute). 1999. Exchanges between Mark Mills and Amory Lovins about the electricity used by the Internet. <http://www.rmi.org/images/other/E-MMABL Internet.pdf>.

RMI (Rocky Mountain Institute). 2001. *The Hypercar Concept.* Old Snowmass, CO: RMI. <http://www.rmi.org/sitepages/pid386.php>.

Robinson, J. 1988. *Yamani: The Inside Story.* London: Simon & Schuster.

Rockwell, T. 1991. *The Rickover Effect: How One Man Made a Difference.* Annapolis, MD: Naval Institute Press.

Rogner, H-H. 1997. An assessment of world hydrocarbon resources. *Annual Review of Energy and the Environment* 22:217–262.

Rogner, H-H. 2000. Energy resources. In *World Energy Assessment: Energy and the Challenge of Sustainability,* ed. J. Goldemberg, 135–171. New York: UNDOP.

Rojstaczer, S., S. M. Sterling, and N. J. Moore. 2001. Human appropriation of photosynthesis products. *Science* 294:2549–2551.

Romm, J. 2000. *The Internet Economy and Global Warming.* Old Snowmass: RMI. <http://www.cool-companies.org>.

Rose, D. J. 1974. Nuclear eclectic power. *Science* 184:351–359.

Rose, D. J. 1979. Views on the U.S. nuclear energy option. Paper prepared for the Conference on the Future of Nuclear Power, Honolulu, October 31–November 3, 1979.

Rose, D. J. 1986. *Learning about Energy.* New York: Plenum Press.

Rosenberg, D. M. et al. 2000. Global-scale environmental effects of hydrological alterations: Introduction. *BioScience* 50:746–751.

Rossin, A. D. 1990. Experience of the U.S. nuclear industry and requirements for a viable nuclear industry in the future. *Annual Review of Energy* 15:153–172.

Rowe, T. et al. 1987. *Santa Clara County Air Pollution Benefit Analysis.* Washington, DC: US EPA.

Rudin, A. 1999. How improved efficiency harms the environment. <http://home.earthlink.net/~andrewrudin/article.html>.

Ruge, G. 1992. *Gorbachev: A Biography.* London: Chatto & Windus.

RWEDP (Regional Wood Energy Development Programme in Asia). 1997. *Regional Study of Wood Energy Today and Tomorrow.* Rome: FAO–RWEDP. <http://www.rwedp.org/fd50.html>.

RWEDP (Regional Wood Energy Development Programme in Asia). 2000. Wood Energy Database. Rome: FAO–RWEDP. <http://www.rwedp.org/d_consumption.html>.

Sadiq, M., and J. C. McCain, eds. 1993. *The Gulf War Aftermath: An Environmental Tragedy.* Boston: Kluwer Academic.

Sagan, S. D. 1988. The origins of the Pacific War. *Journal of Interdisciplinary History* 18: 893–922.

Salter, S. H. 1974. Wave power. *Nature* 249:720–724.

Sanchez, M. C., et al. 1999. *Miscellaneous Electricity Use in the U.S. Residential Sector.* Berkeley, CA: Lawrence Berkeley Laboratory. <http://enduse.lbl.gov/Projects/ResMisc.html>.

Santer, B. D., et al. 2000. Interpreting differential temperature trends at the surface and in the lower troposphere. *Science* 287:1227–1232.

Sassin, W., et al. 1983. *Fueling Europe in the Future.* Laxenburg: IIASA.

Satheesh, S. K., and V. Ramanathan. 2000. Large differences in tropical aerosol forcing at the top of the atmosphere and Earth's surface. *Nature* 405:60–63.

Scheede, G. R. 2001. There's too little power in wind. *Environmental & Climate News* June 2001:1–8. <http://www.heartland.org/environment/jun01/windfarm.htm>.

Schimmoller, B. K. 1999. Advanced coal systems wait in the wings. *Power* 103(7):34–38.

Schimmoller, B. K. 2000. Fluidized bed combustion. *Power* 104(9):36–42.

Schipper, L., and M. Grubb. 2000. On the rebound? Feedback between energy intensities and energy uses in IEA countries. *Energy Policy* 28:367–388.

Schipper, L. J., et al. 1993. Taxation on automobiles, driving, and fuel in OECD countries: Truth and consequences. *LBL Energy Analysis Program 1993 Annual Report,* p. 35.

Schlapbach, L., and A. Züttel. 2001. Hydrogen-storage materials for mobile applications. *Nature* 414:353–358.

Schmidt-Mende, L., et al. 2001. Self-organized discotic liquid crystals for high-efficiency organic photovoltaics. *Science* 293:1119–1122.

Schneider, S. 2001. What is 'dangerous' climate change? *Nature* 411:17–19.

Schrope, M. 2001. Which way to energy utopia? *Nature* 414:682–684.

Schumacher, E. F. 1973. *Small Is Beautiful: A Study of Economics as if People Mattered.* London: Blond and Biggs.

Schurr, S. H. 1984. Energy use, technological change, and productive efficiency: An economic–historical interpretation. *Annual Review of Energy* 9:409–425.

Schurr, S. H., and B. C. Netschert. 1960. *Energy in the American Economy 1850–1975.* Baltimore, MD: Johns Hopkins University Press.

Schwartz, N. N., O. L. Elliot, and G. L. Gower. 1992. Gridded state maps of wind electric potential. Paper presented at Wind Power 1992, Seattle, WA.

Sclater, G. J., et al. 1980. The heat flow through oceanic and continental crust and the heat loss of the Earth. *Reviews of Geophysics and Space Physics* 18:269–311.

Scott, W. G. 1997. Micro-size turbines create market opportunities. *Engineering* 101(9):46–50.

Seaborg, G. T. 1968. Some long-range implications of nuclear energy. *The Futurist* 2(1):12–13.

Seaborg, G. T. 1971. The environment: A global problem, an international challenge. In *Environmental Aspects of Nuclear Power Stations,* 5. Vienna: IAEA, Vienna.

Seaborg, G. T., and W. R. Corliss. 1971. *Man and Atom: Building a New World Through Nuclear Technology.* New York: E. P. Dutton.

Select Committee on Lighting by Electricity of the British House of Commons. 1879. *Hearings on Lighting by Electricity.* London: House of Commons.

Seppa, T. O. 2000. Physical limitations affecting long distance energy sales. Paper presented at IEEE SPM. Seattle, WA: IEEE.

Service, R. F. 1996. New solar cells seem to have power at the right price. *Science* 272:1744–1745.

Service R. F. 1998. Will new catalyst finally tame methane? *Science* 280:525.

Service, R. F. 2001. C_{60} enters the race for the top. *Science* 293:1570.

Service, R. F. 2002. MgB_2 trades performance for a shot at the real world. *Science* 295:786–788.

Shah, A., et al. 1999. Photovoltaic technology: The case for thin-film solar cells. *Science* 285:692–698.

Shaheen, M. 1997. Wind energy transmission. Washington, DC: National Wind Coordinating Committee. <http://www.nationalwind.org/pubs/wes/ibrief09a.htm >.

Shapouri, H., J. A. Duffield, and M. S. Graboski. 1995. *Estimating Net Energy Balance of Corn Ethanol.* Washington, DC: USDA. <http://www.ethanol-gec.org/corn_eth.htm>.

Shell Exploration & Production Company. 1999. URSA deepwater tension leg platform begins production at record setting depth. <http://www.shellus.com/news/press_releases/1999/press_041399.html>.

Shepard, M. 1991. How to improve energy efficiency. *Issues in Science and Technology* 7(2): 85–91.

Shiklomanov, I. A. 1999. *World Water Resources and Water Use.* St. Petersburg: State Hydrological Institute.

Show, I. T., et al. 1979. *Comparative Assessment of Marine Biomass Materials.* Palo Alto, CA: Electric Power Research Institute.

Shukla, J. 1998. Predictability in the midst of chaos: A scientific basis for climate forecasting. *Science* 282:728–731.

Sieferle, R. P. 2001. *The Subterranean Forest: Energy Systems and the Industrial Revolution.* Cambridge: The White Horse Press.

Simakov, S. N. 1986. Forecasting and Estimation of the Petroleum-bearing Subsurface at Great Depths. Leningrad: Nedra.

Simon, J. L. 1981. *The Ultimate Resource.* Princeton, NJ: Princeton University Press.

Simon, J. L. 1996. *The Ultimate Resource 2.* Princeton, NJ: Princeton University Press.

Simon, J., and H. Kahn, eds. 1984. *The Resourceful Earth.* Oxford: Basil Blackwell.

Singer, J. D., and M. Small. 1972. *The Wages of War 1816–1965: A Statistical Handbook.* New York: Wiley.

Singer, S. F. 2001. Who needs higher energy taxes? *Environment & Climate News* December 2001:1–3. <http://www.heartland.org/environment/dec01/singer.htm>.

Smil, V. 1966. Energie, krajina, lide. *Vesmir* 45(5):131–133.

Smil, V. 1974. Energy and the Environment: Scenarios for 1985 and 2000. *The Futurist* 8(1): 4–13.

Smil, V. 1976. *China's Energy.* New York: Praeger.

Smil, V. 1977. China's Future. *Futures* 9:474–489.

Smil, V. 1983. *Biomass Energies.* New York: Plenum Press.

Smil, V. 1984. *The Bad Earth.* Armonk, NY: M. E. Sharpe.

Smil, V. 1985. *Carbon–Nitrogen–Sulfur: Human Interference in Grand Biospheric Cycles.* New York: Plenum Press.

Smil, V. 1987. *Energy, Food, Environment: Realities, Myths, Options.* Oxford: Oxford University Press.

Smil, V. 1988. *Energy in China's Modernization.* Armonk, NY: M. E. Sharpe.

Smil, V. 1991. *General Energetics.* New York: Wiley.

Smil, V. 1992a. Agricultural energy costs: National analyses. In *Energy in Farm Production,* ed. R. C. Fluck, 85–100. Amsterdam: Elsevier.

Smil, V. 1992b. How efficient is Japan's energy use? *Current Politics and Economics of Japan* 2(3/4):315–327.

Smil, V. 1993a. *Global Ecology*. London: Routledge.

Smil, V. 1993b. *China's Environmental Crisis*. Armonk, NY: M. E. Sharpe.

Smil, V. 1994a. *Energy in World History*. Boulder, CO: Westview Press.

Smil, V. 1994b. Energy intensities: Revealing or misleading? *OPEC Review* 18(1):1–23.

Smil, V. 1996. *Environmental Problems in China: Estimates of Economic Costs*. Honolulu: East–West Center.

Smil, V. 1997. *Cycles of Life*. New York: Scientific American Library.

Smil, V. 1998a. China's energy resources and uses: Continuity and change. *The China Quarterly* 156:935–951.

Smil, V. 1998b. Future of oil: Trends and surprises. *OPEC Review* 22(4):253–276.

Smil, V. 1999a. *Energies*. Cambridge, MA: The MIT Press.

Smil, V. 1999b. Crop residues: Agriculture's largest harvest. *BioScience* 49:299–308.

Smil, V. 1999c. China's great famine: 40 years later. *British Journal* 7225:1619–1621.

Smil, V. 2000a. Energy in the 20th century: Resources, conversions, costs, uses, and consequences. *Annual Review of Energy and the Environment* 25:21–51.

Smil, V. 2000b. Jumbo. *Nature* 406:239.

Smil, V. 2000c. *Feeding the World*. Cambridge, MA: The MIT Press.

Smil, V. 2000d. Perils of long-range energy forecasting: Reflections of looking far ahead. *Technological Forecasting and Social Change* 65:251–264.

Smil, V. 2001. *Enriching the Earth*. Cambridge, MA: The MIT Press.

Smil, V. 2002. *The Earth's Biosphere*. Cambridge, MA: The MIT Press.

Smil, V., and W. Knowland, eds. 1980. *Energy in the Developing World*. Oxford: Clarendon Press.

Smil, V., and D. Milton. 1974. Carbon dioxide—Alternative futures. *Atmospheric Environment* 8(12):1213–1232.

Smil, V., P. Nachman, and T. V. Long, II. 1982. *Energy Analysis in Agriculture*. Boulder, CO: Westview Press.

Smith, D. J. 2001. Will the new millennium see the re-birth of cogeneration? *Power* 105(1):41–43.

Smith, D. R. 1987. The wind farms of the Altamont Pass area. *Annual Review of Energy* 12:145–183.

Smith, K. 1993. Fuel combustion, air pollution exposure and health: The situation in developing countries. *Annual Review of Energy and the Environment* 18:529–566.

Smith, K. R. 1988. Energy indexing: The weak link in the energy Weltanschauung. In *Energy Planning*, 113–153. Paris: UNESCO.

Smith, K. R., et al. 1993. One hundred million improved cookstoves in China: How was it done? *World Development* 21:941–961.

Smithsonian Institution. 2001. Energy efficiency: Light sources in the 20th century. <http://americanhistory.si.edu/lighting/chart.htm>.

Society of Automotive Engineers. 1992. *Automotive Emissions and Catalyst Technology.* Warrendale, PA: SAE.

Society of Petroleum Engineers. 1991. *Horizontal Drilling.* Richardson, TX: Society of Petroleum Engineers.

Socolow, R. H. 1977. The coming age of conservation. *Annual Review of Energy* 2:239–289.

Socolow, R. H. 1985. The physicist's role in using energy efficiently: Reflections on the 1974 American Physical Society summer study and on the task ahead. In *Energy Sources: Conservation and Renewables,* eds. D. Hafemeister, H. Kelly and B. Levi, 15–32. New York: American Institute of Physics Press.

SolarPACES. 1999. Status of the technologies. Paris: IEA. <http://www.solarpaces.org/publications/sp99_tec.htm>.

Sørensen, B. 1980. *An American Energy Future.* Golden, CO: Solar Energy Research Institute.

Sørensen, B. 1984. Energy storage. *Annual Review of Energy* 9:9–29.

Sørensen, B. 1995. History of, and recent progress in, wind-energy utilization. *Annual Review of Energy and the Environment* 20:387–424.

Southern California Edison Company. 1988. Planning for uncertainty: A case study. *Technological Forecasting and Social Change* 33:119–148.

Speer, A. 1970. *Inside the Third Reich.* London: Macmillan.

Spinrad, B. I. 1971. The role of nuclear power in meeting world energy needs. In *Environmental Aspects of Nuclear Power Stations,* 57. Vienna: IAEA.

Spreng, D. T. 1978. *On Time, Information, and Energy Conservation.* Oak Ridge, TN: Institute for Energy Analysis.

SRES (Special Report on Emission Scenarios). 2001. *Summary for Policymakers.* Geneva: WMO and UNEP. <http://www.ipcc.ch/pub/SPM_SRES.pdf>.

Srinivasan, S., et al. 1999. Fuel cells: Reaching the ear of clean and efficient power generation in the twenty-first century. *Annual Review of Energy and the Environment* 24:281–328.

Starr, C. 1973. Realities of the energy crisis. *Bulletin of the Atomic Scientists* 29(7):15–20.

Starr, C., and R. Rudman. 1973. Parameters of technological growth. *Science* 182:235–253.

Statistics Bureau. 1970–2001. *Japan Statistical Yearbook.* Tokyo: Statistics Bureau.

Steele, B. C. H., and A. Heinzel. 2001. Materials for fuel-cell technologies. *Nature* 414:345–352.

Stern A. C. 1976–1986. *Air Pollution.* New York: Academic Press.

Stern, D. I., and R. K. Kaufmann. 1998. *Annual Estimates of Global Anthropogenic Methane Emissions: 1860–1994.* Oak Ridge, TN: CDIAC. <http://cdiac.ornl.gov/trends/meth/methane.htm>.

Stoddard, J. C., et al. 1999. Regional trends in aquatic recovery from acidification in North America and Europe. *Nature* 401:575–578.

Stone, R. 2002. Caspian ecology teeters on the brink. *Science* 295:430–433.

Stout, B. A. 1990. *Handbook of Energy for World Agriculture.* New York: Elsevier.

Street, D. G., et al. 1999. Energy consumption and acid deposition in Northeast Asia. *Ambio* 28:135–143.

Sullivan, T. J. 2000. *Aquatic Effects of Acid Deposition.* Boca Raton, FL: Lewis Publishers.

Suess, E. et al. 1999. Flammable ice. *Scientific American* 281(5):76–83.

Suncor Energy. 2001. Oil Sands. <http://www.suncor.com/>.

Swezey, B. G., and Y. Wan. 1995. *The True Cost of Renewables: An Analytic Response to the Coal Industry's Attack on Renewable Energy.* Golden, CO: National Renewable Energy Laboratory.

Szewzyk, U., et al. 1994. Thermophilic, anaerobic bacteria isolated from a deep borehole in granite in Sweden. *Proceedings of the National Academy of Sciences USA* 91:1810–1813.

Taintner, J. A. 1988. *The Collapse of Complex Societies.* Cambridge: Cambridge University Press.

Tavoulareas, E. S. 1991. Fluidized-bed combustion technology. *Annual Review of Energy and the Environment* 16:25–57.

Tavoulareas, E. S. 1995. *Clean Coal Technologies for Developing Countries.* Washington, DC: World Bank.

Taylor, M. J. H., ed. 1989. *Jane's Encyclopedia of Aviation.* New York: Portland House.

Teller, E., et al. 1996. *Completely Automated Nuclear Reactors for Long-Term Operation II: Toward a Concept-Level Point-Design of a High-temperature, Gas-cooled Central Power Station System.* Livermore, CA: Lawrence Livermore National Laboratory.

TGV (Train de grand vitesse). 2002. *TGV: Prenez le temps d'aller vite.* <http://www.tgv.com/homepage/index.htm>.

Thomas, C. 2001. Energy policies haven't worked. *Dallas Business Journal,* March 23, 2001. <http://dallas.bcentral.com/dallas/stories/2001/03/26/editorial3.html>.

Thomas, C. A., et al. 1946. *The Economics of Nuclear Power.* Saint Louis, MO: Monsanto Company.

Thomas, J. 1999. Quantifying the black economy: "Measurement without theory" yet again. *The Economic Journal* 109:381–389.

Thorne, J., and M. Suozzo. 1997. Leaking electricity estimates. *Science News Online* October 25, 1997. <http://www.sciencenews.org/sn_arc97/10_25_97/bob1a.htm>.

TIGR (The Institute for Genomic Research). 2000. TIGR Microbial Database: A listing of published genomes and chromosomes and those in progress. <http://www.tigr.org/tdb/mdb/mdbcomplete.html>.

Tilton, J. E., and B. J. Skinner. 1987. The meaning of resources. In *Resources and World Development,* eds. D. J. McLaren and B. J. Skinner, 13–27. Chichester: Wiley.

Transocean Sedco Forex. 2001. Facts and Firsts. <http://www.deepwater.com/FactsandFirsts.cfm>.

Tunzelmann, G. W. de. 1901. *Electricity in Modern Life.* New York: P. F. Collier.

Turkenburg, W. C. 2000. Renewable energy technologies. In *World Energy Assessment,* ed. J. Goldemberg, 219–272. New York: UNDP.

Turner, B. L., et al., eds. 1990. *The Earth as Transformed by Human Action.* New York: Cambridge University Press.

Turner, J. A. 1999. A realizable renewable energy future. *Science* 285:687–689.

UCS (Union of Concerned Scientists). 2001. Clean Energy Blueprint. Cambridge, MA: UCS. <http://www.ucsusa.org/energy/blueprint.html>.

UK Coal. 2001. *Facts About British Coal Production.* Doncaster: UK Coal. <http://www.rjr.co.uk>.

UKAEA (United Kingdom Atomic Energy Authority). 2001. Decommissioning. <http://www.ukaea.org.uk/decommissioning/>.

UNDP (United Nations Development Programme). 2001. *Human Development Report 2001.* New York: UNDP. <http://www.undp.org/hdr2001/>.

UNO (United Nations Organization). 1956. World energy requirements in 1975 and 2000. In *Proceedings of the International Conference on the Peaceful Uses of Atomic Energy,* Volume 1, 3–33. New York: UNO.

UNO (United Nations Organization). 1976. *World Energy Supplies 1950–1974.* New York: UNO.

UNO (United Nations Organization). 1990. *Global Outlook 2000.* New York: UNO.

UNO. 1991. *World Population Prospects 1990.* New York: UNO.

UNO. 1998. *World Population Prospects: The 1998 Revision.* New York: UNO. <http:/www.un.org/esa/populations/longrange/longrange.htm>.

UNO. 2001. *Yearbook of World Energy Statistics.* New York: UNO.

UNO. 2002. *Long-range World Population Projections.* New York: UNO. <http://www.un.org/esa/population/publications/longrange/longrange.html>.

United Nations Development Programme. 2001. *Human Development Report 2001.* New York: UNDP. <http://www.undp.org/hdr2001/>.

Urbanski, T. 1967. *Chemistry and Technology of Explosives.* Oxford: Pergamon Press.

USBC (U.S. Bureau of the Census). 1975. *Historical Statistics of the United States.* Washington, DC: U.S. Department of Commerce.

USBC (U.S. Bureau of the Census). 2002. Characteristics of new housing. Washington, DC: USCB. <http://www.census.gov/ftp/pub/const/www.charindex.html>.

USGS (United States Geological Survey). 2000. *U.S. Geological Survey World Petroleum Assessment 2000.* Denver, CO: USGS. <http://geology.cr.usgs.gov/energy/WorldEnergy/DDS-60/index.html>.

Valenti, M. 1991. New life from old oil wells. *Mechanical Engineering* 113(2):37–41.

Valk, M., ed. 1995. *Atmospheric Fluidized Bed Coal Combustion.* Amsterdam: Elsevier.

van der Eng, P. 1992. The real domestic product of Indonesia, 1880–1989. *Explorations in Economic History* 29:343–373.

van Gool, W. 1978. *Limits to Energy Conservation in Chemical Processes.* Oak Ridge, TN: Oak Ridge National Laboratory.

Vendryes, G. A. 1977. Superphénix: A full-scale breeder reactor. *Scientific American* 236(3): 26–35.

Vendryes, G. A. 1984. The French liquid-metal fast breeder reactor program. *Annual Review of Energy* 9:263–280.

Vine, E. 1992. *Persistence of Energy Savings: What Do We Know and How Can It be Ensured?* Berkeley, CA: Lawrence Berkeley Laboratory.

Viscusi, W. K., et al. 1994. Environmentally responsible energy pricing. *The Energy Journal* 15:23–42.

Vitousek, P., et al. 1986. Human appropriation of the products of photosynthesis. *BioScience* 36:368–373.

Vogel, E. F. 1979. *Japan as Number One: Lessons for America.* Cambridge, MA: Harvard University Press.

von Braun, W., and F. I. Ordway. 1975. *History of Rocketry and Space Travel.* New York: Thomas Y. Crowell.

Voorhees, A. S., et al. 2001. Cost–benefit analysis methods for assessing air pollution control programs in urban environments—A review. *Environmental Health and Preventive Medicine* 6:63–73.

Vörösmarty, C. J., and D. Sahagian. 2000. Anthropogenic disturbance of the terrestrial water cycle. *BioScience* 50:753–765.

Voss, A. 1979. Waves, currents, tides—Problems and prospects. *Energy* 4:823–831.

WAES (Workshop on Alternative Energy Strategies). 1977. *Energy Supply–Demand Integrations to the Year 2000.* Cambridge, MA: The MIT Press.

Waley, A. 1938. *The Analects of Confucius* (Translation of *Lunyu*). London: George Allen & Unwin.

Walker, B. H., and W. L. Steffen, eds. 1998. *The Terrestrial Biosphere and Global Change: Implications for Natural and Managed Ecosystems.* Cambridge: Cambridge University Press.

Wang, M., and Y. Ding. 1998. Fuel-saving stoves in China. *Wood Energy News* 13(3):9–10.

Ward's Communications. 2000. *2000 Motor Vehicle Facts & Figures.* Southfield, MI: Ward's Communications.

Warman, H. R. 1972. The Future of Oil. *The Geographical Journal* 138:287–297.

Watt, K. F. 1989. Evidence for the role of energy resources in producing long waves in the United States economy. *Ecological Economics* 1:181–195.

Wavegen. 2001. Applications. Inverness, UK: Wavegen. <http://www.wavegen.co.uk/>.

Wayne, W. W. 1977. *Tidal Power Study for the U.S. ERDA.* Washington, DC: US ERDA.

WCD (World Commission on Dams). 2000. *Dams and Development.* London: Earthscan Publishers.

WCED (World Commission on Environment and Development). 1987. *Our Common Future.* Oxford: Oxford University Press.

WCI (World Coal Institute). 2000. *Coal & Steel Facts.* London: WCI. <http://www.wci-coal.com/facts.coal&steel99.htm>.

WCI (World Coal Institute). 2001. *Coal Facts.* London: WCI. <http://www.wci-coal.com/facts.coal99.htm>.

WEC (World Energy Council). 1993. *Energy for Tomorrow's World.* London: Kogan Page.

WEC (World Energy Council). 1998. *Survey of Energy Resources.* London: WEC.

WEC and IIASA. 1995. *Global Energy Perspectives to 2050 and Beyond.* London: World Energy Council.

Weinberg, A. M. 1972. Social institutions and nuclear energy. *Science* 177:27–34.

Weinberg, A. M. 1973. Long-range approaches for resolving the energy crisis. *Mechanical Engineering* 95(6):14–18.

Weinberg, A. M. 1978. Reflections on energy wars. *American Scientist* 66:153–158.

Weinberg, A. M. 1979a. *Limits to Energy Modeling.* Oak Ridge, TN: Institute for Energy Analysis.

Weinberg, A. M. 1979b. Are the alternative energy strategies achievable? *Energy* 4:941–951.

Weinberg, A. M. 1982. From the director's diary. *Institute for Energy Analysis Newsletter* 5:2.

Weinberg, A. M., et al. 1984. *The Second Nuclear Era.* Oak Ridge, TN: Institute of Energy Analysis.

Weinberg, A. M. 1994. *The First Nuclear Era: The Life and Times of a Technological Fixer.* New York: American Institute of Physics.

Weisman, J. 1985. *Modern Power Plant Engineering.* Englewood Cliffs, NJ: Prentice-Hall.

Wells, J. 1992. *Efforts Promoting More Efficient Energy Use.* Washington, DC: GAO.

Whipple, C. G. 1996. Can nuclear waste be stored safely at Yucca Mountain? *Scientific American* 274(6)72–79.

WHO (World Health Organization). 1992. *Indoor Air Pollution from Biomass Fuel.* Geneva: WHO.

Wigley, T., and S. Raper. 2001. Interpretation of high projections for global-mean warming. *Science* 293:451–454.

Wigley, T. M., and D. S. Schimel, eds. 2000. *The Carbon Cycle.* Cambridge: Cambridge University Press.

Williams, R. H. 2000. Advanced energy supply technologies. In *World Energy Assessment,* ed. J. Goldemberg, 273–329. New York: UNDP.

Williams, R. H., and E. D. Larson. 1988. Aeroderivative turbines for stationary power. *Annual Review of Energy* 13:429–489.

Wilshire, H., and D. Prose. 1987. Wind energy development in California, USA. *Environmental Management* 11:13–20.

Wilson, A., and J. Morrill. 1998. *Consumer Guide to Home Energy Savings.* Washington, DC: ACEEE. <http://aceee.org>.

Wilson, C. L., ed. 1980. *Coal: Bridge to the Future.* Cambridge, MA: Ballinger.

Wilson, R. 1998. Accelerator Driven Subcritical Assemblies. Report to Energy, Environment and Economy Committee of the U.S. Global Strategy Council. Cambridge, MA: Harvard University.

Winter, C-J., and J. Nitsch, eds. 1988. *Hydrogen as an Energy Carrier: Technologies, Systems, Economy.* Berlin: Springer-Verlag.

WNA (World Nuclear Association). 2001a. World nuclear power reactors 2000–2001. <http://www/world-nuclear.org/info/reactors.htm>.

WNA (World Nuclear Association). 2001b. Plans for new reactors worldwide. <http://www.world-nuclear.org/info/inf17.htm>.

Womack, J. P., et al. 1991. *The Machine that Changed the World.* New York: Harper.

Wood, C. 1992. *The Bubble Economy: Japan's Extraordinary Speculative Boom of the '80s and the Dramatic Bust of the '90s.* Tokyo: Kodansha.

Workshop on Alternative Energy Strategies. 1977. *Energy Supply—Demand Integrations to the Year 2000.* Cambridge, MA: The MIT Press.

World Bank. 1995. *Investment Strategies for China's Coal and Electricity Delivery System.* Washington, DC: World Bank.

World Bank. 2001. *World Development Report 2001.* Washington, DC: World Bank.

World Energy Conference. 1978. *Study Group Report on World Energy Demand.* Guildford, UK: IPC Science and Technology Press.

World Oil. 2000. *Marine Drilling Rigs 2000/2001.* Houston, TX: Gulf Oil Publishing.

WTO (World Trade Organization). 2001. International Trade Statistics 2000. Geneva: WTO. <http://www.wto.org/english/res_e/statis_e/stat_toc_e.htm>.

Wyman, C. E. 1999. Biomass ethanol: Technical progress, opportunities, and commercial challenges. *Annual Review of Energy and the Environment* 24:189–226.

Yamani, S. A. Z. 2000. Interview, September 5, 2000. *Planet Ark.* <http://www.planetark.org/dailynewsstory.cfm?newsid=8054>.

Yemm, R. 2000. Riding the ocean waves. *International Water Power & Dam Construction* 52(12):41–42.

Zener, C. 1977. The OTEC answer to OPEC: Solar sea power! *Mechanical Engineering* 99(6): 26–29.

Zhang, Z. et al. 1998. Fuelwood forest development strategy. *Wood Energy News* 13(3): 6–8. <http://www.rwedp.org/acrobat/wen13-3.pdf>.

Zonis, M. 1987. *Khomeini, the Islamic Republic of Iran, and the Arab World.* Cambridge, MA: Center for Middle Eastern Studies, Harvard University.

Zorpette, G. 1996. Hanford's nuclear wasteland. *Scientific American* 274(5):88–97.

Zweibel, K. 1993. Thin-film photovoltaic cells. *American Scientist* 81:362–369.

Index